SCIENCE
BY SIMULATION

Volume 2: Models of Classical Physics

SCIENCE
BY SIMULATION

Volume 2: Models of Classical Physics

ANDREW FRENCH

Winchester College, UK

World Scientific

NEW JERSEY • LONDON • SINGAPORE • BEIJING • SHANGHAI • HONG KONG • TAIPEI • CHENNAI • TOKYO

Published by

World Scientific Publishing Europe Ltd.

57 Shelton Street, Covent Garden, London WC2H 9HE

Head office: 5 Toh Tuck Link, Singapore 596224

USA office: 27 Warren Street, Suite 401-402, Hackensack, NJ 07601

Library of Congress Cataloging-in-Publication Data
Names: French, Andrew, author.
Title: Science by simulation / Andrew French, Winchester College, UK.
Description: New Jersey : World Scientific, [2022] | Includes bibliographical references and index. |
      Contents: volume 1. A mezze of mathematical models.
Identifiers: LCCN 2021041907 | ISBN 9781800611078 (hardcover) | ISBN 9781800611214 (paperback) |
      ISBN 9781800611085 (ebook) | ISBN 9781800611092 (ebook other)
Subjects: LCSH: Simulation methods.
Classification: LCC T57.62 .F74 2022 | DDC 658.4/0352--dc23/eng/20211004
LC record available at https://lccn.loc.gov/2021041907

Volume 2: Models of Classical Physics
ISBN   978-1-80061-666-0 (hardcover)
ISBN   978-1-80061-680-6 (paperback)
ISBN   978-1-80061-667-7 (ebook for institutions)
ISBN   978-1-80061-668-4 (ebook for individuals)

British Library Cataloguing-in-Publication Data
A catalogue record for this book is available from the British Library.

For any available supplementary material, please visit
https://www.worldscientific.com/worldscibooks/10.1142/Q0491#t=suppl

Desk Editors: Soundararajan Raghuraman/Gabriel Rawlinson

Typeset by Stallion Press
Email: enquiries@stallionpress.com

To my family, to wise, beautiful and ever tolerant Lucinda, to my friends Matt, Lynne, Walter, Veerle, Greg, and all my students, teachers and colleagues from Carisbrooke, Cambridge, Cowes, Chelmsford, Dorchester, Sherborne and Winchester. And to Sybil, who shall be memorialised in several illustrations throughout this book.

Foreword

An education in physics generally follows a somewhat hierarchical path. The highlights for students may come in several guises: the realisation that one finally understands some small aspect that had previously eluded them, the successful solving of a tricky problem, or even the mere perception of progress in a subject that encompasses a depth of conceptual and mathematical understanding. There is no doubt that students often feel far from the frontiers of the subject, alleviated by momentary exhilaration through lectures on discoveries from researchers in these fields. The step-by-step grinding through a course may never enable them to reach a level of understanding of the exciting topics or help them uncover the complications of an analysis which always seems to be simplified and offers no glimpse of how they could handle a real perspective.

Andrew French's second volume of *Science by Simulation* is a revelation in showing that the opportunity exists for students from school and upwards to engage actively through computational methods in solving problems and analysing examples in physics that lie beyond their purely analytical skills. The calculated results can illuminate thinking whilst revealing the dependence on model parameters which would typically be reduced to a mention at best or even be swept under the carpet. Constructing models with suitable parametrisation is seen as the launch pad, but this is always mirrored by the skill of presenting results to the audience with clarity and in an understandable form, as seen previously in *Science by Simulation – Volume 1: A Mezze of Mathematical Models.*

The range of topics in this volume alone will excite any student wanting to probe the next level in applications they may have come across in introductory courses, such as the Computational Challenge seminars delivered by the British Physics Olympiad. It is a tribute to Dr Andrew French's labour in developing these computational challenges, and indeed the *Science by Simulation* series of books, that many students now have a doorway to access and participate in this wonderful collection of physics examples and applications for themselves.

Robin Hughes
British Physics Olympiad & Isaac Physics
June 2025

Preface

Models of Classical Physics is the second volume of *Science by Simulation*. It is a recipe book of mathematical models that can be enlivened by the transmutation of equations into computer code. The subject matter of this volume includes topics that most students and teachers would associate with the standard canon of classical physics. The chapter titles are: *Mechanics, Thermodynamics, Waves and Ray Optics* and *Electromagnetism.* More contemporary topics (nuclear and quantum, special relativity) will be covered in *Models of Modern Physics* (the third volume of *Science by Simulation*) in addition to models of orbits of stars, planets and moons in systems interacting via gravity. Although I have endeavoured to aim for fairly wide topic coverage (from pre-university to early undergraduate), this is not a textbook and certainly not intended to be comprehensive in scope. It is instead a collection of model systems that can be readily turned into code and then used to make quantitative predictions about the behaviour of certain specific aspects of the physical world. I have chosen a selection of my favourite systems that form the basis of much of my teaching of mathematical modelling and problem-solving skills in physics. I would like to thank Laurent and the editorial and marketing teams from World Scientific for their continued enthusiasm for this project and professionalism in the final production stages. Particular thanks to Anson, Robin and Kimlam from BPhO and Jeremy and Tony from Winchester College for so many inspirational conversations, and to Chas for encouraging me to transform a very speculative idea at a summer track & field match into this series of books. *Science by Simulation* is dedicated to the next generation, particularly my nieces and nephews Dominic, Sophie, Jamie, Charlie, Esme and Sebastian. I am very grateful to the governing body and headmaster of Winchester College for granting a sabbatical in 2021 to complete the final phase of *Science by Simulation – Volume 1* and to commence Volumes 2 and 3.

About the Author

**Andrew French** has taught physics and mathematics at Winchester College since 2011. Previously he taught at Sherborne School, and before that worked as a Systems Engineer for eight years on a wide range of projects, from meteorological sensors, to wind farms, to signal processing algorithms associated with marine and land-based radar systems. While in industry, he completed a PhD at University College London on aspects of novel stepped-frequency waveforms that can be transmitted by a phased-array antenna. He studied at Christ's College, Cambridge, from 1997 to 2002, culminating in a master's degree in experimental and theoretical physics. He also holds a postgraduate master of philosophy in fluid dynamics from the BP Institute in Cambridge and a PGCE in secondary mathematics teaching from Southampton University. Dr French is currently working on a number of educational outreach and research projects, such as the *Eclecticon* resource website (www.eclecticon.info) and the *Science by Simulation* book series. He has also co-authored several recent papers in *Physics Education* relating to epidemiology and numerical methods in introductory calculus teaching. Dr French helps deliver the British Physics Olympiad (BPhO) Round 1 online seminars and writes the annual BPhO Computational Physics Challenge.

Contents

Chapter 1

Introduction

1.1 Problem-Solving, Practical Experiments and Computer Simulations

This book is hopefully more about the *how* of physics rather than the *what*.[1] I believe there are three basic strands, three primary colours, or three legs of a wooden stool (choose your favourite analogy) that are required to yield the full richness and potential for artistry in physics. And yes, I really do mean art. Physics is so much more than a rational processing machine for quantifying the essential physical processes of the Universe and making predictions. Although these attributes are themselves quite impressive, since you are probably scrolling through this book on an almost indestructible self-illuminating ultrahigh-definition screen a few millimetres thick, in a temperature-controlled and weatherproof cafe, sipping exquisite coffee that was harvested at the other side of the world, all powered and financed via a typically reliable global electrical, fibre-optic and wireless network. My proposed concept trio for a complete physics experience is as follows:

(1) **Problem-solving.** Rather than just associate the name of a phenomenon to a *Why* question, we endeavour to calculate something specific and, through this process, infer new information. We begin from a set of basic physical principles that we deem reasonable and, in most situations, have withstood years or perhaps centuries of experimental tests. This is, in essence, the same process as mathematical proof, in the sense that we apply logical reasoning to a set of statements that we assert as fundamental truths, or *axioms*. The difference in physics is that we are always aware that our systems are *approximations* of the real world, whereas in pure mathematics we define an abstract, perfect world in our minds with a finite set of objects and rules that we can assign and change with deity powers. Why is the sky blue? A correct but, I would argue, unsatisfactory answer is the label 'Rayleigh scattering'. A much better

---

[1] Although calculating the rate of flow of energy in standard units is, of course, very important.

answer is to *show*[2] that electromagnetic waves scatter from particles (e.g. air molecules) that are much smaller than the radiation wavelength λ, preferentially as the wavelength is reduced. In fact, a quantitative relationship is that 'amount of scattering' is proportional to $\frac{1}{\lambda^4}$. Since blue (and violet) light are the parts of the visible spectrum which have the smallest wavelengths, blue light is scattered away from a light source more than green, yellow and red. It also explains why a low sun at dusk or dawn may result in a red sky, as, in this case, only these longer wavelengths (of the visible spectrum) are insufficiently scattered by the atmosphere to reach our eyes. Obviously, scientific discourse would be somewhat cumbersome if all discussions began from 'first principles'. We would never get anywhere! My point is that we should be mindful that a phenomenon-label answer should only be used when all parties in a discussion have both experienced the construction of the associated model. A shorthand is appropriate to signpost a shared memory, which could be rich in complexity. The danger is when this hasn't happened and one or all parties don't really understand in a practical sense the meaning of the phenomenon label. This is particularly acute in science education, where there may be pressure to regurgitate a series of magic words to pass exams but insufficient time and/or resources to fully embed the ideas that are labelled. A key message in this series of books is my recommendation that scientific ideas can be much better understood via a creative problem-solving approach. This is at the core of enrichment initiatives such as the courses and competitions of the British Physics Olympiad.[3]

(2) A direct encounter with reality via **practical experiments**, which themselves can be represented by simplified mathematical models. The *scientific method* can be thought of as a cyclical process. At one point in the cycle, an *observation* is made of a physical phenomenon, which ideally is a quantitative measurement that can be repeated by other scientists and, indeed, students in laboratories all over the world. For example, the root-mean-squared voltage induced across a coil placed next to a spinning bar magnet rises in direct proportion to the rate at which the magnet spins. This is a fact that can be observed using appropriate measurement equipment, such as an oscilloscope. Next, a *theory* is proposed which explains the observation in terms of more fundamental ideas. In the case of the magnet and the coil, the magnet produces a magnetic field in

---

[2]The overall concept of *Rayleigh scattering* is as follows: an electromagnetic wave varies sinusoidally with frequency f. Small air molecules are likely to be *polarised* (i.e. charge separated, resulting in *electric fields*) in alignment with the electric field of the waves. The charges in the air molecules therefore oscillate. If the charges oscillate at frequency f, this means their displacement will be proportional to $e^{2\pi i f t}$ and their acceleration (the rate of change of the rate of change of displacement) will therefore be proportional to f^2 since $\frac{d}{dt}\left(e^{2\pi i f t}\right) = 2\pi i f e^{2\pi i f t}$ and, therefore, $\frac{d^2}{dt^2}\left(e^{2\pi i f t}\right) = -4\pi^2 f^2 e^{2\pi i f t}$. *Accelerating charges radiate*, and one can show that radiative power scales as the *square* of acceleration. Therefore, expect EM waves to scatter as f^4. Since the speed of light $c = f\lambda$ and hence $f = \frac{c}{\lambda}$, this means scattering as $\frac{1}{\lambda^4}$.

[3]British Physics Olympiad, https://www.bpho.org.uk/.

the space around it, and the coil experiences a *flux*[4] of this field which changes in a sinusoidal fashion with time. The frequency of the sinusoid is the same as the rate of spin of the magnet. Now, *Faraday's*[5] *law of electromagnetism* states that 'the rate of change of magnetic flux linked is proportional to the voltage induced'. Note if we can assume a fixed resistance of the coil, we might also expect the induced current to be proportional to the voltage induced, and therefore the rate of change of magnetic flux linked, and hence the rate at which the magnet spins.[6] A theory begins its life as an abstract mathematical model. It only becomes physics if it survives repeated encounters with the real world. Many more observations must be made, inspected rigorously for bias or unaccounted-for influencing factors and cross-checked against theoretical predictions. The scientific method loop continues to run, refining our models that make good predictions and rejecting the ones that don't. If a poor correlation between model and experiment remains after all the method verification checks have been completed, then *it is the theory which needs to be modified.* The alternative, where experimental results that don't align with our beliefs, opinions, ideological stances or cultural norms *are instead rejected,* is a manifestation of confirmation bias and flawed thinking. Put more simply, it is a wilful departure from the truth. Practical experiments are our encounter with the real world and arbitrate the efficacy of our models. Therefore, they are at the root of the physics experience.

(3) The creation of **computer simulations** that turn mathematical models into something that can be directly compared to what can be measured in the real world. From a mathematical perspective, computer models effectively define the entire *solution space.* If you specify the relationships between all sensible inputs and all computable outputs of a planet-and-star orbit scenario, then you have a mechanism for solving any astrophysics problem that might be posed. I think this confers a deep and practically very useful conceptual understanding. The creation of a computer simulation also perhaps illustrates a balance between human thinking and machine processes. A human defines the model, hopefully through a nice, clear diagram and associated equations. When suitably programmed into a computer, the equations can be evaluated repeatedly over a large range of inputs in a tiny fraction of the time it would take a human to do it. The fundamental purpose of this type of computer, perhaps all computers, is to perform pre-programmed repeated calculations rapidly without getting bored

---

[4]The magnetic flux linked Φ is the magnetic flux density B projected onto the normal to the coil, multiplied by its cross-sectional area A, multiplied by the number of coils N, i.e. $\Phi = NBA$.

[5]Michael Faraday (1791–1867).

[6]As discussed in the *Toroidal Inductor* and *Transformer* sections of the chapter on Electromagnetism, the transfer of energy from magnet to coil may be rather more subtle than this simplistic argument implies. Induced current (i.e. the flow of electric charge) in the coil will *also* create a magnetic field, which will *oppose the change that produced it* (Lenz's law). If our coil has a significant *inductance*, then it will manifest a 'back EMF', i.e. reducing the overall voltage induced and, indeed, the efficiency of energy transfer from magnet to coil.

or distracted. Humans can't do this very well at all by comparison. I think we should view computers as a tool to augment our human powers, but with us at the centre of things, rather than a potentially higher-functioning lifeform that may consider us fleshy beings obsolete in the near future. As the late Steve Jobs of Apple Computers has surely demonstrated a thousand-billion-fold in dollar units, placing human interaction at the very heart of the computing experience can also be a highly lucrative business.

1.2   The Style and Structure of *Science by Simulation* (Volume 2)

1.2.1   *This isn't really a textbook*

As per *A Mezze of Mathematical Models* [9], this book is not intended to be a coursebook of examinable syllabus content.[7] I have chosen a collection of example systems that I personally find interesting, which I hope provides suitable context and opportunity for creating rich mathematical models that can be readily evaluated in computer code to produce graphs, animations, coloured surfaces, etc. Many of the illustrations in this book have been created in this way. The intention is to inspire the reader to *create their own simulations* and actively learn physics in the process. Revision of previously encountered topics is not test preparation or postmortem, but instead multiple dives into the models and recipes of this book to help fix problems that will inevitably occur when writing code from scratch yourself. Perhaps, in a small way, this approach could be a counter to my fear that the goal of physics education is becoming solely the ability to regurgitate just enough information in order to pass standardised tests.[8] If we wish students to become useful as physicists when they graduate, we ought to coach them to experience what 'being useful' means, well in advance of their departure from school and university. I think it is to be able to create models, run them error-free in software and check their predictive power against experimental reality. Passing written examinations is, of course, a vital element of the learning process, but it should not be the only reason why we choose to study.

Although this book (and the next volume in the series) follows quite a traditional journey through the topics of physics, the content is by no means comprehensive. I would highly recommend a good textbook (or indeed several) to pair with each chapter. A list of my personal favourites is provided in the Bibliography, and the majority will be referenced as appropriate in the text.

---

[7]Although if you *do* make a course out of it, and you can convince yourselves, your students and the people who audit you that it is successful, do please get in contact!

[8]In UK secondary schools, we have (at the time of writing) an additional problem of multiple examination boards, all competing for students. This is surely a creeping mechanism for reducing content and difficulty over time, as the board which offers the greatest chance of a good grade is probably the one most schools will go for. It baffles me why something like an A-Level examination shouldn't be a national standard.

1.2.2 *Chapter synopsis for Models of Classical Physics*

(1) **Mechanics.** *Conservation of energy and momentum* ideas to describe the motion of bouncing balls, based on the surprisingly high-velocity recoil of a tennis ball dropped vertically on top of a basket ball. Can we extend this idea to create an (albeit eccentrically impractical) launch system for the transfer of material from the Earth to the Moon? Everything you wanted to know (but were perhaps too afraid of the mathematics to ask) about *projectile motion.* From the parabolic trajectory equation, to an elegant geometric solution to the maximum-range problem, to incorporating air resistance into a numerical scheme based upon Newton's second law: *mass $\times$ acceleration = vector sum of force.* A progression of classic *statics* problems (ladder, drawbridge, stack of drums), which require the application of force vector addition, moments, Newton's third law, friction models and, indeed, *geometric* as well as *algebraic* mathematical methods. Then, wobbling *weebles* (that won't fall down), the full range of *inclined plane scenarios* and *rolling and sliding* along linear and loop-the-loop tracks. The concluding example of a rolling sphere falling off a vertically circular track and landing precisely at the lowest point of the track is a rather beautiful hybrid of non-uniform rotational dynamics and projectile motion. A few examples of *Atwood machines* (pulley system curios) and then a recipe for the calculation of the *forces on the hinge* of a swinging door, which brings together pretty much all the techniques of rigid body dynamics about a fixed axis. To finish, *Lagrangian mechanics* to the rescue of a surprisingly tricky sliding block problem and then similar magic tricks involving the *Euler–Lagrange equation* to solve the *brachistochrone* and an asymmetric generalisation of the *catenary* curve, which describes the shape of a weighted chain held between two fixed points.

(2) **Thermodynamics.** *Ideal gases* and the relationships between pressure, temperature, volume, heat exchanged, work done and entropy change for *isothermal, isochoric, isobaric* and *isentropic* processes. These rules are then put together to develop models of *Carnot, Otto* and *Diesel heat engines* by considering the corresponding trajectories in pressure, volume (p, V) and absolute temperature, entropy (T, S) spaces. Finally, *Fourier's law of heat conduction* simplified as *Newton's law of cooling.* Then, a change of concept from a continuum/fluidic theory of heat and gases to some of the ideas of statistical thermodynamics. We begin with *kinetic theory, Knudsen's number* for ideal gases and learn coding recipes for *random walks* that visually represent our kinetic theory predictions. Finally, theoretical and computer simulation justifications for *Boltzmann's exponential statistics*, which we then apply to determine the *speed distribution of molecules in an ideal gas.*

(3) **Waves and ray optics.** A *mathematical anatomy of waves* involving the spatial and temporal variations of the phase of cosine functions. Music technology examples follow, involving *standing waves on guitar strings and in organ pipes.* We then move to optics, starting with *Fermat's minimum travel time principle* applied to derive the laws of *reflection* and *refraction.* Then, a coordinate

transformation interpretation of the behaviour of *thin lenses* and the really strange *real* images one can obtain using *concave spherical mirrors*. Next, a classic mathematical journey involving a laser-illuminated grating which produces (within certain geometrical limits) *Fraunhofer* and *Fresnel* diffraction patterns. We then go supersonic with *Mach's construction* and create a geometric visualisation of the *Doppler effect* and *shock fronts*. We will follow this with an algebraic derivation of the Doppler shift formula and the cosmologically useful concept of *redshift*. The penultimate models involve water waves and the *Kelvin wedge,* i.e. why ducks on a river, giant cruise ships and, indeed, an alien hydrocarbon ocean skiff would all produce the same wake geometry. To conclude, *simple harmonic motion* and *resonance* in a driven mass–spring system.

(4) **Electromagnetism.** *Resistors in series and parallel,* and how these can be combined to produce *golden circuits*, and indeed the *circuit of Deep Thought*. We then make excellent use of quadratic equations and basic calculus to derive the *maximum power theorem* and make sure our microphones and cables have a nicely matched impedance. We then revise electrical resistance essentials via an 'electric sausage' concept and discuss how this enables us to design cost-effective long-distance *power cables*. And then to magnetism! We start with the *Ising model* of *ferromagnetism* and the associated *Metropolis algorithm*, which can be used to model magnetisation effects using a computer program. We then look at *Ampère's theorem* and its application to *toroidal inductors*, and apply wave ideas to analyse oscillating currents in circuits. We will model the frequency response of *transformers* and consider the essentials of analogue radio by considering *resonance* in a circuit comprising an *inductor, a capacitor and a resistor*. The penultimate model is a detailed look at reflection and refraction at a boundary by deriving the *Fresnel equations of S and P electric and magnetic field polarisations* and also the *Brewster angle*. Finally, an introduction to *Maxwell's equations* and various general properties we can infer about electromagnetic waves: their speed, transverse nature and Poynting's relationship between energy flow and electric and magnetic field components.

(5) **Appendices** on the *calculus of variations* (*Euler–Lagrange equation, Beltrami identity* and *Lagrangian mechanics*) and a *general dispersion relationship* for waves on the interface of two fluid layers of different density, with the lower layer being a fixed depth above a solid surface.

1.3 Mathematics: The Language of Physics

The title of this short section is also the title of an excellent book called the *Language of Physics* [4], co-authored by one of my colleagues at Winchester, Dr. John Cullerne. The title of the book is essentially its thesis: if you wish to master physics and use it in any meaningful analytical way for quantitative prediction, you need to express your ideas mathematically. I will assume familiarity with mathematical notation common to most pre-university courses. Scalar quantities such as time t, energy E and charge Q will be represented by *italic* symbols, whereas vector quantities such as velocity $\mathbf{v}$ will be in **bold** font. I will use accent 'hats' to denote *unit vectors*. For example, a displacement vector $\mathbf{r}$ in a Cartesian coordinate x, y, z system (see Fig. 1.1) would be written as

$$\mathbf{r} = x\hat{\mathbf{x}} + y\hat{\mathbf{y}} + z\hat{\mathbf{z}}.$$

The set of unit vectors $\{\hat{\mathbf{x}}, \hat{\mathbf{y}}, \hat{\mathbf{z}}\}$ are all of unit length, indeed define the Cartesian axes set, and are mutually perpendicular.

To multiply vectors, I will use *dot products*, $\mathbf{a} \cdot \mathbf{b}$, or *cross products*, $\mathbf{a} \times \mathbf{b}$. The dot product $\mathbf{a} \cdot \mathbf{b} = |\mathbf{a}|\,|\mathbf{b}|\cos\theta$, where angle θ is the angle between the vectors $\mathbf{a}$ and $\mathbf{b}$, which have respective lengths $|\mathbf{a}|$ and $|\mathbf{b}|$. This means $\frac{\mathbf{a} \cdot \mathbf{b}}{|\mathbf{b}|}$ is the *projection* of $\mathbf{a}$ in the direction of $\mathbf{b}$. The cross product magnitude $|\mathbf{a} \times \mathbf{b}| = |\mathbf{a}|\,|\mathbf{b}|\sin\theta$, and the direction of $\mathbf{a} \times \mathbf{b}$ is *perpendicular to the plane which contains* $\mathbf{a}$ and $\mathbf{b}$. In Cartesian coordinates,

$$\mathbf{a} \times \mathbf{b} = \hat{\mathbf{x}}(a_y b_z - a_z b_y) + \hat{\mathbf{y}}(a_z b_x - a_x b_z) + \hat{\mathbf{z}}(a_x b_y - a_y b_x),$$

where $\mathbf{a} = a_x\hat{\mathbf{x}} + a_y\hat{\mathbf{y}} + a_z\hat{\mathbf{z}}$ and $\mathbf{b} = b_x\hat{\mathbf{x}} + b_y\hat{\mathbf{y}} + b_z\hat{\mathbf{z}}$.

The instantaneous gradient of a curve x versus t (i.e. evaluated at a particular value of t) is defined by a *derivative* $\frac{dx}{dt}$, and if the curve is smooth enough such that the derivative is continuous, a *second derivative* $\frac{d^2x}{dt^2}$ might also be defined. If t is time, then these derivatives can be written as $\dot{x} = \frac{dx}{dt}$ and $\ddot{x} = \frac{d^2x}{dt^2}$, i.e. *velocity* and *acceleration*.[9] The area under a curve $v(t)$ is given by the *integral* $\int v\,dt$, which is the *anti-derivative* of *integrand* v, evaluated between the starting and stopping limits of t, which define the area. The use of derivatives, i.e. *differentiation*, and its inverse process, *integration*, are the two aspects of the *calculus* subset of mathematics. This is a fundamental linguistic component of physics, simply because many of the key equations (e.g. Newton's second law: mass $\times$ acceleration = vector sum of force) involve a derivative.

In later sections of this book (e.g. in the section on *Maxwell's equations* and in discussions of the *calculus of variations* and *Lagrangian mechanics*), I will use *partial derivatives* and *vector calculus* definitions (*grad, div, curl*). A partial derivative, such as $\frac{\partial}{\partial x}\left(2x^2 y + z\right)$, treats all other variables in an expression as constants, apart from the variable at the 'bottom' of the derivative expression (x in this case). So, $\frac{\partial}{\partial x}\left(2x^2 y + z\right) = 4xy$. The *gradient* operator ∇ produces a *vector* which represents the direction toward which a slope is increasing. Consider a function $\phi(x, y)$ which represents the height of a surface $z = \phi(x, y)$ mapped over a Cartesian x, y horizontal grid. In this coordinate system,

$$\nabla\phi = \frac{\partial\phi}{\partial x}\hat{\mathbf{x}} + \frac{\partial\phi}{\partial y}\hat{\mathbf{y}}.$$

---

[9] ... in one dimension. **Bold font** the quantities to extend to vectors in two or three dimensions, e.g. $\dot{\mathbf{x}} = \frac{d\mathbf{x}}{dt}$.

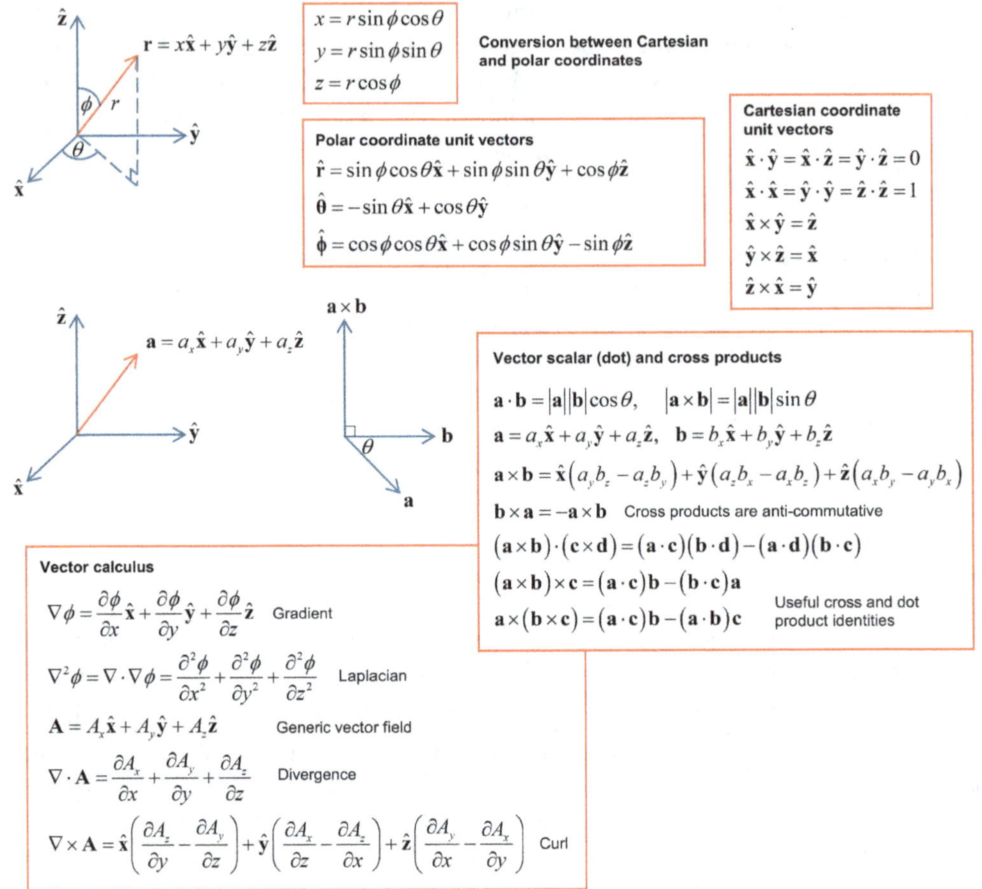

Fig. 1.1. The Cartesian $\{\hat{\mathbf{x}}, \hat{\mathbf{y}}, \hat{\mathbf{z}}\}$ set of unit vectors are all mutually perpendicular. We can use a Cartesian coordinate system to describe 2D (x, y) or 3D (x, y, z) trajectories, defined at an instant in time by a displacement vector (from the coordinate origin) $\mathbf{r} = x\hat{\mathbf{x}} + y\hat{\mathbf{y}} + z\hat{\mathbf{z}}$. In certain situations, such as rotational motion, it is often more convenient to use polar coordinates (r, θ, ϕ) rather than Cartesians. Dot, cross and vector calculus operations are used throughout this book to describe equations involving physical laws which involve 2D or 3D quantities, e.g. electric and magnetic fields.

If we have a *vector field* (i.e. a spatially varying grid of vectors), such as electric field strength $\mathbf{E}$, then *vector operators* such as $\nabla \cdot \mathbf{E}$, (the *divergence* or 'div') and $\nabla \times \mathbf{E}$ (the *curl*) can yield useful properties of the vector field. In the case of Maxwell's equations, vector calculus operators applied to electric and magnetic fields relate to their source quantities, i.e. charges and their movement.

Chapter 2

Mechanics

2.1 Beginning with Classical Mechanics

Classical mechanics is usually the starting point for most physics courses. There is a good reason for this. Fundamentally, our experimental understanding of mechanics begins when we are born and is refined as we learn to walk, learn to feed and dress ourselves, run, jump and fall over and have lots of fun with the dynamics of air-filled spheres of various sizes. Eventually, we learn to master wheels and pedals, although at some point, we may have to let hydrocarbon thermodynamics or electric motors do the hard work for us. A close study of the rules of motion, and indeed the conditions for equilibrium,[1] is a sensible first step in making sense of the world we experience. Unlike electricity, and certainly the quantum and special relativistic topics of modern physics, which can appear rather exotic and abstract and require specialised equipment to reveal their mysteries, mechanics is rooted in everyday experience. In my teaching, I like to begin a course for 13-year-olds with measurement and density concepts and then follow this with an introduction to mechanical work, kinetic energy and gravitational potential energy. For sixth-formers (16 to 17-year-olds), I always begin with *kinematics* (i.e. the relationship between *displacement*, *velocity* and *acceleration*), starting with one-dimensional motion and then extending a constant-acceleration special case to the two dimensions of projectile motion of objects in uniform gravitational fields. I then move on to forces and Newton's laws, develop energy ideas, use momentum conservation to determine the result of a collision and then apply all the tools experienced so far to objects that are connected in some way through surfaces, rods, chains, string etc. Things that stretch, compress and oscillate are investigated a few months later.

---

[1] For a rigid object to be in *mechanical equilibrium*, the vector sum of forces must be zero (i.e. the force vectors connect tip to tail to form a closed polygon) and the sum of turning moments (about any axis) must also be zero. Equilibrium means no acceleration of the centre of mass and no angular acceleration in a rotational sense. In other words, a rigid object need not be static when in equilibrium, its centre of mass could move with constant velocity and it could rotate with constant *angular momentum* $\Sigma_i m_i \mathbf{r}_i \times \mathbf{v}_i$, where $\mathbf{r}_i$ is a position vector of a point mass m_i component of a system, moving with velocity $\mathbf{v}_i$.

In most courses I have taught, circular motion (leading to rigid body dynamics) usually occurs about a term after linear mechanics. However, the concept of *no net turning moments about any axis in a rigid body being an additional condition for mechanical equilibrium* is often employed in the discussion of statics problems (i.e. blocks on rough inclined planes, ladders resting on walls, conditions for toppling instead of sliding etc.) and is therefore often sneaked into the first tranche of sixth-form mechanics. The study of rotational motion, certainly in the UK, is, in my opinion, a particularly shameful casualty resulting from a sustained erosion of mathematical content from A-Level courses over recent decades.[2] At the time of writing, a potentially top-grade student may only encounter basic circular motion, i.e. no angular momentum, moments of inertia, the concept of torque, rolling etc., even if they study *both* physics and further mathematics. I'm not sure if this is progress, and it places even more pressure on universities to make up the shortfall in knowledge. In my personal opinion, having a selection of examination boards in the UK is not helpful. If you are a college interested in maximising your number of A-Level A* grades, why opt for a harder and broader syllabus? And if you are the exam board, you are financially incentivised to make an attractive offer to as many schools as possible. Both of these are potentially drivers for a diminution in content. I'm all for competition in industry, but when it comes to standards, it makes no sense to have alternatives. Recall the bad old days before USB ports, and imagine the frustration of different plug sockets, voltage standards, or even *time* standards. I'm in no position to hold back the teach-to-the-test tide, but I do urge both students and teachers to consider what should really be fundamental physics knowledge. If the latter is not on your syllabus specification, teach it anyway, but make sure you have a bombproof counter to the death-of-learning phrase: 'Is this going to be on the test?' If you are a student and you don't have much of an opportunity to use calculus, torque = moment of inertia × angular acceleration, and you lack a suitably mystical appreciation of gyroscopes, then I urge you to teach yourself these things. This book won't offer you a complete story, but I hope it may be a start. And let me be entirely clear: the language will be mathematical.

2.2   Bouncing Balls

We will start with one-dimensional motion and discuss how using conservation of momentum and energy can help us make predictions. A simulation of the displacement and velocity of a bouncing ball is quite interesting, but dropping a tennis ball on top of a basket ball is delightful. You will make small children smile and mathematically curious older children (and hopefully some adults too) fascinated when you tell them that the upper ball has a fundamental limit of reaching *nine times the height from which it was dropped*. In practice, it is more like about a factor of three or four, but impressive nonetheless.

---

[2]You could probably say the same for *geometric optics*.

2.2.1 *Two balls dropped together*

Consider a pair of balls dropped from rest, one above the other (see (c) and (d) in Fig. 2.1). The drop distance h is that of the centre of mass of (either) ball. The upper ball has mass m, and the lower ball has mass M, such that $M > m$. Assuming that the strength of gravity g is constant over height h and that the balls don't move fast enough for air resistance to become important, they will hit a solid floor at speed u such that (by conservation of energy)

$$\tfrac{1}{2}Mu^2 = Mgh \tag{2.1}$$

and

$$\tfrac{1}{2}mu^2 = mgh.$$

Therefore (using either ball),

$$u^2 = 2gh. \tag{2.2}$$

Assume that the lower ball collides with the floor with *coefficient of restitution*[3] C_1. The recoil speed of the mass M is therefore C_1u.

This ball now collides with the mass m, which is moving downwards with speed u. By conservation of momentum (and taking upward velocity as positive),

$$MC_1u - mu = MV + mv, \tag{2.3}$$

where V is the resulting upward velocity of mass M and v is the (upward) recoil velocity of mass m. If the ball–ball collision has coefficient of restitution C_2,

$$C_2 = \frac{v - V}{C_1u + u}. \tag{2.4}$$

So, velocity V is

$$V = v - C_2(C_1u + u). \tag{2.5}$$

Substituting for V into the conservation of momentum equation $MC_1u - mu = MV + mv$,

$$C_1u - \frac{m}{M}u = v - C_2\left(C_1u + u\right) + \frac{m}{M}v \tag{2.6}$$

$$u\left(C_1 - \frac{m}{M} + C_2C_1 + C_2\right) = v\left(1 + \frac{m}{M}\right). \tag{2.7}$$

---

[3]*Coefficient of restitution* $C = \frac{\text{speed of separation}}{\text{speed of approach}}$. A *perfectly inelastic* collision corresponds to $C = 0$, i.e. the colliding objects stick together and hence their speed of separation is 0. In a *perfectly elastic* collision, the kinetic energy of the colliding objects is unchanged, which implies $C = 1$. Unless some form of explosion occurs which converts stored energy into extra kinetic energy, real-world collisions are modelled by C in the range $0 < C < 1$.

Therefore,

$$v = \frac{C_1 - \frac{m}{M} + C_2 C_1 + C_2}{1 + \frac{m}{M}} u. \tag{2.8}$$

If the recoiling ball of mass m subsequently rises to a height H metres, conservation of energy implies

$$\tfrac{1}{2}mv^2 = mgH.$$

$$\therefore H = \frac{v^2}{2g}. \tag{2.9}$$

Hence, since $h = \frac{u^2}{2g}$,

$$\frac{H}{h} = \frac{v^2}{u^2} = \left(\frac{C_1 - \frac{m}{M} + C_2 C_1 + C_2}{1 + \frac{m}{M}} \right)^2. \tag{2.10}$$

The most extreme case is when both ground and inter-ball collisions are both *perfectly elastic.* i.e. $C_1 = C_2 = 1$, so the height ratio

$$\frac{H}{h} < \left(\frac{3 - \frac{m}{M}}{1 + \frac{m}{M}} \right)^2. \tag{2.11}$$

The absolute limit for $\frac{H}{h}$ is when $M \gg m$, which means $\frac{m}{M} \to 0$. In this case,

$$\frac{H}{h} \to 9. \tag{2.12}$$

So, the upper ball can rise up to nine times the original height! This is nicely demonstrated by dropping a tennis ball and a basket ball together from between 50 cm and 1 m. If conducted inside a classroom, the tennis ball will probably hit the ceiling, i.e. $\frac{H}{h} \approx 4$. The inter-ball coefficient of restitution C_2 can be computed by fixing the basket ball (mass M) to a solid horizontal surface, like a laboratory bench, and then dropping the tennis ball onto it from a height of up to about $h = 1.00\,\text{m}$. To prevent the tennis ball recoiling at at angle, perhaps the tennis ball could be dropped inside a wide, transparent tube. If the tennis ball recoils and reaches height h', by conservation of energy,

$$mgh' = \tfrac{1}{2}m \left(C_2 u \right)^2, \tag{2.13}$$

$$mgh = \tfrac{1}{2}mu^2. \tag{2.14}$$

Therefore,

$$C_2 = \sqrt{\frac{h'}{h}}. \tag{2.15}$$

This measurement will need plenty of repeats. For a more accurate measure of h', the ball could be filmed in slow motion. 60 frames per second (or higher) is now typical for most smartphones.

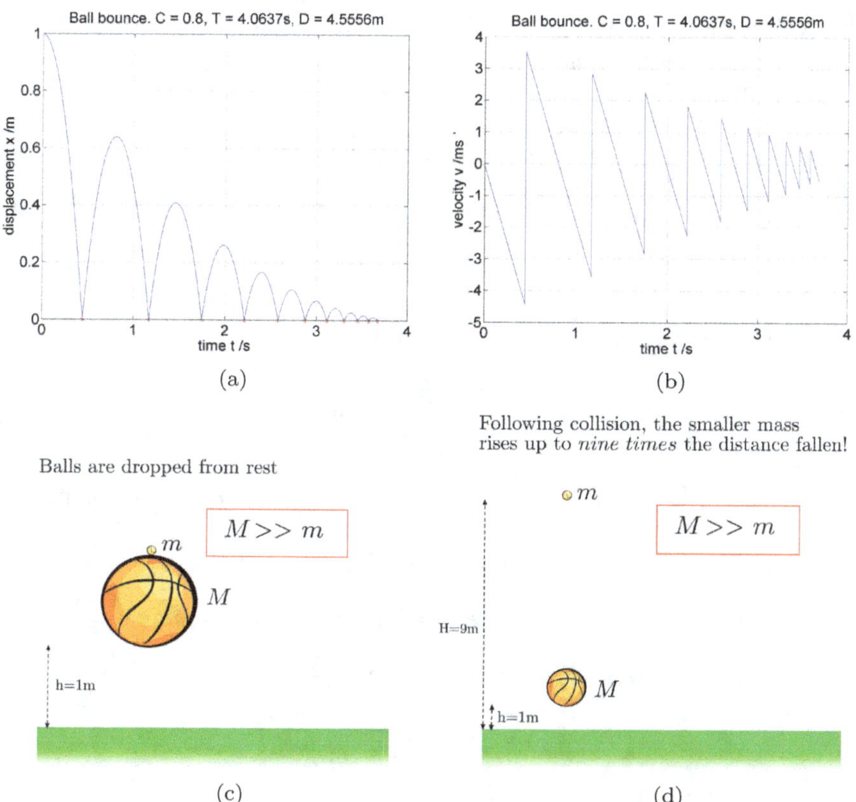

Fig. 2.1. (a) and (b) model of the vertical motion of a bouncing ball dropped from rest from 1.00 m. Air resistance is ignored. $g = 9.81$ ms^{-2}, and the coefficient of restitution $C = \frac{\text{speed of separation}}{\text{speed of approach}} = 0.8$. The displacement vs. time curves in (a) are inverted parabolas due to constant acceleration motion when the ball is in the air. This is even more evident in (b), which tracks the *upward* velocity vs. time. The time between bounces decreases geometrically with powers of C, leading to a finite bounce time of $T = 4.06$ s and an overall distance travelled of $D = 4.56$ m. (c) and (d) illustrate the limiting case of a *double ball bounce* in which all collisions are perfectly elastic. If the ball-to-ball coefficient of restitution is C_2 and lower ball to ground is C_1, then the apogee to original height ratio is $\frac{H}{h} = \left(\frac{C_1 - \frac{m}{M} + C_2 C_1 + C_2}{1 + \frac{m}{M}} \right)^2$. In the limit when $M \gg m$ and $C_1 = C_2 = 1$, then $\frac{H}{h} \to 9$.

To determine C_1, the easiest way of doing this is to drop the basket ball from height h and time how long it takes to stop bouncing.[4] The displacement and velocity versus time graphs for such a vertical ball bounce are illustrated in (a) and (b) of Fig. 2.1. The time to drop height h is given by

$$h = \tfrac{1}{2}gt_1^2. \tag{2.16}$$

$$\therefore t_1 = \sqrt{\frac{2h}{g}}.$$

---

[4]This method could also be applied to find the ball-ball coefficient of restitution, C_2 as long as a method of preventing the tennis ball from rolling off is implemented!

The recoil speed is $C_1\sqrt{2gh}$, so the ball will then rise to height h_1. We can calculate what this is by equating the gravitational potential energy (GPE) gain with the initial kinetic energy (KE):

$$mgh_1 = \tfrac{1}{2}m(C_1\sqrt{2gh})^2, \tag{2.17}$$

i.e. h_1 in terms of h and C_1 is

$$h_1 = C_1^2 h. \tag{2.18}$$

So, the time between the first and second bounces will be

$$\Delta t_{12} = 2 \times \sqrt{\frac{2h_1}{g}} = 2C_1 t_1. \tag{2.19}$$

Generalising to the time interval Δt_n between bounce $n-1$ and bounce n,

$$\Delta t_n = 2C_1^{n-1} t_1. \tag{2.20}$$

Therefore, the total bouncing time t is a *geometric progression*[5]:

$$t = t_1(1 + 2C_1 + 2C_1^2 + 2C_1^3 + \cdots). \tag{2.21}$$

$$\therefore \tfrac{1}{2}\left(\frac{t}{t_1} + 1\right) = 1 + C_1 + C_1^2 + C_1^3 + \cdots \tag{2.22}$$

$$\therefore \tfrac{1}{2}\left(\frac{t}{t_1} + 1\right) = \frac{1}{1 - C_1}, \tag{2.23}$$

which we can evaluate as $n \to \infty$ since $C_1 < 1$ (i.e. it is physically improbable to be exactly unity).

Hence, since $t_1 = \sqrt{\frac{2h}{g}}$, we can calculate C_1 using the formula

$$C_1 = 1 - \frac{1}{\tfrac{1}{2}\left(\frac{t}{t_1} + 1\right)}. \tag{2.24}$$

$$\therefore C_1 = 1 - \frac{2}{\sqrt{\frac{g}{2h}}t + 1}. \tag{2.25}$$

Let's use some sensible numbers. If $t = 4.06\,\text{s}$ and $h = 1.00\,\text{m}$, then

$$C_1 = 1 - \frac{2}{\sqrt{\frac{9.81}{2 \times 1.00}} \times 4.06 + 1}, \tag{2.26}$$

$$C_1 \approx 0.800. \tag{2.27}$$

---

[5]The sum of a *geometric progression* has a closed form expression, $\sum_{n=1}^{N} ar^{n-1} = a\frac{1-r^N}{1-r}$. If $|r| < 1$, then $\sum_{n=1}^{\infty} ar^{n-1} = \frac{a}{1-r}$.

2.2.2 *Ball-bounce moonshot*

A rather silly but intriguing extension of the problem is to consider how many balls (N) are required such that the top ball (of mass m_N) *can escape the Earth*. The ball labelled $n = 1$ is the bottom ball. Assume, in this case, that all collisions are elastic, and the mass m_n of each ball is k times the mass of the ball above it, m_{n+1}.

$$\therefore m_{n+1} = \frac{m_n}{k}. \tag{2.28}$$

To conserve the momentum of mass m_n moving upwards with speed v_n and mass m_{n+1} moving downwards with speed u,

$$m_n v_n - m_{n+1} u = m_n V + m_{n+1} v_{n+1}. \tag{2.29}$$

By restitution, since elastic collisions are assumed,

$$\frac{v_{n+1} - V}{v_n + u} = 1. \tag{2.30}$$

Therefore, the recoil speed V of mass m_n is

$$V = v_{n+1} - v_n - u. \tag{2.31}$$

Substitution for m_{n+1} and V in the conservation of momentum expression yields a recurrence relation for v_n:

$$m_n v_n - \frac{m_n}{k} u = m_n \left(v_{n+1} - v_n - u \right) + \frac{m_n}{k} v_{n+1} \tag{2.32}$$

$$v_{n+1} = \frac{2 v_n + \left(1 - \frac{1}{k} \right) u}{1 + \frac{1}{k}} \tag{2.33}$$

$$v_{n+1} = \frac{2k}{k+1} v_n + \frac{k-1}{k+1} u \tag{2.34}$$

$$v_{n+1} = a v_n + b. \tag{2.35}$$

Now,

$$v_1 = u, \tag{2.36}$$

$$v_2 = au + b, \tag{2.37}$$

$$v_3 = a(au + b) + b = a^2 u + ba + b, \tag{2.38}$$

$$v_4 = a(a^2 u + ba + b) + b = a^3 u + ba^2 + ba + b, \tag{2.39}$$

$$v_5 = a(a^3 u + ba^2 + ba + b) + b = a^4 u + ba^3 + ba^2 + ba + b. \tag{2.40}$$

Therefore,

$$v_N = a^{N-1} u + b \sum_{i=0}^{N-2} a^i \tag{2.41}$$

$$v_N = a^{N-1}u + b\frac{a^{N-1} - 1}{a - 1},$$ (2.42)

where we have again used the formula for the sum of a geometric progression in the final step.[6] Substituting for a and b,

$$\frac{v_N}{u} = \left(\frac{2k}{k+1}\right)^{N-1} + \frac{k-1}{k+1}\frac{\left(\frac{2k}{k+1}\right)^{N-1} - 1}{\frac{2k}{k+1} - 1}$$ (2.43)

$$\frac{v_N}{u} = \left(\frac{2k}{k+1}\right)^{N-1} + \left(\frac{2k}{k+1}\right)^{N-1} - 1$$ (2.44)

$$\frac{v_N}{u} = 2\left(\frac{2k}{k+1}\right)^{N-1} - 1.$$ (2.45)

Given that we have just performed quite a bit of algebra, let's now run a quick sanity check that our general result matches the double-ball bounce system, i.e. set $N = 2$ and $k \to \infty$.

$$\therefore \frac{v_2}{u} = 2\left(\frac{2}{1 + \frac{1}{k}}\right) - 1.$$ (2.46)

So, if $k \to \infty$,

$$\frac{v_2}{u} \to 3,$$ (2.47)

which is consistent with our previous discussion.

Now, to escape the Earth, the launch kinetic energy must exceed the work done[7] (against gravity) to move a mass m from the surface of the Earth (of mass $M_\oplus$ and radius $R_\oplus$) to an infinite distance away. Hence,

$$\tfrac{1}{2}mv_N^2 > \frac{GM_\oplus m}{R_\oplus},$$ (2.48)

i.e. the launch speed is subject to the inequality

$$v_N > \sqrt{\frac{2GM_\oplus}{R_\oplus}}.$$ (2.49)

Earth's mass and radius are $M_\oplus = 5.97 \times 10^{24}$ kg and $R_\oplus = 6.38 \times 10^6$ m, respectively. The *universal gravitational constant* is $G = 6.67 \times 10^{-11}$ Nm^{2}kg^{-2}. Therefore, the *escape velocity* $\sqrt{\frac{2GM_\oplus}{R_\oplus}}$ of the Earth is about 11.2 km/s.

---

[6] For a geometric progression, $\sum_{n=1}^{N} a^{n-1} = \frac{1-a^N}{1-a}$. Let term index $n \to i+1$. $\therefore \sum_{n=1}^{N} a^{n-1} \to \sum_{i=0}^{N-1} a^i = \frac{1-a^N}{1-a}$, which implies $b\sum_{i=0}^{N-2} a^i = b\frac{a^{N-1}-1}{a-1}$, as stated.

[7] $-\frac{GMm}{R}$ is the *gravitational potential energy* (GPE) of a mass m at distance R from the centre of another mass M. The work done to take m from R to ∞ is therefore $\frac{GMm}{R}$.

If the masses are dropped from height h, then conservation of energy implies $u = \sqrt{2gh}$. Therefore, if $\frac{v_N}{u} = 2\left(\frac{2k}{k+1}\right)^{N-1} - 1$ and $v_N > \sqrt{\frac{2GM_\oplus}{R_\oplus}}$ to escape,

$$\sqrt{2gh}\left(2\left(\frac{2k}{k+1}\right)^{N-1} - 1\right) > \sqrt{\frac{2GM_\oplus}{R_\oplus}}. \tag{2.50}$$

$$\therefore \log 2 + (N-1)\log\left(\frac{2k}{k+1}\right) > \log\left(\sqrt{\frac{GM_\oplus}{ghR_\oplus}} + 1\right).$$

Hence,

$$N > \frac{\log\left(\sqrt{\frac{GM_\oplus}{ghR_\oplus}} + 1\right) - \log 2}{\log\left(\frac{2k}{k+1}\right)} + 1. \tag{2.51}$$

Now, the gravitational field strength at the surface of the Earth is

$$g = \frac{GM_\oplus}{R_\oplus^2}. \tag{2.52}$$

Therefore,

$$\frac{GM_\oplus}{ghR_\oplus} = \frac{GM_\oplus R_\oplus}{ghR_\oplus^2} = \frac{gR_\oplus}{gh} = \frac{R_\oplus}{h}. \tag{2.53}$$

Hence,

$$N > \frac{\log\left(\sqrt{\frac{R_\oplus}{h}} + 1\right) - \log 2}{\log\left(\frac{2k}{k+1}\right)} + 1. \tag{2.54}$$

Let the mass multiplier be $k = 2$ and the initial drop height be $h = 1.00\,\text{m}$. Therefore, in order for the top ball to escape the Earth, the number of balls N must be subject to the inequality

$$N > \frac{\log\left(\sqrt{6.38 \times 10^6} + 1\right) - \log 2}{\log\left(\frac{4}{3}\right)} + 1, \tag{2.55}$$

$$N > 25.8. \tag{2.56}$$

So, if you have an elastic ball tower of $N = 26$ balls, then the top ball should be able to reach the moon! Is this a practical lunar taxi service? Well, to launch a top mass $m_{26} = 1,000$ kg, the bottom mass $m_1 = 2^{25} \times 1,000\,\text{kg} = 3.355 \times 10^7 \times 1,000\,\text{kg}$. To lift this (and the other balls) by a metre would probably require far more energy than launching a rocket in the conventional way. The assumption of elastic collisions, particularly with the ground, given this mass, is also going to be somewhat flawed. Furthermore, the initial acceleration of the top ball would almost certainly be harmful to the occupants of the space capsule! If the balls were made from iron of density $\rho = 7{,}870\,\mathrm{kgm}^{-3}$, then this means the lower ball would have radius r, where $\frac{4}{3}\pi r^3 \rho = m_1$.

$$\therefore r = \sqrt[3]{\frac{3m_1}{4\pi\rho}}. \tag{2.57}$$

$$\therefore r = \sqrt[3]{\frac{3 \times 3.355 \times 10^7 \times 1{,}000}{4\pi \times 7{,}870}} = 100.6\,\mathrm{m}, \tag{2.58}$$

so the moon-launch contraption is going to be quite a megastructure.

2.3   Projectiles and Constant Acceleration Motion

2.3.1   *Equations for position and velocity vs. time*

Projectiles are typically modelled as a point mass m (i.e. a 'particle') falling under gravity of fixed strength $g\,\mathrm{Nkg}^{-1}$, as illustrated in Fig. 2.2. In other words, internal motion and rotation are ignored, and only the centre of mass of the projectile is considered. Air resistance is often ignored to enable analysis to proceed

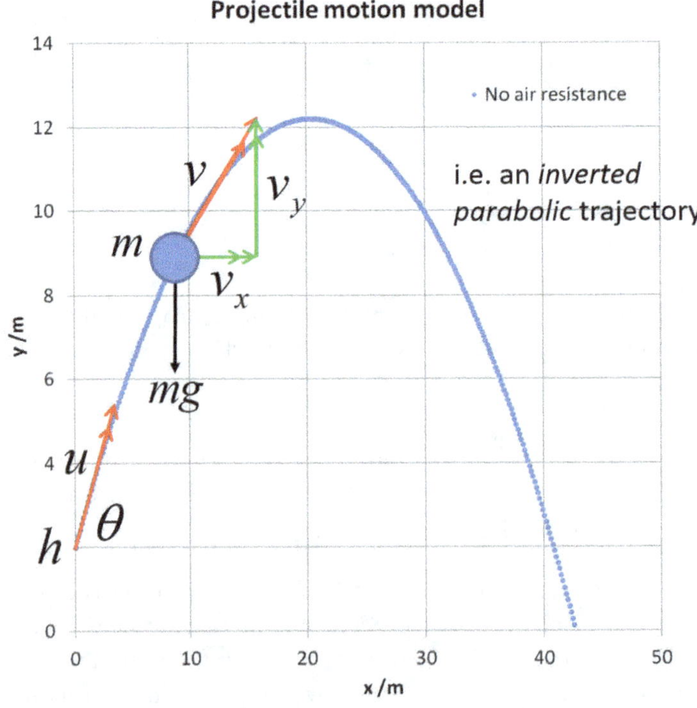

Fig. 2.2. A projectile of mass m is launched at speed u from $(0, h)$ at angle θ from the horizontal. The projectile feels a constant gravitational force mg acting downwards, and we shall (at least for the moment) ignore the effect of air resistance. The velocity, at a tangent to the *parabolic* trajectory, has magnitude $v = \sqrt{v_x^2 + v_y^2}$, where v_x and v_y are the respective velocity components in the x and y directions.

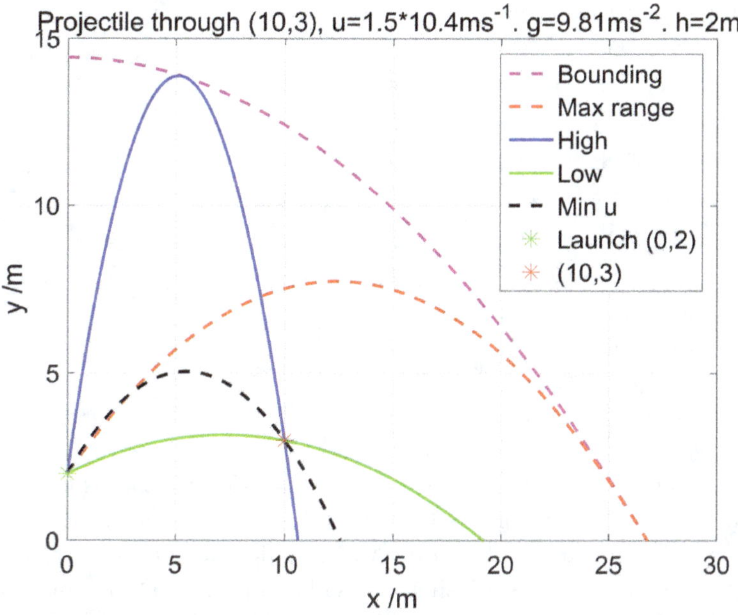

Fig. 2.3. Drag-free projectile motion using $g = 9.81$ ms^{-2}. A particle is projected from $(0, 2)$ with speed $u = 15.6$ ms^{-1}. The blue and green inverted parabolas pass through a defined target $(10, 3)$, using (respective) elevation angles of $\theta = 77.9°$ and $\theta = 17.8°$ from the horizontal. The black curve corresponds to the minimum u $(10.4$ ms$^{-1})$ to pass through $(10, 3)$. The red curve results in a maximum range of $(26.8$ m$)$, following a launch elevation of $\theta = 42.9°$. The magenta curve is the *bounding parabola* for launch speed $u = 15.6$ ms^{-1}. No (x, y) targets could be reached for y values in excess of the bounding parabola, regardless of the elevation θ chosen.

without any numerical iteration and approximation. Note that this assumption may be significantly invalid for many real projectiles, and a numerical recipe will be described at the end of this section. Figure 2.3 describes the various inverted parabolic trajectories that are associated with projectile motion, which we shall derive. Calculations particular to Fig. 2.3 are as follows:

Target X, Y coordinates in metres $= (10,3)$
$g = 9.81$ ms^{-2}
Launch height $h = 2$ m
Minimum launch speed $u_{\min} = 10.4$ m/s
Launch speed $u = 1.5 \times 10.4$ m/s $= 15.6$ m/s

Maximum range $R_{\max} = 26.8$ m
Maximum range elevation $= 42.9°$
Maximum range time of flight $= 2.34$ s
Maximum range y apogee $= 7.75$ m
Maximum range x apogee $= 12.4$ m

Minimum u range $R = 12.6$ m
Minimum u elevation $= 47.9°$
Minimum u time of flight $= 1.8$ s

Minimum uy apogee $= 5.04\,\mathrm{m}$
Minimum ux apogee $= 5.5\,\mathrm{m}$

High ball range $R = 10.6\,\mathrm{m}$
High ball elevation $= 77.9°$
High ball time of flight $= 3.24$ s
High ball y apogee $= 13.9\,\mathrm{m}$
High ball x apogee $= 5.11\,\mathrm{m}$

Low ball range $R = 19.2\,\mathrm{m}$
Low ball elevation $= 17.8°$
Low ball time of flight $= 1.29$ s
Low ball y apogee $= 3.17\,\mathrm{m}$
Low ball x apogee $= 7.25\,\mathrm{m}$

The idea, which is a general theme of *Science by Simulation*, is that one should use the mathematical recipes presented here to construct your own computer code to perform calculations and make plots. All the different parabolic trajectories discussed are plotted automatically using a MATLAB program, and this also exports a text file with the information above. The great thing about this is not only the speed and precision that the use of a computer program provides, it is also the idea that you can program the computer to give you *exactly* the information you want, in a way that is optimal to aid *your* understanding. From a teaching and learning perspective, coding up the projectile problem in this way is a very efficient way to understanding *everything* about projectile motion. Your computer model effectively represents almost every problem that could be set, and coding up the simulation is effectively an exploration of this scenario space.

The projectile motion system reduces to a two-dimensional kinematics problem, where acceleration $\mathbf{a} = -g\hat{\mathbf{y}}$ is constant. Let the coordinates of the projectile be $\mathbf{r} = x\hat{\mathbf{x}} + y\hat{\mathbf{x}}$ on a Cartesian grid at time t since launch. The velocity is $\mathbf{v} = v_x\hat{\mathbf{x}} + v_y\hat{\mathbf{y}}$, where $\hat{\mathbf{x}}, \hat{\mathbf{y}}$ are unit vectors in the Cartesian x, y directions. Let the initial velocity be u at an elevation of θ, and let the projectile be launched from $(0, h)$.

Since acceleration is constant,

$$v_x = u\cos\theta, \tag{2.59}$$

$$v_y = u\sin\theta - gt, \tag{2.60}$$

$$x = ut\cos\theta, \tag{2.61}$$

$$y = h + ut\sin\theta - \tfrac{1}{2}gt^2. \tag{2.62}$$

By conservation of energy,

$$\tfrac{1}{2}mu^2 + mgh = \tfrac{1}{2}m\left(v_x^2 + v_y^2\right) + mgy \tag{2.63}$$

$$u^2 + 2gh = v_y^2 + u^2\cos^2\theta + 2gy \tag{2.64}$$

$$u^2\left(1 - \cos^2\theta\right) - 2g(y - h) = v_y^2. \tag{2.65}$$

Therefore,

$$v_y^2 = u^2 \sin^2 \theta - 2g(y - h).$$ (2.66)

The first line of the conservation of energy equation also allows us to directly compute the projectile speed v:

$$u^2 + 2gh = v^2 + 2gy.$$ (2.67)

$$\therefore v = \sqrt{u^2 - 2g(y - h)}.$$ (2.68)

The angle ϕ of the velocity with respect to the x-axis can be found by setting

$$\mathbf{v} = v_x\hat{\mathbf{x}} + v_y\hat{\mathbf{x}} = v\left(\cos\phi\hat{\mathbf{x}} + \sin\phi\hat{\mathbf{y}}\right).$$ (2.69)

Hence,

$$\phi = \tan^{-1}\left(\frac{v_y}{v_x}\right) = \tan^{-1}\left(\frac{u\sin\theta - gt}{u\cos\theta}\right) = \tan^{-1}\left(\tan\theta - \frac{gt}{u}\sec\theta\right).$$ (2.70)

2.3.2 *Apogee*

An immediate result is to determine the time t_a and coordinates (x_a, y_a) of the highest point on the trajectory. This occurs when $v_y = 0$, and this is called the *apogee*:

$$t_a = \frac{u\sin\theta}{g},$$ (2.71)

$$x_a = \frac{u^2\sin\theta\cos\theta}{g} = \frac{u^2\sin 2\theta}{2g},$$ (2.72)

$$y_a = h + \frac{u^2}{g}\sin^2\theta - \frac{1}{2}g\frac{u^2}{g^2}\sin^2\theta = h + \frac{u^2}{2g}\sin^2\theta.$$ (2.73)

2.3.3 *Time and height relationships*

An elegant set of relationships can be formed by considering the times t_1 and t_2, such that $t_2 > t_1$, which correspond to a projectile being at height H, where $h < H < y_a$. Using the expression for $y(t)$,

$$H = h + ut_1\sin\theta - \tfrac{1}{2}gt_1^2,$$ (2.74)

$$H = h + ut_2\sin\theta - \tfrac{1}{2}gt_2^2.$$ (2.75)

Adding these equations,

$$2H = 2h + u\left(t_1 + t_2\right)\sin\theta - \tfrac{1}{2}g\left(t_1^2 + t_2^2\right).$$ (2.76)

We can also find t_1 and t_2 by solving $H = h + ut\sin\theta - \frac{1}{2}gt^2$ for t:

$$t^2 - \frac{2ut}{g}\sin\theta + \frac{2}{g}(H-h) = 0 \tag{2.77}$$

$$\left(t - \frac{u}{g}\sin\theta\right)^2 - \frac{u^2}{g^2}\sin^2\theta + \frac{2}{g}(H-h) = 0. \tag{2.78}$$

$$\therefore t_1 = \frac{u}{g}\sin\theta - \sqrt{\frac{u^2}{g^2}\sin^2\theta - \frac{2}{g}(H-h)}. \tag{2.79}$$

$$\therefore t_2 = \frac{u}{g}\sin\theta + \sqrt{\frac{u^2}{g^2}\sin^2\theta - \frac{2}{g}(H-h)}. \tag{2.80}$$

By adding these expressions for t_1 and t_2,

$$t_1 + t_2 = \frac{2u}{g}\sin\theta. \tag{2.81}$$

Hence,

$$u = \frac{g}{2\sin\theta}(t_1 + t_2). \tag{2.82}$$

Substituting $u = \frac{g}{2\sin\theta}(t_1 + t_2)$ into $2H = 2h + u(t_1 + t_2)\sin\theta - \frac{1}{2}g(t_1^2 + t_2^2)$,

$$2H = 2h + \frac{g}{2\sin\theta}(t_1 + t_2)(t_1 + t_2)\sin\theta - \frac{1}{2}g(t_1^2 + t_2^2) \tag{2.83}$$

$$\frac{4(H-h)}{g} = (t_1 + t_2)(t_1 + t_2) - t_1^2 - t_2^2 \tag{2.84}$$

$$\frac{4(H-h)}{g} = t_1^2 + t_2^2 + 2t_1t_2 - t_1^2 - t_2^2 \tag{2.85}$$

$$\frac{4(H-h)}{g} = 2t_1t_2. \tag{2.86}$$

$$\therefore H = h + \frac{1}{2}gt_1t_2. \tag{2.87}$$

The last result provides a neat mechanism for measuring g experimentally since the equation

$$g = \frac{2(H-h)}{t_1t_2} \tag{2.88}$$

neither depends on the launch speed u nor the launch angle θ. The idea is to launch a projectile from beneath two horizontal laser beams $H - h$ apart. A timing system is triggered when the laser beam comprising the *lower* light gate is cut. The time t_1 later is when the second (higher) light gate is cut for the *first* time. The time t_2

corresponds to when the higher light gate is cut for the *second* time, i.e. on the way down.

2.3.4 *The inverted parabola trajectory equation*

The constant x-direction velocity v_x is a key feature of drag-free projectile motion, and the resulting direct proportion between x and t is a very useful relationship:

$$t = \frac{x}{u\cos\theta}. \tag{2.89}$$

Substituting for t in $y = h + ut\sin\theta - \frac{1}{2}gt^2$ yields the *trajectory equation* $y(x)$, which is an *inverted parabola* (or 'unhappy mathematician'):

$$y = h + \frac{xu\sin\theta}{u\cos\theta} - \frac{1}{2}\frac{g}{u^2}x^2\frac{1}{\cos^2\theta}. \tag{2.90}$$

Using trigonometric identities $\tan\theta = \frac{\sin\theta}{\cos\theta}$ and $\frac{1}{\cos^2\theta} = 1 + \tan^2\theta$, we can write the trajectory equation as

$$y = h + x\tan\theta - \frac{1}{2}\frac{g}{u^2}x^2(1 + \tan^2\theta). \tag{2.91}$$

2.3.5 *Find the angle(s) to pass through (X, Y) given u*

If a projectile is required to pass through (or collide with) a particular coordinate (X, Y), we can solve the trajectory equation to determine the elevation angle θ, given that speed u is known. This calculation underpins the models of all ball sports and, indeed, rather more serious problems of gunnery, ballistics etc.,

$$Y = h + X\tan\theta - \frac{1}{2}\frac{g}{u^2}X^2(1 + \tan^2\theta) \tag{2.92}$$

$$\frac{gX^2}{2u^2}\tan^2\theta - X\tan\theta + Y - h + \frac{gX^2}{2u^2} = 0 \tag{2.93}$$

$$\tan^2\theta - \frac{2u^2}{gX}\tan\theta + \frac{2u^2}{gX^2}(Y - h) + 1 = 0 \tag{2.94}$$

$$\left(\tan\theta - \frac{u^2}{gX}\right)^2 - \frac{u^4}{g^2X^2} + \frac{2u^2}{gX^2}(Y - h) + 1 = 0 \tag{2.95}$$

$$\left(\tan\theta - \frac{u^2}{gX}\right)^2 - \frac{u^4}{g^2X^2}\left(1 - \frac{g^2X^2}{u^4}\frac{2u^2}{gX^2}(Y - h) - \frac{g^2X^2}{u^4}\right) = 0 \tag{2.96}$$

$$\left(\tan\theta - \frac{u^2}{gX}\right)^2 - \frac{u^4}{g^2X^2}\left(1 - \frac{2g}{u^2}(Y - h) - \frac{g^2X^2}{u^4}\right) = 0. \tag{2.97}$$

$$\therefore \tan\theta = \frac{u^2}{gX}\left(1 \pm \sqrt{1 - \frac{2g}{u^2}(Y - h) - \frac{g^2X^2}{u^4}}\right). \tag{2.98}$$

So, in general, there are three scenarios, depending on whether the quantity $\Delta = 1 - \frac{2g}{u^2}(Y - h) - \frac{g^2 X^2}{u^4}$ is negative, zero or positive. $\Delta < 0$ means no possible trajectories, i.e. u is not sufficient to reach (X, Y) regardless of the launch angle θ. $\Delta = 0$ implies a single *minimum u parabola*. If $\Delta > 0$, this implies two possible trajectories that will pass through (X, Y). Let's consider the latter case first. The $\pm$ in the expression for $\tan\theta$ means *two* possible trajectories, with two different values of θ. The time of flight from $(0, h)$ to (X, Y) in each case is

$$t = \frac{X}{u \cos\theta}. \tag{2.99}$$

'High ball', longer time of flight:

$$\tan\theta_+ = \frac{u^2}{gX}\left(1 + \sqrt{1 - \frac{2g}{u^2}(Y - h) - \frac{g^2 X^2}{u^4}}\right). \tag{2.100}$$

'More direct', shorter time of flight:

$$\tan\theta_- = \frac{u^2}{gX}\left(1 - \sqrt{1 - \frac{2g}{u^2}(Y - h) - \frac{g^2 X^2}{u^4}}\right). \tag{2.101}$$

For the two possible trajectories to correspond to a real value of θ, the square-rooted quantity $1 - \frac{2g}{u^2}(Y - h) - \frac{g^2 X^2}{u^4}$ must be positive (or zero), i.e.

$$1 - \frac{2g}{u^2}(Y - h) - \frac{g^2 X^2}{u^4} \geq 0 \tag{2.102}$$

$$u^4 - 2g(Y - h)u^2 - g^2 X^2 \geq 0 \tag{2.103}$$

$$\left(u^2 - g(Y - h)\right)^2 - g^2(Y - h)^2 - g^2 X^2 \geq 0. \tag{2.104}$$

Since u^2 is positive, we can take the positive root

$$u^2 \geq g(Y - h) + g\sqrt{(Y - h)^2 + X^2}. \tag{2.105}$$

Therefore, in order for a trajectory to pass though (X, Y),

$$u \geq \sqrt{g}\sqrt{(Y - h) + \sqrt{(Y - h)^2 + X^2}}. \tag{2.106}$$

2.3.6   *Minimum u parabola*

A single possible trajectory, i.e. θ can only be one value, occurs when $\tan\theta = \frac{u^2}{gX}$. This is for the minimum possible value of u, which we showed in the previous subsection to be

$$u = \sqrt{g}\sqrt{(Y-h) + \sqrt{(Y-h)^2 + X^2}}. \tag{2.107}$$

Since $\tan\theta = \frac{u^2}{gX}$,

$$\tan\theta = \frac{(Y-h) + \sqrt{(Y-h)^2 + X^2}}{X}. \tag{2.108}$$

Now, since

$$\frac{g}{u^2} = \frac{1}{X\tan\theta} = \frac{1}{(Y-h) + \sqrt{(Y-h)^2 + X^2}}, \tag{2.109}$$

it means that the minimum u trajectory $y = h + x\tan\theta - \frac{g}{2u^2}x^2\left(1 + \tan^2\theta\right)$ is *independent* of both u and g. However, the required minimum speed itself $u = \sqrt{g}\sqrt{(Y-h) + \sqrt{(Y-h)^2 + X^2}}$ is, of course, *not* independent of g, as it is $\propto \sqrt{g}$.

The minimum u parabola, defined in terms of X, Y only, and illustrated in Fig. 2.3, is

$$y = x\left(\frac{Y + \sqrt{X^2 + Y^2}}{X}\right) - \frac{\sqrt{X^2 + Y^2}}{X^2}x^2. \tag{2.110}$$

2.3.7 *Bounding parabola*

A slightly different perspective on the projectile scenario is the idea of the *bounding parabola*, i.e. the set of X, Y positions reachable given a value of u. Let's start with

$$u^4 - 2g\left(Y-h\right)u^2 - g^2X^2 \geq 0, \tag{2.111}$$

which satisfies real values of $\tan\theta$, for the possible trajectories through (X, Y). Rearranging this inequality,

$$u^4 + 2ghu^2 - g^2X^2 \geq 2gYu^2 \tag{2.112}$$

$$\frac{1}{2}\frac{u^2}{g} + h - \frac{1}{2}\frac{gX^2}{u^2} \geq Y. \tag{2.113}$$

So, the equation

$$y = h + \frac{u^2}{2g} - \frac{1}{2}\frac{g}{u^2}x^2 \tag{2.114}$$

defines the bounding parabola. Beyond this, no coordinates (X, Y) can be reached given the launch speed u. Note that when $y = 0$ and $x > 0$,

$$x = \sqrt{\frac{2u^2}{g}\left(h + \frac{u^2}{2g}\right)}. \tag{2.115}$$

2.3.8   *The maximum range problem*

Define $x = R$ when $y = 0$ for a projectile trajectory. First, let's calculate the time of flight t from the expression for $y(t)$:

$$h + ut\sin\theta - \frac{1}{2}gt^2 = 0 \tag{2.116}$$

$$t^2 - \frac{2u\sin\theta}{g}t - \frac{2h}{g} = 0 \tag{2.117}$$

$$\left(t - \frac{u\sin\theta}{g}\right)^2 - \frac{u^2\sin^2\theta}{g^2} - \frac{2h}{g} = 0. \tag{2.118}$$

$$\therefore t = \frac{u\sin\theta}{g} + \frac{u}{g}\sqrt{\sin^2\theta + \frac{2gh}{u^2}}. \tag{2.119}$$

Note that, in the last step, we must use the positive root since $t > 0$.
Now, since

$$R = ut\cos\theta, \tag{2.120}$$

the range R is

$$R = \frac{u^2}{g}\left(\sin\theta\cos\theta + \cos\theta\sqrt{\sin^2\theta + \frac{2gh}{u^2}}\right). \tag{2.121}$$

In many practical problems involving projectiles, it is useful to find the maximum possible R given a fixed u. In a gunnery or athletics scenario, the key goal is to calculate what elevation angle θ maximises R. This can be achieved by solving $\frac{dR}{d\theta} = 0$. The expression above is a little unwieldy, so let us first simplify it by looking at the special case of $h = 0$, i.e. the projectile is launched from ground level at $(0, 0)$:

$$R = \frac{2u^2}{g}\sin\theta\cos\theta = \frac{u^2}{g}\sin 2\theta, \tag{2.122}$$

$$\frac{dR}{d\theta} = \frac{2u^2}{g}\cos 2\theta, \tag{2.123}$$

and therefore $\frac{dR}{d\theta} = 0$ when $\theta = 45°$.

Although it is possible to find $\frac{dR}{d\theta}$ for the $h \neq 0$ general case and then solve for $\frac{dR}{d\theta} = 0$, there is an alternative geometric argument (see Fig. 2.4) which requires much less algebra and perhaps offers deeper insights into the problem. In this case, consider the velocity $\mathbf{v}$ of the projectile when $x = R$ and $y = 0$ at time t. If the initial velocity was $\mathbf{u} = u(\cos\theta\hat{\mathbf{x}} + \sin\theta\hat{\mathbf{y}})$, then

$$\mathbf{v} = \mathbf{u} - gt\hat{\mathbf{y}}. \tag{2.124}$$

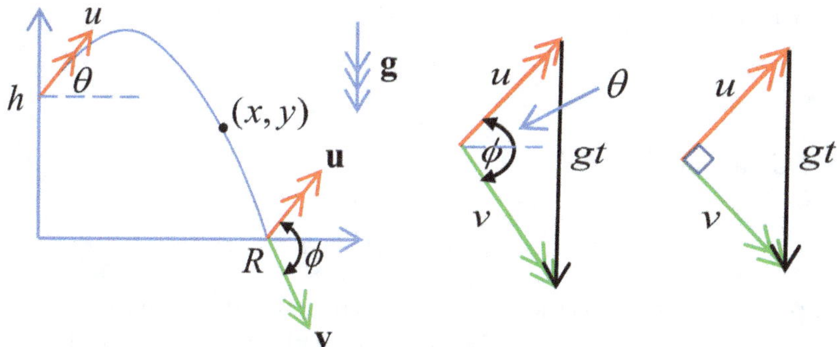

Fig. 2.4. A projectile is launched with speed u from $(0, h)$ at elevation angle θ from the horizontal. By using a geometric argument involving the vector triangle formed by launch velocity **u**, impact velocity **v** and velocity change due to gravity $-gt\hat{\mathbf{y}}$, it is possible to show that range $R = \frac{u^2}{g}\sqrt{1 + \frac{2gh}{u^2}} \sin\phi$ where ϕ is the angle between the vectors **v** and **u**. The maximum range is when $\phi = 90°$, which means $R_{\max} = \frac{u^2}{g}\sqrt{1 + \frac{2gh}{u^2}}$ and $\theta = \sin^{-1}\sqrt{\frac{1}{2 + \frac{2gh}{u^2}}}$.

The vectors **v**, **u**, $-gt\hat{\mathbf{y}}$ form a vector triangle of sides v, u and gt. The area A of the triangle can be evaluated in two ways:

$$A = \tfrac{1}{2} \times gt \times u\cos\theta, \tag{2.125}$$

$$A = \tfrac{1}{2} \times u\sin\phi \times v, \tag{2.126}$$

where ϕ is the angle between the vectors **v** and **u**. Equating these expressions,

$$gt\cos\theta = v\sin\phi. \tag{2.127}$$

$$\therefore ut\cos\theta = \frac{uv}{g}\sin\phi. \tag{2.128}$$

Now, since $R = ut\cos\theta$,

$$R = \frac{uv}{g}\sin\phi. \tag{2.129}$$

By conservation of energy and noting that $y = 0$ at $x = R$,

$$mgh + \tfrac{1}{2}mu^2 = \tfrac{1}{2}mv^2. \tag{2.130}$$

$$\therefore v = \sqrt{2gh + u^2}. \tag{2.131}$$

Hence,

$$R = \frac{u}{g}\sqrt{2gh + u^2}\sin\phi \tag{2.132}$$

$$R = \frac{u^2}{g}\sqrt{1 + \frac{2gh}{u^2}}\sin\phi. \tag{2.133}$$

R is clearly maximised when $\phi = 90°$ given the $\sin\phi$ factor, so

$$R_{\max} = \frac{u^2}{g}\sqrt{1 + \frac{2gh}{u^2}}. \tag{2.134}$$

Note that the ratio

$$\alpha = \frac{2gh}{u^2} = \frac{mgh}{\frac{1}{2}mu^2} = \frac{\text{GPE at } t = 0}{\text{KE at } t = 0}. \tag{2.135}$$

When $\phi = 90°$, this means the $\mathbf{v}$, $\mathbf{u}$, $-gt\hat{\mathbf{y}}$ vector triangle is *right-angled* (between $\mathbf{v}$ and $\mathbf{u}$), so we can use Pythagoras' theorem to determine the time of flight:

$$g^2 t^2 = u^2 + v^2. \tag{2.136}$$

Now, $v = \sqrt{2gh + u^2}$, so

$$g^2 t^2 = u^2 + 2gh + u^2. \tag{2.137}$$

Therefore, the time of flight t for maximum range is

$$t = \frac{u}{g}\sqrt{2 + \frac{2gh}{u^2}}. \tag{2.138}$$

$R = ut\cos\theta$; therefore, for maximum range $R_{\max} = \frac{u^2}{g}\sqrt{1 + \frac{2gh}{u^2}}$, we can use these expressions for R to determine the angle θ, which corresponds to the maximum range $R_{\max}$:

$$\frac{u^2}{g}\sqrt{1 + \frac{2gh}{u^2}} = u\frac{u}{g}\sqrt{2 + \frac{2gh}{u^2}}\cos\theta \tag{2.139}$$

$$1 + \frac{2gh}{u^2} = \left(2 + \frac{2gh}{u^2}\right)\cos^2\theta \tag{2.140}$$

$$1 + \frac{2gh}{u^2} = \left(2 + \frac{2gh}{u^2}\right)(1 - \sin^2\theta) \tag{2.141}$$

$$1 + \frac{2gh}{u^2} = 2 + \frac{2gh}{u^2} - \left(2 + \frac{2gh}{u^2}\right)\sin^2\theta \tag{2.142}$$

$$\sin^2\theta = \frac{1}{2 + \frac{2gh}{u^2}}. \tag{2.143}$$

To maximise projectile range given a fixed launch speed u, the elevation angle must be

$$\theta = \sin^{-1}\sqrt{\frac{1}{2 + \frac{2gh}{u^2}}}. \tag{2.144}$$

Note that when $h = 0$, this reduces to $\theta = \sin^{-1}\left(\frac{1}{\sqrt{2}}\right) = 45°$, which is consistent with our previous calculation for this ground-launch scenario.

2.3.9 *Modelling projectile motion with air resistance*

Newton's second law ('mass $\times$ acceleration $=$ vector sum of force') describing the rate of change of velocity $\mathbf{v}$ for a projectile of mass m and cross-sectional area A, being launched at speed u at elevation angle θ from $(0, h)$ into air of density ρ, is

$$m\frac{d\mathbf{v}}{dt} = -mg\hat{\mathbf{y}} - \tfrac{1}{2}c_D A\rho \mathbf{v}\,|\mathbf{v}|\,, \tag{2.145}$$

where c_D is the *aerodynamic drag coefficient* for the projectile. For a sphere, the drag coefficient[8] is $c_D \approx 0.47$. The $\mathbf{v}\,|\mathbf{v}|$ term ensures that the drag proportional to v^2 always *opposes* the motion. Unfortunately, this formulation makes it difficult to make analytical progress. In other words, to solve for the trajectory, one requires a *iterative* numerical method based on a small, finite time step Δt. An idea which develops from the previous part of this projectile motion section is to assume constant acceleration motion between time steps. This is known as the *Verlet* method.[9]

Initial conditions:

$$t_0 = 0, \tag{2.146}$$

$$\mathbf{r}_0 = \begin{pmatrix} 0 \\ h \end{pmatrix}, \tag{2.147}$$

$$\mathbf{v}_0 = \begin{pmatrix} u\cos\theta \\ u\sin\theta \end{pmatrix}. \tag{2.148}$$

Iteration:

$$t_{n+1} = t_n + \Delta t, \tag{2.149}$$

$$\mathbf{a} = -g\hat{\mathbf{y}} - \tfrac{1}{2}\frac{c_D A\rho}{m}\mathbf{v}_n|\mathbf{v}_n|, \tag{2.150}$$

$$\mathbf{r}_{n+1} = \mathbf{r}_n + \mathbf{v}_n\Delta t + \tfrac{1}{2}\mathbf{a}\Delta t^2, \tag{2.151}$$

$$\mathbf{V} = \mathbf{v}_n + \mathbf{a}\Delta t, \tag{2.152}$$

$$\mathbf{A} = -g\hat{\mathbf{y}} - \tfrac{1}{2}\frac{c_D A\rho}{m}\mathbf{V}|\mathbf{V}|, \tag{2.153}$$

$$\mathbf{v}_{n+1} = \mathbf{v}_n + \tfrac{1}{2}\left(\mathbf{a} + \mathbf{A}\right)\Delta t. \tag{2.154}$$

Note that, in this *velocity-dependent acceleration Verlet* (VDAV)[10] incarnation, we avoid a circular reference by computing an intermediate velocity $\mathbf{V}$ and subsequent acceleration $\mathbf{A}$. An average between acceleration $\mathbf{a}$ evaluated at iteration n and $\mathbf{A}$ is used to determine the updated velocity $\mathbf{v}_{n+1}$.

A sensible stopping criteria after N iterations might be when the y component of $\mathbf{r}_{N+1}$ becomes negative, i.e. when $\mathbf{r}_{N+1}\cdot\hat{\mathbf{y}} < 0$. As long as Δt is small, this means

---

[8] https://en.wikipedia.org/wiki/Drag_coefficient accessed 26 April 2021.

[9] There are more precise numeric schemes, such as *Runge–Kutta*, but Verlet will be good enough to serve as example.

[10] *Velocity-dependent acceleration Verlet* (VDAV) is explored for various classic dynamics problems in French, Cullerne, Kanchanasakdichai (2019) [10].

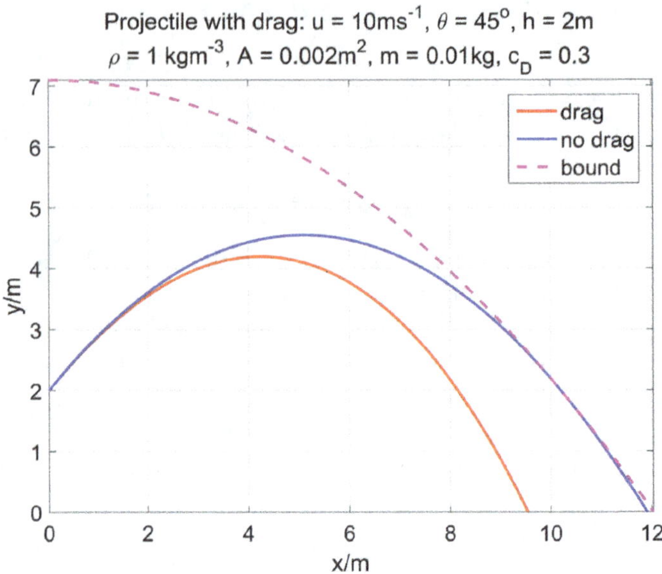

Fig. 2.5. Comparison of a drag-free parabolic trajectory of a projectile (blue) with an air resistance model implemented using a *Verlet* iterative numeric method (red). This assumes constant acceleration motion during (small) finite time steps Δt. Since work is done against drag forces, both the apogee and maximum range of the drag-inclusive curve are smaller than those of the drag-free trajectory. The dashed magenta curve is the *bounding parabola* for the $u = 10$ ms^{-1} launch speed, from $(0, 2)$ (metres). Regardless of the initial launch angle θ, it is impossible to reach a target (x, y) location beyond the bounding parabola, given a fixed launch speed u.

the range R can be calculated from the x component of $\mathbf{r}_{N+1}$, or perhaps better, an average:

$$R = \tfrac{1}{2}\left(\mathbf{r}_N + \mathbf{r}_{N+1}\right) \cdot \hat{\mathbf{x}}. \tag{2.155}$$

The time of flight can be found in a similar fashion:

$$t_{\text{flight}} = t_N + \tfrac{1}{2}\Delta t. \tag{2.156}$$

Figure 2.5 illustrates the effect of drag compared to a drag-free inverted-parabolic trajectory. In this simulation, $u = 10$ ms^{-1}, $g = 9.81$ ms^{-2}, $\theta = 45°$ and $h = 2.00$ m. The blue curve is for the no drag case, and the magenta dashed curve is the bounding parabola. The red curve is a drag-inclusive model, computed using the Verlet method with time step $\Delta t = 0.01$ s. The projectile mass $m = 0.01$ kg, cross-sectional area $A = 0.002$ m^2, air density 1.0 kgm^{-3} and drag coefficient $c_D = 0.3$. Unsurprisingly, the effect of drag is to reduce both the apogee and maximum range.

2.4   The Ladder Problem

We have commenced our exploration of classical mechanics via one-dimensional kinematics and collisions and then (two-dimensional) projectile motion. Now, let's consider some problems in *statics*.

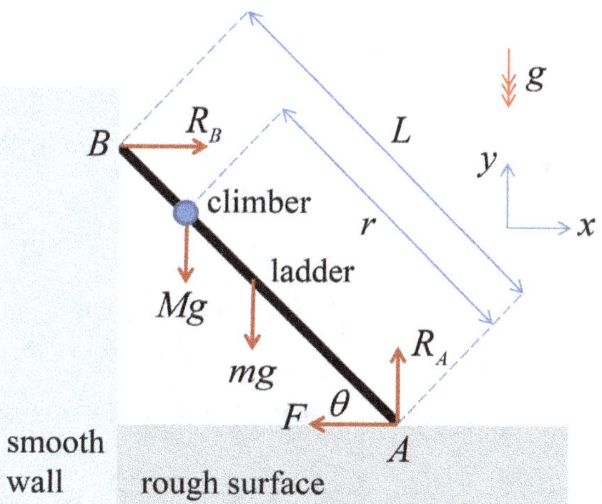

Fig. 2.6. Force diagram for the ladder problem. A rigid ladder of mass m is placed at an angle of θ against a smooth ice wall. The ends of the ladder extend from point A on a rough horizontal surface to point B on the ice wall. The coefficient of friction between the ladder and the rough surface at A is μ. If a climber of mass M ascends the ladder, $\mu > \frac{1}{2}\left(\frac{m+2M}{m+M}\right)\frac{1}{\tan\theta}$ in order for the ladder not to slip.

A rigid ladder of length L is placed on rough ground with *coefficient of static friction* μ next to a frictionless ice wall. The ladder has mass m, and a climber of mass M ascends the ladder. A diagram of the situation, including arrows to represent all forces, is shown in Fig. 2.6. What would be a lower limit on μ such that the ladder doesn't slip?

Assuming equilibrium, we can write down Newton's second law for the ladder-and-climber system in the x, y directions.

In the x direction,

$$0 = R_B - F. \tag{2.157}$$

$$\therefore R_B = F. \tag{2.158}$$

In the y direction,

$$0 = R_A - (m + M)g. \tag{2.159}$$

$$\therefore R_A = (m + M)g, \tag{2.160}$$

where R_A and R_B are the respective normal contact forces at A and B, and F is the frictional force at A, the base of the ladder.

For equilibrium, we can also assume that the net turning moment *about any axis* is zero. If the climber is at a distance of r up the ladder ($0 \leq r \leq L$), taking clockwise moments about point A,

$$0 = R_B L \sin\theta - \tfrac{1}{2}mgL\cos\theta - Mgr\cos\theta \tag{2.161}$$

$$0 = R_B \tan\theta - \left(\tfrac{1}{2}m + \frac{Mr}{L}\right)g. \tag{2.162}$$

$$\therefore R_B = \left(\tfrac{1}{2}m + \frac{Mr}{L} \right) \frac{g}{\tan\theta}. \tag{2.163}$$

If there is no slip at A,

$$F < \mu R_A. \tag{2.164}$$

Therefore, since

$$F = R_B = \left(\tfrac{1}{2}m + \frac{Mr}{L} \right) \frac{g}{\tan\theta} \tag{2.165}$$

and no slip must occur for the entire range $0 \le r \le L$, the largest F is when $r = L$, i.e.

$$F_{\mathrm{max}} = \left(\tfrac{1}{2}m + M \right) \frac{g}{\tan\theta}. \tag{2.166}$$

Therefore, for the ladder not to slip,

$$\left(\tfrac{1}{2}m + M \right) \frac{g}{\tan\theta} < \mu(m + M)g. \tag{2.167}$$

$$\therefore \mu > \tfrac{1}{2} \left(\frac{m + 2M}{m + M} \right) \frac{1}{\tan\theta}. \tag{2.168}$$

Note that, in the limit when $M \gg m$,

$$\mu > \frac{1}{\tan\theta}. \tag{2.169}$$

This relationship is explored in Fig. 2.7, which plots $\mu = \tfrac{1}{2} \left(\frac{1 + \frac{2M}{m}}{1 + \frac{M}{m}} \right) \frac{1}{\tan\theta}$ vs $\frac{M}{m}$ for θ values of 30°, 40°, 50°, 60° and 70°. Interestingly, the minimum μ increases less steeply as $\frac{M}{m}$ is increased. In other words, if μ is beyond $\frac{1}{\tan\theta}$, it doesn't really matter how much heavier the climber is than the ladder.

2.5   The Drawbridge Problem

A rigid drawbridge of weight W and length L is lowered from a vertical position via fixed chains at height L above the hinge of the drawbridge. Assume that the drawbridge is lowered slowly enough for it to be considered to be in equilibrium. Let the total chain tension be T (in practice, $\tfrac{1}{2}T$ for each of two chains), and define the angle β to be the angle between the chain and the drawbridge, as illustrated in Fig. 2.8.

If the elevation of the drawbridge is θ, the isosceles triangle formed from the bridge, the initial vertical bridge position and the chain mean that

$$180° = 2\beta + 90° - \theta. \tag{2.170}$$

$$\therefore \beta = 45° + \tfrac{1}{2}\theta. \tag{2.171}$$

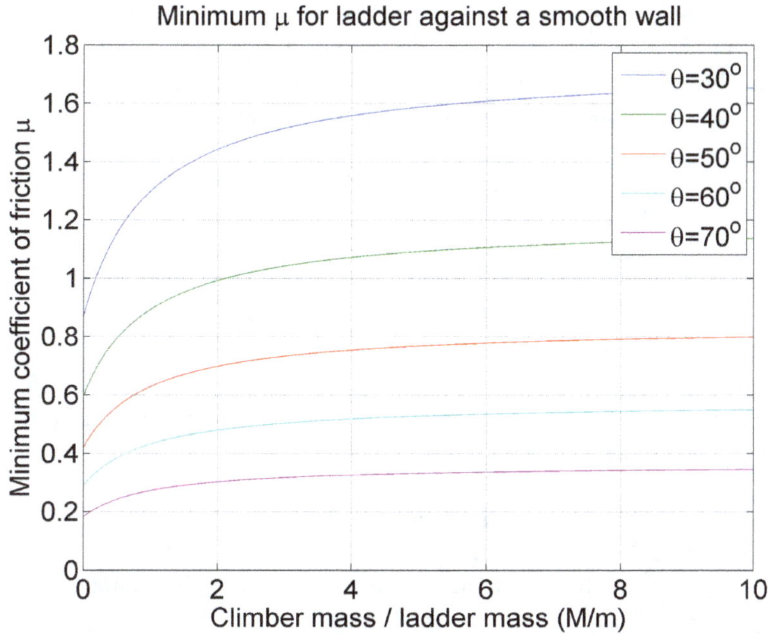

Fig. 2.7. Coefficient of friction $\mu > \frac{1}{2}\left(\frac{m+2M}{m+M}\right)\frac{1}{\tan\theta}$ in order for a ladder of mass m, bearing a climber of mass M, to not slip. θ is the angle of the ladder from the horizontal. The ladder is placed on a rough horizontal surface and is resting against a frictionless vertical wall.

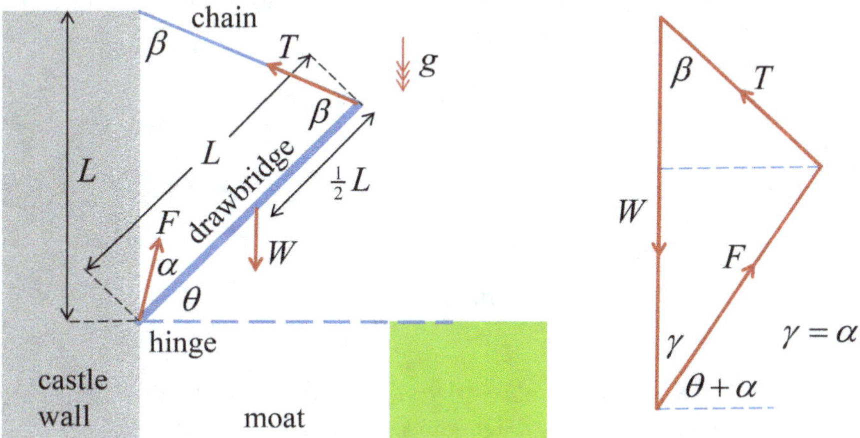

Fig. 2.8. Diagram representing forces on a model of a rigid drawbridge of length L, which is lowered (in equilibrium) by inextensible chains providing tension T. The drawbridge weight is W, and the hinge at the base of the drawbridge provides a force F at an angle α above the drawbridge. The goal is to develop a model of how α, F and T vary with angle θ, given inputs of L and W. This particular problem can be tricky to solve without a *geometric* discussion of why $2\alpha + \theta = 90°$. The latter can be shown by drawing a force triangle of W, F and T and using the result $F\sin\alpha = T\sin\beta$ (which we derive from taking moments about the centre of the drawbridge) to show that the angle γ between F and W must equal α.

Let F be the force exerted on the drawbridge by the hinge (and by Newton's third law, the force exerted on the hinge by the drawbridge), at angle α from the elevation of the drawbridge. Given that the system is assumed to be in equilibrium, taking clockwise moments from the *centre* of the drawbridge,

$$0 = F \sin\alpha \times \tfrac{1}{2}L - T \sin\beta \times \tfrac{1}{2}L, \tag{2.172}$$

i.e.

$$F \sin\alpha = T \sin\beta. \tag{2.173}$$

If we consider a triangle of forces of drawbridge weight, total chain tension and hinge force, we can define the angle between the hinge force and the weight vector to be γ (see Fig. 2.8). From the force triangle,

$$F \sin\gamma = T \sin\beta. \tag{2.174}$$

Comparing this to the moments expression above, this means $\gamma = \alpha$, which implies $90° = 2\alpha + \theta$ and therefore

$$\alpha = 45° - \tfrac{1}{2}\theta. \tag{2.175}$$

This is perhaps the key step in the problem and indicates the benefits of a geometric perspective *in addition to* purely algebraic approaches.

Taking clockwise moments about the hinge,

$$0 = W \times \tfrac{1}{2}L \cos\theta - T \sin\beta \times L. \tag{2.176}$$

Hence,

$$T = \tfrac{1}{2}\frac{W \cos\theta}{\sin\beta}, \tag{2.177}$$

$$F = \frac{T \sin\beta}{\sin\alpha} = \tfrac{1}{2}\frac{W \cos\theta}{\sin\alpha}. \tag{2.178}$$

Now, $\beta = 45° + \tfrac{1}{2}\theta$ and $\alpha = 45° - \tfrac{1}{2}\theta$ suggest that we should investigate the possibility of simplification using trigonometric addition formulae:

$$\cos\theta = \cos\left(\tfrac{1}{2}\theta + \tfrac{1}{2}\theta\right) = \cos^2\left(\tfrac{1}{2}\theta\right) - \sin^2\left(\tfrac{1}{2}\theta\right).$$

$$\therefore \sin\beta = \sin\left(45° + \tfrac{1}{2}\theta\right) = \sin 45° \cos\left(\tfrac{1}{2}\theta\right) + \cos 45° \sin\left(\tfrac{1}{2}\theta\right)$$

$$= \tfrac{1}{\sqrt{2}}\left(\cos\left(\tfrac{1}{2}\theta\right) + \sin\left(\tfrac{1}{2}\theta\right)\right). \tag{2.179}$$

$$\therefore \sin\alpha = \sin\left(45° - \tfrac{1}{2}\theta\right) = \sin 45° \cos\left(\tfrac{1}{2}\theta\right) - \cos 45° \sin\left(\tfrac{1}{2}\theta\right)$$

$$= \tfrac{1}{\sqrt{2}}\left(\cos\left(\tfrac{1}{2}\theta\right) - \sin\left(\tfrac{1}{2}\theta\right)\right). \tag{2.180}$$

Therefore,

$$T = \tfrac{1}{2}W\,\frac{\cos^2\left(\tfrac{1}{2}\theta\right) - \sin^2\left(\tfrac{1}{2}\theta\right)}{\frac{1}{\sqrt{2}}\left(\cos\left(\tfrac{1}{2}\theta\right) + \sin\left(\tfrac{1}{2}\theta\right)\right)} \tag{2.181}$$

$$= \tfrac{1}{\sqrt{2}}W\,\frac{\left(\cos\left(\tfrac{1}{2}\theta\right) + \sin\left(\tfrac{1}{2}\theta\right)\right)\left(\cos\left(\tfrac{1}{2}\theta\right) - \sin\left(\tfrac{1}{2}\theta\right)\right)}{\cos\left(\tfrac{1}{2}\theta\right) + \sin\left(\tfrac{1}{2}\theta\right)}. \tag{2.182}$$

$$\therefore T = \tfrac{1}{\sqrt{2}}W\left(\cos\left(\tfrac{1}{2}\theta\right) - \sin\left(\tfrac{1}{2}\theta\right)\right). \tag{2.183}$$

Similarly,

$$F = \tfrac{1}{2}W\,\frac{\cos^2\left(\tfrac{1}{2}\theta\right) - \sin^2\left(\tfrac{1}{2}\theta\right)}{\frac{1}{\sqrt{2}}\left(\cos\left(\tfrac{1}{2}\theta\right) - \sin\left(\tfrac{1}{2}\theta\right)\right)} \tag{2.184}$$

$$= \tfrac{1}{\sqrt{2}}W\,\frac{\left(\cos\left(\tfrac{1}{2}\theta\right) + \sin\left(\tfrac{1}{2}\theta\right)\right)\left(\cos\left(\tfrac{1}{2}\theta\right) - \sin\left(\tfrac{1}{2}\theta\right)\right)}{\cos\left(\tfrac{1}{2}\theta\right) - \sin\left(\tfrac{1}{2}\theta\right)}. \tag{2.185}$$

$$\therefore F = \tfrac{1}{\sqrt{2}}W\left(\cos\left(\tfrac{1}{2}\theta\right) + \sin\left(\tfrac{1}{2}\theta\right)\right). \tag{2.186}$$

Note that we have a sensible limiting case when the drawbridge is fully open in the horizontal position, i.e. $\theta = 0°$. In this case, $\alpha = \beta = 45°$ (which means the hinge force and tension are at right angles) and $F = T = \tfrac{1}{\sqrt{2}}W$, which can be easily verified by applying Pythagoras' theorem to this special case of the force triangle. A plot of chain tension T and hinge force F versus drawbridge elevation angle θ is provided in Fig. 2.9. Unsurprisingly, the chain tension drops to zero as the drawbridge closes. At $\theta = 90°$, all the weight (22 kN in this example) is balanced by the hinge force, which will act vertically upwards since $\alpha = 45° - \tfrac{1}{2}\theta = 0$.

2.6 The Three Barrels Problem

Two cylinders of weight W and radius a are placed next to each other on a rough horizontal surface with coefficient of friction μ_A. A cylinder of weight w and radius b is placed carefully between and above the cylinders. What are the equilibrium requirements of μ_A, and indeed μ_B too, the latter being the coefficient of friction at the interface between the cylinders weighing w and W? Let F be the magnitude of the friction force between the lower and upper cylinders and R be the corresponding normal contact force, i.e. a *total contact force* of magnitude $\sqrt{R^2 + F^2}$. For the lower cylinder, let the friction force with the horizontal surface be G and normal contact be S. To solve this interesting statics problem adapted from Quadling, *Mechanics M3&4* pp. 90–93 [29], first draw three diagrams depicting the following: (a) the overall geometry; (b) the forces on one of the lower cylinders; and (c) the forces on the upper cylinder. Note that the symmetry of the problem means you can choose either the left or right cylinders since they are identical. This force-diagram triptych is illustrated in Fig. 2.10.

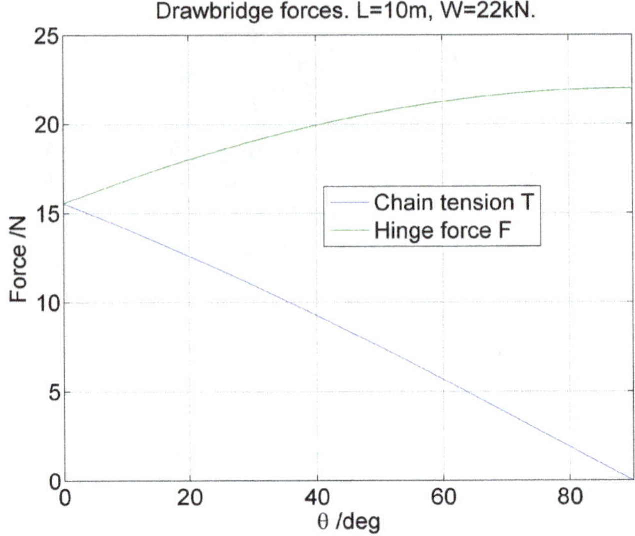

Fig. 2.9. A plot of chain tension T and hinge force F vs. drawbridge elevation angle θ, given drawbridge weight W. $T = \frac{1}{\sqrt{2}}W\left(\cos\left(\frac{1}{2}\theta\right) - \sin\left(\frac{1}{2}\theta\right)\right)$ and $F = \frac{1}{\sqrt{2}}W\left(\cos\left(\frac{1}{2}\theta\right) + \sin\left(\frac{1}{2}\theta\right)\right)$.

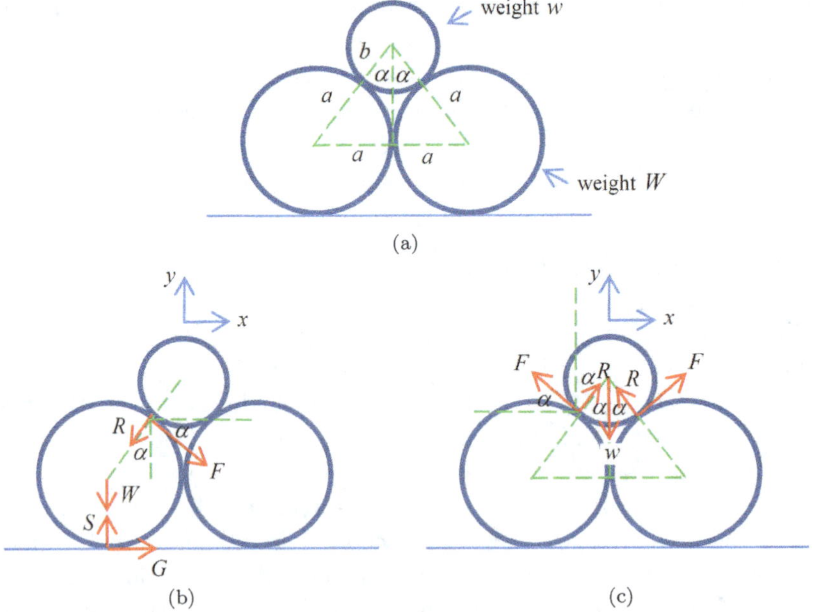

Fig. 2.10. Two cylinders of weight W and radius a are placed next to each other on a rough horizontal surface. A cylinder of weight w and radius b is placed carefully between and above the cylinders. Let F be the magnitude of the friction force between the lower and upper cylinders and R be the corresponding normal contact force (i.e. a total contact force of magnitude $\sqrt{R^2 + F^2}$). For the lower cylinder, let the friction force with the horizontal surface be G and normal contact be S. The goal of this problem is to determine expressions for the minimum coefficients of friction between the lower balls and the horizontal surface, and also between the upper and lower balls, to maintain static equilibrium. (a) Describes the overall geometry, (b) defines forces on the lower cylinder and (c) defines forces on the upper cylinder.

In Fig. 2.10, the two lower cylinders are touching and the centre of the upper cylinder is directly above this point. Hence:

$$(a + b) \sin \alpha = a. \tag{2.187}$$

$$\therefore \alpha = \sin^{-1}\left(\frac{a}{a+b}\right). \tag{2.188}$$

Resolving the forces in the y direction for the upper cylinder,

$$0 = 2R \cos \alpha + 2F \sin \alpha - w. \tag{2.189}$$

$$\therefore w = 2R \cos \alpha + 2F \sin \alpha. \tag{2.190}$$

Resolving the forces in the x direction for the lower cylinder,

$$0 = G - R \sin \alpha + F \cos \alpha. \tag{2.191}$$

$$\therefore G = R \sin \alpha - F \cos \alpha. \tag{2.192}$$

Resolving the forces in the y direction for the lower cylinder,

$$0 = S - R \cos \alpha - F \sin \alpha - W. \tag{2.193}$$

$$\therefore S = R \cos \alpha + F \sin \alpha + W. \tag{2.194}$$

Taking clockwise moments about the centre of the lower cylinder,

$$0 = F \times a - G \times a. \tag{2.195}$$

$$\therefore F = G. \tag{2.196}$$

Hence,

$$F = R \sin \alpha - F \cos \alpha \tag{2.197}$$

$$F(1 + \cos \alpha) = R \sin \alpha. \tag{2.198}$$

$$\therefore 2F(1 + \cos \alpha) \cos \alpha = 2R \sin \alpha \cos \alpha. \tag{2.199}$$

Also,

$$w = 2R \cos \alpha + 2F \sin \alpha. \tag{2.200}$$

$$\therefore w \sin \alpha - 2F \sin^2 \alpha = 2R \sin \alpha \cos \alpha. \tag{2.201}$$

By equating $2R \sin \alpha \cos \alpha$ between these expressions,

$$w \sin \alpha - 2F \sin^2 \alpha = 2F(1 + \cos \alpha) \cos \alpha. \tag{2.202}$$

Therefore,

$$F = \tfrac{1}{2} w \frac{\sin \alpha}{(1 + \cos \alpha) \cos \alpha + \sin^2 \alpha} \tag{2.203}$$

$$F = \tfrac{1}{2} w \frac{\sin \alpha}{\cos \alpha + \cos^2 \alpha + \sin^2 \alpha}, \tag{2.204}$$

$$F = \tfrac{1}{2}w\frac{\sin\alpha}{\cos\alpha + 1}. \tag{2.205}$$

Hence, using $F = G$ and $F(1 + \cos\alpha) = R\sin\alpha$,

$$G = \tfrac{1}{2}w\frac{\sin\alpha}{\cos\alpha + 1}, \tag{2.206}$$

$$R = \frac{1 + \cos\alpha}{\sin\alpha}F \tag{2.207}$$

$$= \tfrac{1}{2}w \tag{2.208}$$

and

$$S = R\cos\alpha + F\sin\alpha + W \tag{2.209}$$

$$= \tfrac{1}{2}w\cos\alpha + \tfrac{1}{2}w\frac{\sin\alpha}{\cos\alpha + 1}\sin\alpha + W \tag{2.210}$$

$$= \tfrac{1}{2}w\left(\frac{\cos^2\alpha + \cos\alpha + \sin^2\theta}{\cos\alpha + 1}\right) + W \tag{2.211}$$

$$= \tfrac{1}{2}w + W. \tag{2.212}$$

For no slip at the interface between the upper and lower cylinders,

$$F < \mu_A R. \tag{2.213}$$

Therefore, using $F = \tfrac{1}{2}w\frac{\sin\alpha}{\cos\alpha + 1}$ and $R = \tfrac{1}{2}w$,

$$\tfrac{1}{2}w\frac{\sin\alpha}{\cos\alpha + 1} < \mu_A\tfrac{1}{2}w. \tag{2.214}$$

$$\therefore \mu_A > \frac{\sin\alpha}{\cos\alpha + 1}. \tag{2.215}$$

For no slip at the interface between the lower cylinder and horizontal ground interface,

$$G < \mu_B S. \tag{2.216}$$

Therefore, using $G = \tfrac{1}{2}w\frac{\sin\alpha}{\cos\alpha + 1}$ and $S = \tfrac{1}{2}w + W$,

$$\tfrac{1}{2}w\frac{\sin\alpha}{\cos\alpha + 1} < \mu_B\left(\tfrac{1}{2}w + W\right). \tag{2.217}$$

$$\therefore \mu_B > \frac{w}{w + 2W}\frac{\sin\alpha}{\cos\alpha + 1}. \tag{2.218}$$

In summary, we can express all the forces in terms of weights W and w and the cylinder radii a and b:

$$R = \tfrac{1}{2}w, \tag{2.219}$$

$$S = \tfrac{1}{2}w + W, \tag{2.220}$$

$$\alpha = \sin^{-1}\left(\frac{a}{a+b}\right),\tag{2.221}$$

$$F = \tfrac{1}{2}w\frac{\sin\alpha}{\cos\alpha+1},\tag{2.222}$$

$$G = \tfrac{1}{2}w\frac{\sin\alpha}{\cos\alpha+1}.\tag{2.223}$$

The coefficients of friction for no slip are (note that A means 'between the cylinders' and B means 'between the lower cylinder and the ground')

$$\mu_A > \frac{\sin\alpha}{\cos\alpha+1},\tag{2.224}$$

$$\mu_B > \frac{w}{w+2W}\frac{\sin\alpha}{\cos\alpha+1}.\tag{2.225}$$

Consider the special case of a nice Pythagorean triple, $5^2 = 3^2 + 4^2$ (i.e. the cylinder ratios described by Quadling [29]):

$$a + b = 5r,\tag{2.226}$$

$$a = 3r.\tag{2.227}$$

$$\therefore b = 2r.\tag{2.228}$$

$$\therefore \sin\alpha = \tfrac{3}{5}.\tag{2.229}$$

$$\therefore \cos\alpha = \tfrac{4}{5}.\tag{2.230}$$

This means

$$\frac{\sin\alpha}{\cos\alpha+1} = \frac{\tfrac{3}{5}}{\tfrac{4}{5}+1} = \tfrac{3}{9} = \tfrac{1}{3}.\tag{2.231}$$

Hence, the conditions for no slip are

$$\mu_A > \tfrac{1}{3},\tag{2.232}$$

$$\mu_B > \frac{w}{3w+6W}.\tag{2.233}$$

For problem-setters, note that any Pythagorean triple involving a and $a+b$ will yield nice integer ratios for $\sin\alpha$ and $\cos\alpha$. For example, using $17^2 = 15^2 + 8^2$,

$$a + b = 17r,\tag{2.234}$$

$$a = 15r.\tag{2.235}$$

$$\therefore b = 17r - 15r = 2r.\tag{2.236}$$

$$\therefore \cos\alpha = \tfrac{8}{17}.\tag{2.237}$$

$$\therefore \sin\alpha = \tfrac{15}{17}.\tag{2.238}$$

This means

$$\frac{\sin\alpha}{\cos\alpha + 1} = \frac{\frac{15}{17}}{\frac{8}{17} + 1} = \tfrac{3}{5}. \tag{2.239}$$

The conditions for no slip in this case are

$$\mu_A > \tfrac{3}{5}, \tag{2.240}$$

$$\mu_B > \frac{3w}{5w + 10W}. \tag{2.241}$$

2.7   Weebles Wobble But Don't Fall Down

A *weeble* is a curious toy, popular in the 1970s, typically based on a caricature figurine mounted on a solid hemisphere. The toy is designed to always right itself when tipped over. To achieve this feat, *the centre of mass of the weeble must be below the centre of the hemisphere*. Since the normal contact force between the weeble and a horizontal surface will always act radially inwards, i.e. towards the centre of the hemisphere,[11] this means the weight of the weeble will always result in a turning moment about the point of contact with the surface which will serve to restore the weeble to an upright position. To enable some illustrative mathematical analysis of appropriate brevity, consider a suitably modernist weeble to be constructed from a uniform solid hemisphere of radius r and density ρ_1, topped by a solid cone of density ρ_2 (see Fig. 2.11). The upper cone has radius r and height h. The question is, what are the limits on h such that the assembly works as a weeble?

The first challenge is to work out the position of the centre of mass of the hemisphere. Turning it on its side and imagining it to be consisting of thin cylinders

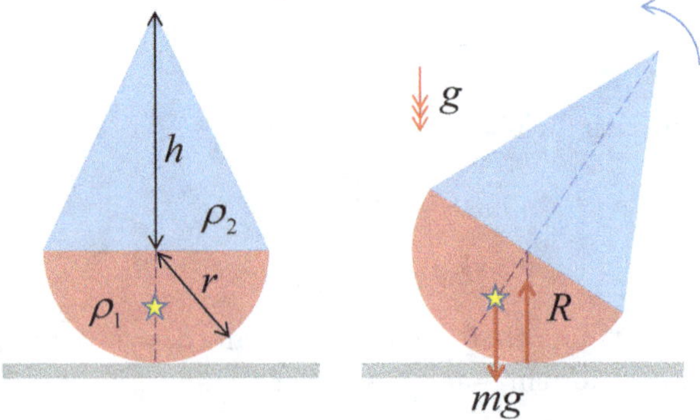

Fig. 2.11. A weeble, constructed from a solid hemisphere of radius r and density ρ_1, topped with a solid cone of density ρ_2, height h and base radius r. If the centre of mass of the combined system (the yellow star) is *below* the centre of the hemispherical circle, the turning moment of the weeble's weight about the point of contact with a horizontal surface will always cause the weeble to right itself. Therefore, a 'weeble may wobble but won't fall down'.

---

[11] The normal contact force of magnitude R is perpendicular to the hemispherical surface, and since the radials of a sphere are all tangential to the surface, the normal contact force must point towards the centre of the hemisphere.

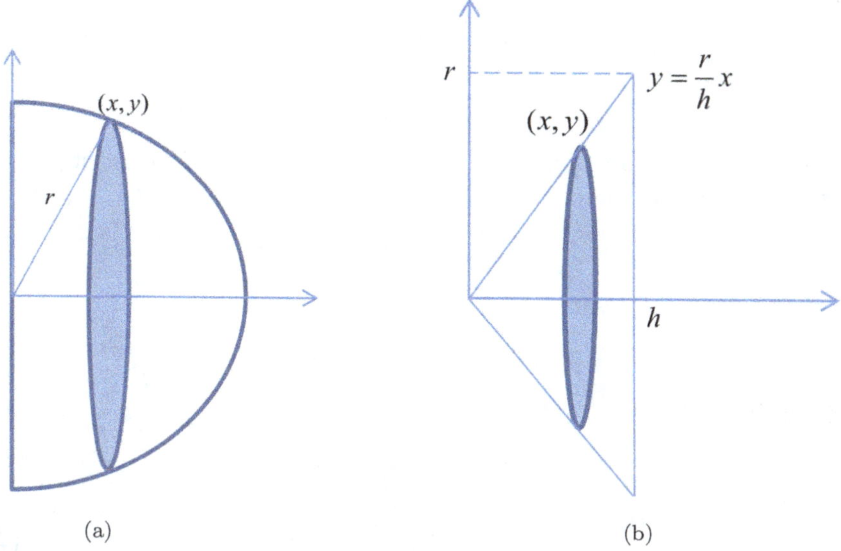

Fig. 2.12. The centre of mass of (a) a uniform solid *hemisphere* of density ρ_1 and (b) a *cone* of density ρ_2 can be determined by considering a series of infinitely thin cylindrical slices of mass $dm = \rho_{1,2} 2\pi y^2 dx$. For the hemisphere, the total mass is $m = \frac{2}{3}\pi r^3 \rho_1$. Since $r^2 = x^2 + y^2$, the centre of mass is $(\bar{x}, 0)$, where $\bar{x} = \frac{\int x\,dm}{\frac{2}{3}\pi r^3 \rho_1} = \frac{3}{8}r$. For the cone, the total mass is $m = \frac{1}{3}\pi r^2 h \rho_2$. Since $y = h - \frac{h}{r}x$, the centre of mass is $(\bar{x}, 0)$, where $\bar{x} = \frac{\int x\,dm}{\frac{1}{3}\pi r^2 h \rho_2} = \frac{1}{4}h$.

of mass $dm = \rho_1 \pi y^2 dx$, as illustrated in Fig. 2.12(a), the position $\bar{x}$ of the centre of mass is

$$\bar{x} = \frac{\int x\,dm}{\frac{2}{3}\pi r^3 \rho_1}. \tag{2.242}$$

Note that $\frac{2}{3}\pi r^3 \rho_1$ is the mass of the hemisphere, i.e. half the mass of a sphere of radius r and density ρ_1.

Now, the curve $y(x)$ of the hemisphere satisfies

$$r^2 = x^2 + y^2. \tag{2.243}$$

So, using $dm = \rho_1 \pi y^2 dx$ and $y^2 = r^2 - x^2$,

$$\bar{x} = \frac{\rho_1 \pi \int_0^r x\left(r^2 - x^2\right) dx}{\frac{2}{3}\pi r^3 \rho_1} \tag{2.244}$$

$$= \frac{3}{2r^3}\left[\frac{1}{2}x^2 r^2 - \frac{1}{4}x^4\right]_0^r \tag{2.245}$$

$$= \frac{3}{8}r. \tag{2.246}$$

In other words, the centre of mass of a hemisphere balanced on its round side is $r - \frac{3}{8}r = \frac{5}{8}r$ vertically upwards from the point of contact.

Now, consider a cone to be constructed from thin cylinders of mass $dm = \rho_2 \pi y^2 dx$, as illustrated in Fig. 2.12(b). The centre of mass is

$$\bar{x} = \frac{\int x\, dm}{\frac{1}{3}\pi r^2 h \rho_2}, \tag{2.247}$$

where $\frac{1}{3}\pi r^2 h \rho_2$ is the mass of the cone. Now, the sides of the cone are straight lines of gradient $\frac{h}{r}$ from the origin. Hence,

$$y = \frac{r}{h}x \tag{2.248}$$

and therefore

$$dm = \rho_2 \pi y^2 dx = \rho_2 \pi \frac{r^2}{h^2} x^2 dx.$$

The centre of mass of the cone is therefore

$$\bar{x} = \frac{\rho_2 \pi \frac{r^2}{h^2} \int_0^h x^3 dx}{\frac{1}{3}\pi r^2 h \rho_2} \tag{2.249}$$

$$= \frac{3}{h^3} \left[\frac{1}{4} x^4 \right]_0^h \tag{2.250}$$

$$= \tfrac{3}{4} h \tag{2.251}$$

from the apex, or $\frac{1}{4}h$ from the base.

The centre of mass of the weeble formed from the hemisphere and cone is now (measured from the hemispherical end)

$$\bar{y} = \frac{\underbrace{\frac{5}{8}r \times \frac{2}{3}\pi r^3 \rho_1}_{\text{hemisphere}} + \underbrace{\left(r + \frac{1}{4}h\right) \times \frac{1}{3}\pi r^2 h \rho_2}_{\text{cone}}}{\underbrace{\frac{2}{3}\pi r^3 \rho_1 + \frac{1}{3}\pi r^2 h \rho_2}_{\text{mass of hemisphere + cone}}} \tag{2.252}$$

$$= \frac{\frac{5}{8}r + \frac{1}{2}\frac{h}{r}\frac{\rho_2}{\rho_1}\left(r + \frac{1}{4}h\right)}{1 + \frac{1}{2}\frac{h}{r}\frac{\rho_2}{\rho_1}} \tag{2.253}$$

$$= \frac{5r + \frac{h}{r}\frac{\rho_2}{\rho_1}\left(4r + h\right)}{8 + \frac{4h}{r}\frac{\rho_2}{\rho_1}}. \tag{2.254}$$

$$\therefore \bar{y} = \frac{5 + \frac{h}{r}\frac{\rho_2}{\rho_1}\left(4 + \frac{h}{r}\right)}{8 + \frac{4h}{r}\frac{\rho_2}{\rho_1}} r. \tag{2.255}$$

For the weeble to 'wobble but not fall down',

$$\bar{y} < r, \tag{2.256}$$

$$5 + \frac{h}{r}\frac{\rho_2}{\rho_1}\left(4 + \frac{h}{r}\right) < 8 + \frac{4h}{r}\frac{\rho_2}{\rho_1}, \tag{2.257}$$

$$\frac{h^2}{r^2}\frac{\rho_2}{\rho_1} < 3. \tag{2.258}$$

$$\therefore h < r\sqrt{\frac{3\rho_1}{\rho_2}}. \tag{2.259}$$

So, if $\rho_1 = \rho_2$,

$$h < r\sqrt{3}, \tag{2.260}$$

$$\frac{h}{r} \lesssim 1.732. \tag{2.261}$$

The whole weeble could be CNC-machined out of wood—this would be an excellent project! Perhaps set $\frac{h}{r}$ as 1.73, i.e. just a bit lower than $\sqrt{3}$. Hopefully, subject to the amount of varnishing, it may be possible for the weeble to maintain a tilted position and, with a small disturbance, slowly right itself.

2.8   The Inclined Plane

Consider a skier[12] of mass m on a fixed slope of inclination θ from the horizontal. A tow rope of tension T acts uphill at an angle of ϕ from the slope, as indicated in Fig. 2.13. Depending on the value of T, the angle of the slope, coefficients of static friction etc., the skier can be:

(1) accelerating downhill with acceleration a;
(2) sliding downhill at constant speed (dynamic equilibrium);
(3) in static equilibrium with friction F acting uphill;
(4) in static equilibrium with friction F acting downhill;
(5) sliding uphill at constant speed (dynamic equilibrium);
(6) accelerating uphill with acceleration a.

In order to build a computer model of the situation, as illustrated in Fig. 2.13, we will need to consider the conditions for each scenario.

We shall ignore air resistance and any tumbling/rolling that may result during the non-static situations, i.e. a *particle* model. Model inputs are m, g, θ, T, ϕ and the coefficient of friction μ between the skier and the slope. We will assume that μ is the same for both static and dynamic situations.[13] In all scenarios, we can apply Newton's second law in the y direction perpendicular to the slope:

$$0 = R + T\sin\phi - mg\cos\theta. \tag{2.262}$$

$$\therefore R = mg\cos\theta - T\sin\phi, \tag{2.263}$$

---

[12] In Fig. 2.13, the 'skier' is Sybil the cat, who regularly features in the mechanics problems I set for my students. No cats, real or imagined, were perturbed during the construction of this simulation.

[13] In reality, once something starts to slide, a constant friction model $F = \mu_D R$ may require a different coefficient of friction μ than used to model static friction, i.e. when $F \leq \mu R$.

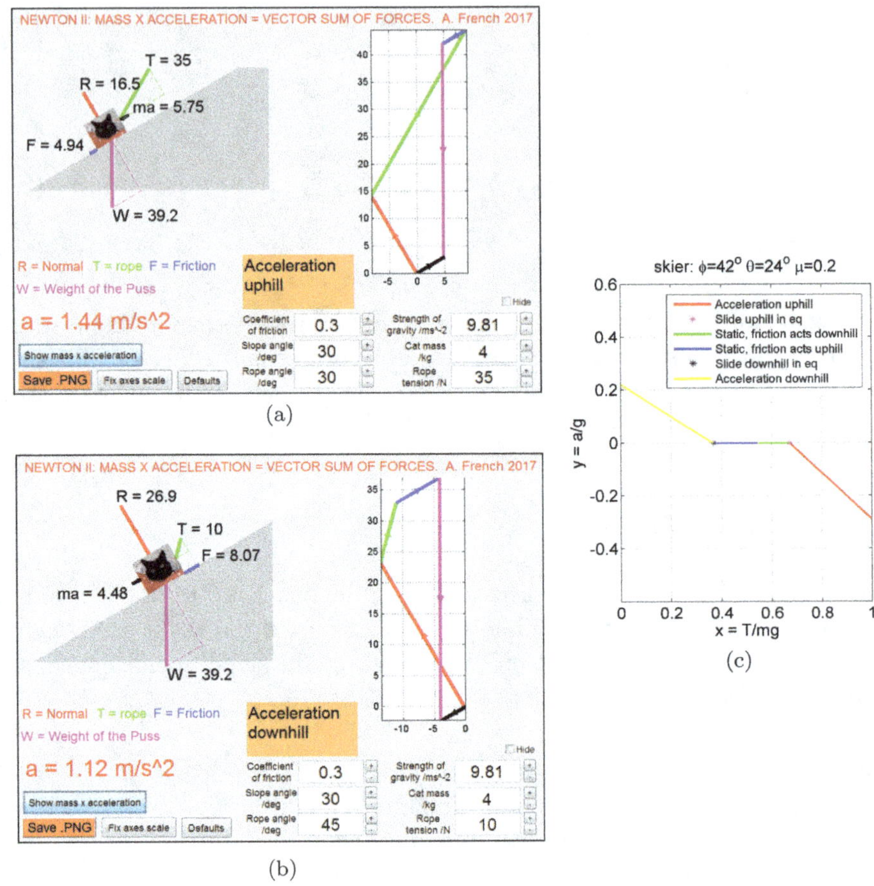

(a)

(b)

(c)

Fig. 2.13. Model of the dynamics of a particle on an inclined plane, such as a ski slope. The vector sum of four forces determines whether the particle (in this case, Sybil the cat grudgingly adopts this moniker) is either in equilibrium or accelerating up or down the slope. The four forces are (1) normal contact force R with slope, (2) rope tension T, (3) friction force F and (4) weight W. (a) and (b) are screenshots of a MATLAB 'app' constructed to illustrate the scenario and compute the forces as well as any acceleration that might arise. The force diagram on the left is compared to a vector sum representation on the right. Whereas the former is easier to draw and understand, the latter is a geometric visualisation of Newton's second law and provides an accurate description of how mass × acceleration equates to the vector sum of forces. (c) plots the *six* possible distinct scenarios resulting from a varied magnitude of rope tension T, from downhill acceleration through various equilibrium states (but where the direction of the friction force vector will change) to uphill acceleration.

where R is the normal contact force. For no slip,

$$F \leq \mu R, \tag{2.264}$$

and when slipping occurs,

$$F = \mu R. \tag{2.265}$$

We can set a simple rule to determine whether the friction force F acts uphill or downhill:

$$\begin{aligned} T\cos\phi > mg\sin\theta: \;\; & F \text{ acts downhill;} \\ T\cos\phi \leq mg\sin\theta: \;\; & F \text{ acts uphill.} \end{aligned} \tag{2.266}$$

2.8.1 *Friction acts downhill*

If F acts *downhill*, i.e. $T\cos\phi > mg\sin\theta$, apply Newton's second law in the x direction *up* the slope:

$$ma = -mg\sin\theta - F + T\cos\phi. \tag{2.267}$$

Assume equilibrium ($a = 0$) and then test it with $F \leq \mu R$. Assuming equilibrium, $ma = \ldots$ becomes

$$F = T\cos\phi - mg\sin\theta. \tag{2.268}$$

Since $F \leq \mu R$ and $R = mg\cos\theta - T\sin\phi$,

$$F \leq \mu R \tag{2.269}$$

$$T\cos\phi - mg\sin\theta \leq \mu(mg\cos\theta - T\sin\phi) \tag{2.270}$$

$$T(\cos\phi + \mu\sin\phi) \leq mg(\mu\cos\theta + \sin\theta). \tag{2.271}$$

$$\therefore T \leq \frac{\mu\cos\theta + \sin\theta}{\cos\phi + \mu\sin\phi}mg. \tag{2.272}$$

So, if

$$T > \frac{\mu\cos\theta + \sin\theta}{\cos\phi + \mu\sin\phi}mg, \tag{2.273}$$

the acceleration is *uphill* with a magnitude of a, where

$$a = -g\sin\theta - \mu\left(g\cos\theta - \frac{T}{m}\sin\phi\right) + \frac{T}{m}\cos\phi. \tag{2.274}$$

$$\therefore a = \frac{T}{m}(\cos\phi + \mu\sin\phi) - g(\sin\theta + \mu\cos\theta). \tag{2.275}$$

If $T\cos\phi > mg\sin\theta$ and $T < \frac{\mu\cos\theta+\sin\theta}{\cos\phi+\mu\sin\phi}mg$, then the skier is *static, with friction acting downhill*. If $T\cos\phi > mg\sin\theta$ and $T = \frac{\mu\cos\theta+\sin\theta}{\cos\phi+\mu\sin\phi}mg$, the skier is *on the point of sliding uphill* and may well do so at constant speed.

2.8.2 *Friction acts uphill*

If F acts *uphill*, i.e. $T\cos\phi \leq mg\sin\theta$, apply Newton's second law in the x direction *down* the slope:

$$ma = mg\sin\theta - F - T\cos\phi. \tag{2.276}$$

Assume equilibrium ($a = 0$) and then test it with $F \leq \mu R$. Assuming equilibrium, $ma = \ldots$ becomes

$$F = mg\sin\theta - T\cos\phi. \tag{2.277}$$

Since $F \leq \mu R$ and $R = mg\cos\theta - T\sin\phi$,

$$F \leq \mu R \tag{2.278}$$

$$mg\sin\theta - T\cos\phi \leq \mu(mg\cos\theta - T\sin\phi) \tag{2.279}$$

$$mg\left(\sin\theta - \mu\cos\theta\right) \leq T\left(\cos\phi - \mu\sin\phi\right). \tag{2.280}$$

$$\therefore T \geq \frac{\sin\theta - \mu\cos\theta}{\cos\phi - \mu\sin\phi}mg. \tag{2.281}$$

So, if

$$T < \frac{\sin\theta - \mu\cos\theta}{\cos\phi - \mu\sin\phi}mg, \tag{2.282}$$

the acceleration is *downhill* with a magnitude of a, where

$$a = g\sin\theta - \mu\left(g\cos\theta - \frac{T}{m}\sin\phi\right) - \frac{T}{m}\cos\phi. \tag{2.283}$$

$$\therefore a = g\left(\sin\theta - \mu\cos\theta\right) - \frac{T}{m}\left(\cos\phi - \mu\sin\phi\right). \tag{2.284}$$

If $T\cos\phi \leq mg\sin\theta$ and $T > \frac{\sin\theta - \mu\cos\theta}{\cos\phi - \mu\sin\phi}mg$, then the skier is *static, with friction acting uphill*. If $T\cos\phi \leq mg\sin\theta$ and $T = \frac{\sin\theta - \mu\cos\theta}{\cos\phi - \mu\sin\phi}mg$, the skier is *on the point of sliding downhill* and may well do so at constant speed.

In summary, as plotted in Fig. 2.13(c), let us first define the dimensionless variables of rope tension/weight and acceleration/free-fall acceleration:

$$x = \frac{T}{mg}, \; y = \frac{a}{g}. \tag{2.285}$$

If $\left(x > \frac{\sin\theta}{\cos\phi}\right)$ and $\left(x > \frac{\mu\cos\theta + \sin\theta}{\cos\phi + \mu\sin\phi}\right)$, then $y = x\left(\cos\phi + \mu\sin\phi\right) - \sin\theta - \mu\cos\theta$, and acceleration a is uphill.

If $x = \frac{\mu\cos\theta + \sin\theta}{\cos\phi + \mu\sin\phi}$, then $y = 0$. The skier can slide uphill at constant speed, i.e. in dynamic equilibrium.

If $\left(x > \frac{\sin\theta}{\cos\phi}\right)$ and $\left(x < \frac{\mu\cos\theta + \sin\theta}{\cos\phi + \mu\sin\phi}\right)$, then $y = 0$. The skier is in static equilibrium, and F acts downhill.

If $\left(x \leq \frac{\sin\theta}{\cos\phi}\right)$ and $\left(x > \frac{\sin\theta - \mu\cos\theta}{\cos\phi - \mu\sin\phi}\right)$, then $y = 0$. The skier is in static equilibrium, and F acts uphill.

If $x = \frac{\sin\theta - \mu\cos\theta}{\cos\phi - \mu\sin\phi}$, then $y = 0$. The skier can slide downhill at constant speed, i.e. in dynamic equilibrium.

If $\left(x \leq \frac{\sin\theta}{\cos\phi}\right)$ and $\left(x < \frac{\sin\theta - \mu\cos\theta}{\cos\phi - \mu\sin\phi}\right)$, then $y = \sin\theta - \mu\cos\theta - x(\cos\phi - \mu\sin\phi)$, and acceleration a is downhill.

These scenarios are used to colour-code the y versus x plot in Fig. 2.13(c). Example scenarios (accelerating uphill and accelerating downhill) are represented using a MATLAB app, which calculates everything about the problem and, importantly, plots a vector sum diagram (which might be tricky to draw accurately without a prior calculation). In both scenarios, the vector polygon is not closed. The gap (the black arrow) is the resulting mass $\times$ acceleration. Sybil the cat was of course very willing to participate in this simulation.

2.9 Rolling Down a Slope

If a cylinder or sphere of radius r and mass m is rolling down a slope of elevation θ without slipping (see Fig. 2.14), the speed v of the centre of mass must equal $r\dot{\theta}$, where $\dot{\theta} = \frac{d\theta}{dt}$ is the rate of change of rotation angle, or 'angular speed'. Therefore, the acceleration down the slope is

$$a = r\ddot{\theta}. \tag{2.286}$$

Newton's second law in the x direction (down the slope) implies

$$ma = mg\sin\theta - F, \tag{2.287}$$

where F is the friction force between the rolling object and the slope. There *must* be friction to prevent it from sliding.

Evaluating Newton's second law in the y direction (perpendicular to slope),

$$0 = R - mg\cos\theta. \tag{2.288}$$

$$\therefore\ R = mg\cos\theta, \tag{2.289}$$

where R is the normal contact force.

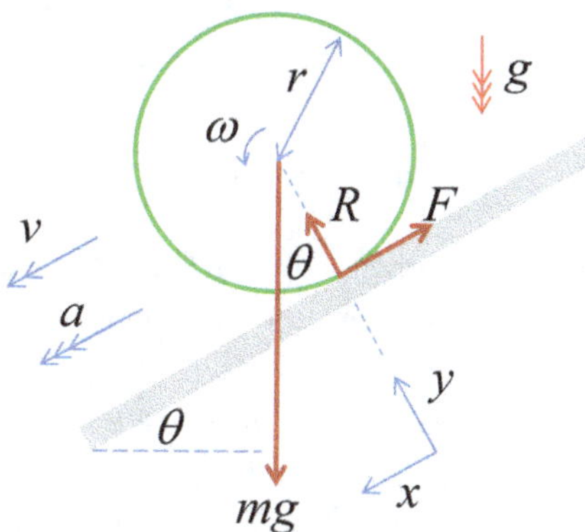

Fig. 2.14. Diagram of forces acting on a sphere or cylinder of radius r that cause it to roll down a slope of elevation θ. The normal contact force is R, and this will always act perpendicular to the slope and point towards the centre of the rolling circle. It therefore does no work and has no turning effect about the centre of mass. The weight mg, where g is the strength of gravity, acts through the centre of the sphere or cylinder and also provides no turning moment about the centre of mass. Therefore, to cause the system to roll, there must be a friction force F parallel to the slope. $\omega = \frac{v}{r}$ is the angular speed of rolling, which implies a rolling condition that the velocity v of the centre of mass down the slope equals $r\omega$, which is the tangential velocity of a point on the circular boundary. The equality of these velocities implies that no slip occurs at the point of contact, and indeed the cylinder surface is *instantaneously stationary* at this point. $a = \dfrac{g\sin\theta}{1+\frac{I}{mr^2}}$ is the acceleration down the slope, which is a constant, given slope angle θ, moment of inertia I, mass m and strength of gravity g. A cylinder of mass m and radius r will have a moment of inertia of $I = \frac{1}{2}mr^2$, whereas a solid sphere will have $I = \frac{2}{5}mr^2$.

If we use 'torque = moment of inertia $\times$ angular acceleration',

$$Fr = I\ddot{\theta}. \tag{2.290}$$

Now, since $\ddot{\theta} = \frac{a}{r}$,

$$F = \frac{Ia}{r^2}. \tag{2.291}$$

Therefore,

$$ma = mg\sin\theta - \frac{Ia}{r^2} \tag{2.292}$$

$$a = \frac{g\sin\theta}{1 + \frac{I}{mr^2}}. \tag{2.293}$$

Let $I = \sigma mr^2$, where $\sigma = \frac{1}{2}$ for a solid cylinder and $\sigma = \frac{2}{5}$ for a solid sphere.

$$\therefore a = \frac{g\sin\theta}{1 + \sigma}. \tag{2.294}$$

For a solid cylinder ($\sigma = \frac{1}{2}$),

$$a = \tfrac{2}{3}g\sin\theta, \tag{2.295}$$

and for a solid sphere ($\sigma = \frac{2}{5}$),

$$a = \tfrac{5}{7}g\sin\theta. \tag{2.296}$$

We can also use the no-slip condition $F \leq \mu R$ to determine the minimum coefficient of friction μ for both the cylinder and sphere scenarios:

$$\frac{Ia}{r^2} \leq \mu mg\cos\theta \tag{2.297}$$

$$\frac{\sigma mr^2}{r^2}\frac{g\sin\theta}{1+\sigma} \leq \mu mg\cos\theta. \tag{2.298}$$

$$\therefore \mu \geq \frac{\sigma}{\sigma + 1}\tan\theta. \tag{2.299}$$

For a solid cylinder ($\sigma = \frac{1}{2}$),

$$\mu \geq \tfrac{1}{3}\tan\theta, \tag{2.300}$$

and for a solid sphere ($\sigma = \frac{2}{5}$),

$$\mu \geq \tfrac{2}{7}\tan\theta. \tag{2.301}$$

Compare this to a rectangular object not sliding down a slope:

$$\mu \geq \tan\theta. \tag{2.302}$$

2.10 Slide and Roll the Loop

2.10.1 *Scope*

A classic dynamics problem is to *slide* or *roll* a small sphere (or cylinder) of radius k and mass m along a track, starting from rest with its centre of mass at height h. Once height h is lost, the sphere slides or rolls around a circular track section of radius r. Two interesting scenarios will now be considered (see Fig. 2.15). First, what is the maximum ratio $\beta = \frac{r-k}{h}$ such that the sphere completes an entire loop-the-loop without falling off the track? The second scenario is to work out β such that the sphere falls off the track at some angle Θ, becomes a projectile and then lands *exactly* at the base of the loop-the-loop. In both cases, we ignore friction and air resistance.

We will show that for a sphere to *slide all the way round*, $h > 2.5(r - k)$, whereas to roll, $h > 2.7(r - k)$. For a sphere to *fall off and land at the base of the loop*, $h = 1.75(r - k)$ for a sliding sphere and $h = 1.85(r - k)$ for a rolling sphere.

2.10.2 *Speed from conservation of energy*

The sphere will enter the loop-the-loop (horizontally) with speed u. By conservation of energy,

$$mgh = \tfrac{1}{2}mu^2 + \rho\tfrac{1}{2}I\Omega^2. \tag{2.303}$$

$I = \tfrac{2}{5}mk^2$ is the moment of inertia of a solid sphere of mass m and radius k, whereas $I = \tfrac{1}{2}mk^2$ is the moment of inertia of a solid cylinder of radius k and mass m. Ω is its angular speed (at the bottom of the loop), and ρ is a binary parameter that determines whether the sphere is sliding or rolling. $\rho = 1$ means it is rolling, whereas $\rho = 0$ means it is sliding. For no slip,[14] the rolling condition is

$$u = k\Omega. \tag{2.304}$$

Define

$$\sigma = \frac{I}{mk^2}, \tag{2.305}$$

---

[14]Let us assume in the 'pure sliding' scenario that there is no rotation. Clearly, some hybrid of the two might be possible.

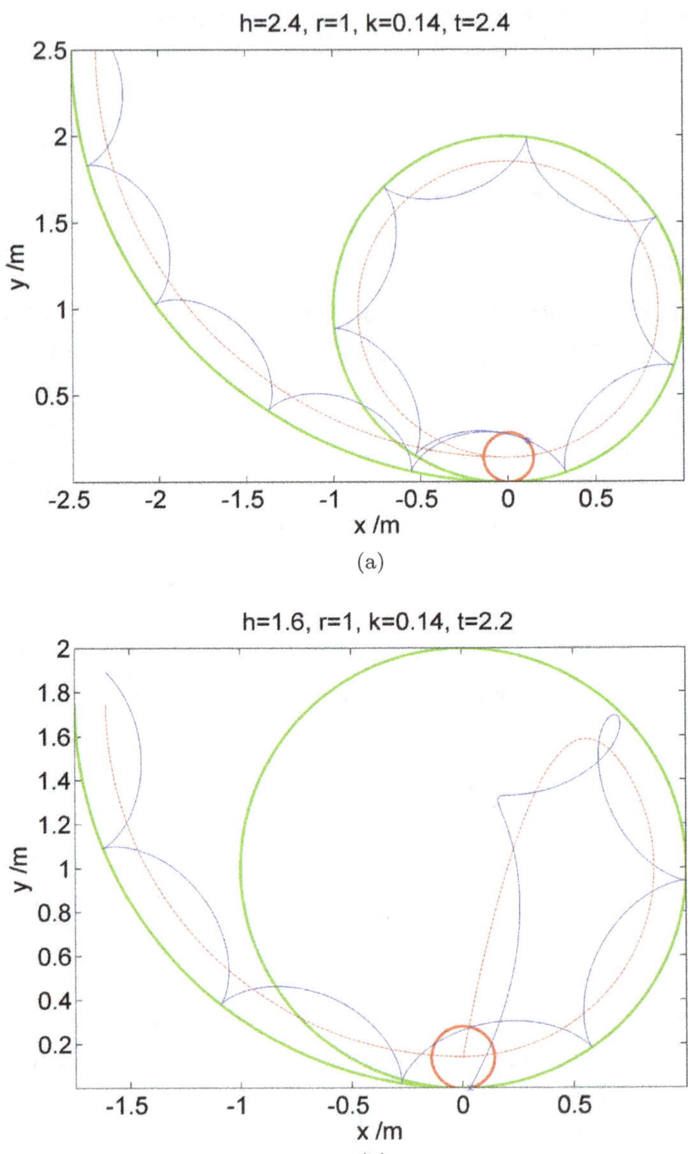

Fig. 2.15. An interesting rotational dynamics challenge is to model the rolling (or sliding) of a sphere or cylinder of radius k around a vertical 'loop-the-loop' consisting of a circle of radius r. The sphere or cylinder is released from rest from a quarter-circle track of radius h (as plotted is a quarter circle, but the geometry doesn't matter; the purpose is to convert GPE into translational and rotational KE at the bottom of the circle). Two scenarios are of particular interest. (a) To find the starting height h such that the sphere or cylinder rolls all the way round the loop-the-loop without falling off. The criteria is $h > \frac{1}{2}(5 + \rho\sigma)(r - k)$, where $\rho = 1$ means rolling and $\rho = 0$ means sliding. $\sigma = \frac{1}{2}$ for a cylinder and $\sigma = \frac{2}{5}$ for a sphere. (b) To determine h such that the sphere or cylinder falls off and lands at exactly the base of the loop-the-loop. In this situation, $h = \frac{1}{4}(7 + \rho\sigma)(r - k)$. For all motion, air resistance and rolling resistance are ignored, i.e. at all times, we expect GPE + linear KE + rotational KE to sum to the same value.

where $\sigma = \frac{2}{5}$ is for a sphere and $\sigma = \frac{1}{2}$ is for a cylinder. We consider a sphere in this problem but keep the parameter σ, so we can recalculate easily for a cylinder (or, indeed, a hollow sphere or hollow cylinder) without having to redo the analysis. Hence, the energy conservation expression becomes

$$mgh = \tfrac{1}{2}mu^2 + \tfrac{1}{2}\rho\sigma mk^2 \frac{u^2}{k^2} \tag{2.306}$$

$$2gh = u^2(1 + \rho\sigma) \tag{2.307}$$

$$u = \sqrt{\frac{2gh}{1 + \rho\sigma}}. \tag{2.308}$$

$$\therefore u = \alpha\sqrt{gh}, \tag{2.309}$$

where

$$\alpha = \sqrt{\frac{2}{1 + \rho\sigma}}. \tag{2.310}$$

This means for the sliding case of $\rho = 0$,

$$\alpha = \sqrt{2} \approx 1.414, \tag{2.311}$$

and for the sphere rolling case of $\rho = 1$,

$$\alpha = \sqrt{\frac{2}{1 + \frac{2}{5}}} = \sqrt{\frac{10}{7}} \approx 1.195. \tag{2.312}$$

When the sphere reaches angle θ (as measured from the downward vertical) around the circular track, assume it is still in contact and has translational (tangential) speed v and angular speed ω. Again using conservation of energy, noting the height gained by the sphere is $(r - k)(1 - \cos\theta)$,

$$mgh = \tfrac{1}{2}mv^2 + \rho\tfrac{1}{2}\sigma mk^2\omega^2 + mg(r - k)(1 - \cos\theta). \tag{2.313}$$

For no slip, the rolling condition is

$$v = k\omega. \tag{2.314}$$

Therefore, using $(r - k) = \beta h$,

$$2gh = v^2 + \rho\sigma v^2 + 2g\beta h(1 - \cos\theta) \tag{2.315}$$

$$v = \sqrt{\frac{2gh - 2g\beta h(1 - \cos\theta)}{1 + \rho\sigma}}. \tag{2.316}$$

$$\therefore v = \alpha\sqrt{gh}\sqrt{1 - \beta(1 - \cos\theta)}. \tag{2.317}$$

2.10.3 *The normal contact force and condition for contact*

Newton's second law applied to the sphere implies that the radially inward (centripetal) mass × acceleration must equate to the normal contact force R minus the radially outward component of the weight of the sphere:

$$\frac{mv^2}{r-k} = R - mg\cos\theta. \tag{2.318}$$

Therefore, again using $(r - k) = \beta h$ for brevity,

$$R = \frac{mv^2}{\beta h} + mg\cos\theta. \tag{2.319}$$

For the sphere to remain in contact with the loop-the-loop track, $R > 0$. Using the expressions for R and v above,

$$\frac{m\alpha^2 gh}{\beta h}\left(1 - \beta(1 - \cos\theta)\right) + mg\cos\theta > 0 \tag{2.320}$$

$$\frac{\alpha^2}{\beta}\left(1 - \beta(1 - \cos\theta)\right) + \cos\theta > 0 \tag{2.321}$$

$$(\alpha^2 + 1)\cos\theta > \alpha^2 - \frac{\alpha^2}{\beta}. \tag{2.322}$$

$$\therefore \cos\theta > -\frac{\alpha^2\left(\frac{1}{\beta} - 1\right)}{\alpha^2 + 1}. \tag{2.323}$$

Note that $h > (r - k)$, $\beta = \frac{r-k}{h}$ and, therefore, $1 > \beta$ and hence $\frac{1}{\beta} > 1$. This explains the purpose of the last step: to clearly define that the lower limit of $\cos\theta$ is *negative*.

Recall loop-entering speed $u = \alpha\sqrt{gh}$ and $\alpha = \sqrt{\frac{2}{1+\rho\sigma}}$. Therefore,

$$\alpha^2 = \frac{2}{1 + \rho\sigma} \tag{2.324}$$

$$\alpha^2 + 1 = \frac{3 + \rho\sigma}{1 + \rho\sigma} \tag{2.325}$$

$$\frac{\alpha^2}{\alpha^2 + 1} = \frac{2}{3 + \rho\sigma}. \tag{2.326}$$

$\beta = \frac{r-k}{h}$ also implies

$$\frac{1}{\beta} - 1 = \frac{h}{r-k} - 1 = \frac{h - r + k}{r - k}. \tag{2.327}$$

So, the critical angle Θ (in radians), where the sphere falls off the track, can be computed from $\cos\Theta = -\frac{\alpha^2\left(\frac{1}{\beta}-1\right)}{\alpha^2+1}$:

$$\Theta = \pi - \cos^{-1}\left(\frac{2}{3+\rho\sigma}\,\frac{h-r+k}{r-k}\right). \tag{2.328}$$

At what speed does the sphere leave the track? The speed is $v = \alpha\sqrt{gh}\sqrt{1-\beta(1-\cos\theta)}$ and $\theta = \Theta$. Hence,

$$v = \alpha\sqrt{gh}\sqrt{1-\beta(1-\cos\Theta)} \tag{2.329}$$

$$= \alpha\sqrt{gh}\sqrt{1-\beta\left(1+\frac{\alpha^2}{\alpha^2+1}\left(\frac{1}{\beta}-1\right)\right)} \tag{2.330}$$

$$= \alpha\sqrt{gh}\sqrt{1-\beta-\frac{\alpha^2}{\alpha^2+1}(1-\beta)} \tag{2.331}$$

$$= \alpha\sqrt{gh}\sqrt{(1-\beta)\left(1-\frac{\alpha^2}{\alpha^2+1}\right)}. \tag{2.332}$$

$$\therefore v = \frac{\alpha}{\sqrt{\alpha^2+1}}\sqrt{gh}\sqrt{(1-\beta)}. \tag{2.333}$$

Now, the minimum value of $\cos\Theta$ is -1, i.e. $\Theta = \pi$ (which corresponds to the top of the loop-the-loop). Therefore, in order for the sphere to make it all the way round,

$$\frac{\alpha^2}{\alpha^2+1}\left(\frac{1}{\beta}-1\right) > 1. \tag{2.334}$$

Using $\frac{1}{\beta}-1 = \frac{h-r+k}{r-k}$ and $\frac{\alpha^2}{\alpha^2+1} = \frac{2}{3+\rho\sigma}$,

$$\frac{2}{3+\rho\sigma}\,\frac{h-r+k}{r-k} > 1 \tag{2.335}$$

$$h-r+k > \tfrac{3+\rho\sigma}{2}(r-k). \tag{2.336}$$

$$\therefore h > \tfrac{1}{2}(5+\rho\sigma)(r-k). \tag{2.337}$$

For sliding ($\rho = 0$),

$$h > \tfrac{5}{2}(r-k). \tag{2.338}$$

For a rolling sphere ($\rho = 1$, $\sigma = \tfrac{2}{5}$),

$$h > \tfrac{27}{10}(r-k). \tag{2.339}$$

2.10.4 *For how long is the sphere on the track?*

If the sphere is executing vertical circular motion of radius R,

$$v = R\frac{d\theta}{dt}. \qquad (2.340)$$

$$\therefore t = R \int_0^\theta \frac{1}{v(\theta')}\,d\theta'. \qquad (2.341)$$

From above, if we consider the first part of the motion to be a vertical circle of radius h, then $\beta = \frac{h-k}{h}$ and

$$v = \alpha\sqrt{gh}\sqrt{1 - \beta(1 - \cos\theta)}, \qquad (2.342)$$

whereas for a loop-the-loop of radius $r - k$, $\beta = \frac{r-k}{h}$.

The integral $t = R \int_0^\theta \frac{1}{v(\theta')}\,d\theta'$ will need to be evaluated numerically. A popular technique is to fit a *piecewise cubic spline* to $\frac{1}{v(\theta)}$ over the required range of θ, using a linearly spaced vector of θ values. The cubic sections can then be integrated analytically (i.e. each bit is a quartic). For our example, consider 1,000 values for the initial arc of radius $h - k$ from $\theta = -\frac{1}{2}\pi$ to 0 and then another set of values separated by $\Delta\theta = \frac{\pi}{2000}$ from $\theta = 0$ to Θ for the loop-the-loop track.

2.10.5 *Calculate h such that subsequent projectile motion passes through $(0, 0)$*

Now, consider a projectile motion such that the centre of mass of a sphere that falls off the track passes through $(0, 0)$, i.e. the bottom of the loop-the-loop. The sphere can only fall off when $\theta > \frac{1}{2}\pi$, so define angle ϕ such that $\theta = \frac{1}{2}\pi + \phi$.

Now,

$$\cos\theta = \cos\left(\tfrac{1}{2}\pi + \phi\right) = \cos\tfrac{1}{2}\pi\cos\phi - \sin\tfrac{1}{2}\pi\sin\phi = -\sin\phi. \qquad (2.343)$$

For brevity, define

$$s = \sin\phi = \frac{\alpha^2}{\alpha^2 + 1}\left(\frac{1}{\beta} - 1\right) = \frac{2}{3 + \rho\sigma}\frac{h - r + k}{r - k}, \qquad (2.344)$$

noting the fall-off angle $\cos\theta = -\frac{\alpha^2}{\alpha^2+1}\left(\frac{1}{\beta} - 1\right)$ calculated in the previous section.

Projectile motion from the fall-off point (where we set time $t = 0$) has an x, y trajectory

$$x = (r - k)\cos\phi - vt\sin\phi, \qquad (2.345)$$

$$y = r - k + (r - k)\sin\phi + vt\cos\phi - \tfrac{1}{2}gt^2, \qquad (2.346)$$

where v is the speed at the fall-off point. When $x = 0$,

$$(r - k)\cos\phi = vt\sin\phi. \tag{2.347}$$

Therefore,

$$t = \frac{(r - k)\cos\phi}{v\sin\phi}. \tag{2.348}$$

When $y = 0$,

$$(r - k)(1 + \sin\phi) + vt\cos\phi = \tfrac{1}{2}gt^2. \tag{2.349}$$

Therefore,

$$(r - k)(1 + \sin\phi) + v\frac{(r - k)\cos\phi}{v\sin\phi}\cos\phi = \tfrac{1}{2}g\frac{(r - k)^2\cos^2\phi}{v^2\sin^2\phi} \tag{2.350}$$

$$1 + \sin\phi + \frac{\cos^2\phi}{\sin\phi} = \tfrac{1}{2}g\frac{(r - k)}{v^2\tan^2\phi} \tag{2.351}$$

$$\frac{\sin\phi + \sin^2\phi + \cos^2\phi}{\sin\phi} = \tfrac{1}{2}g\frac{(r - k)}{v^2\tan^2\phi}. \tag{2.352}$$

$$\therefore v^2 = \tfrac{1}{2}g\frac{(r - k)\sin\phi}{(1 + \sin\phi)\tan^2\phi}. \tag{2.353}$$

Now, from the trigonometric identity $\sin^2\phi + \cos^2\phi = 1$,

$$\frac{1}{\sin^2\phi} = 1 + \frac{1}{\tan^2\phi} \tag{2.354}$$

$$\frac{1}{\tan^2\phi} = \frac{1}{\sin^2\phi} - 1 \tag{2.355}$$

$$\frac{1}{\tan^2\phi} = \frac{1 - \sin^2\phi}{\sin^2\phi}. \tag{2.356}$$

Therefore, if $\sin\phi = s$,

$$\frac{1}{\tan^2\phi} = \frac{1 - s^2}{s^2}. \tag{2.357}$$

We can now write $v^2 = \tfrac{1}{2}g\frac{(r-k)\sin\phi}{(1+\sin\phi)\tan^2\phi}$ in terms of $\beta = \frac{r-k}{h}$, s and h:

$$v^2 = \tfrac{1}{2}g\beta h\frac{s}{1+s}\frac{1-s^2}{s^2} = \tfrac{1}{2}g\beta h\frac{s}{1+s}\frac{(1-s)(1+s)}{s^2} \tag{2.358}$$

$$= \tfrac{1}{2}g\beta h\frac{1-s}{s}. \tag{2.359}$$

Now,

$$s = \frac{\alpha^2}{\alpha^2 + 1}\left(\frac{1}{\beta} - 1\right) = \frac{\alpha^2}{\alpha^2 + 1}\left(\frac{1-\beta}{\beta}\right). \tag{2.360}$$

$$\therefore \frac{\alpha^2}{\alpha^2 + 1} = \frac{\beta s}{1 - \beta}. \tag{2.361}$$

From the analysis of the previous section, the expression for v^2 was derived to be

$$v^2 = \frac{\alpha^2}{\alpha^2 + 1} gh(1 - \beta). \tag{2.362}$$

Using $\frac{\alpha^2}{\alpha^2+1} = \frac{\beta s}{1-\beta}$,

$$v^2 = \frac{\beta s}{1 - \beta} gh\,(1 - \beta) \tag{2.363}$$

$$= \beta sgh. \tag{2.364}$$

Equating $v^2 = \beta sgh$ with $v^2 = \frac{1}{2}g\beta h\frac{1-s}{s}$,

$$\frac{1}{2}\frac{1 - s}{s} = s,$$

$$2s^2 + s - 1 = 0,$$

$$(2s - 1)(s + 1) = 0,$$

i.e.

$$s = -1 \text{ or } s = \tfrac{1}{2}. \tag{2.365}$$

Rather beautifully, since $\sin\phi = s$, if we take $s = \frac{1}{2}$, this means the sliding sphere and the rolling sphere will both fall off the track at $\phi = 30°$.

Now,

$$s = \frac{\alpha^2}{\alpha^2 + 1}\left(\frac{1}{\beta} - 1\right) \tag{2.366}$$

$$\frac{\alpha^2 + 1}{\alpha^2}s + 1 = \frac{1}{\beta}. \tag{2.367}$$

$$\therefore \beta = \frac{1}{\frac{\alpha^2+1}{\alpha^2}s + 1}. \tag{2.368}$$

$\beta = \frac{r-k}{h} > 0$ so $s = -1$ is unphysical. So, using $s = \frac{1}{2}$, $\frac{\alpha^2}{\alpha^2+1} = \frac{2}{3+\rho\sigma}$ and $\beta = \frac{r-k}{h}$,

$$\frac{r - k}{h} = \frac{1}{\frac{3+\rho\sigma}{2}\frac{1}{2} + 1}. \tag{2.369}$$

The drop distance h such that the sphere will fall off the track at $\theta = 120°$ and land at $(0,0)$ (i.e. the base of the loop-the-loop) is therefore

$$h = \tfrac{1}{4}(7 + \rho\sigma)(r - k). \tag{2.370}$$

So, for sliding ($\rho = 0$) cylinders or spheres, $h = \frac{7}{4}(r - k)$. For a rolling cylinder, ($\rho = 1$, $\sigma = \frac{1}{2}$), $h = \frac{15}{8}(r-k)$, and for a rolling sphere, ($\rho = 1$, $\sigma = \frac{2}{5}$), $h = \frac{37}{20}(r-k)$.

2.11 Atwood Machine

An Atwood machine is a system of masses connected by (ideally) 'light and inextensible strings' and 'light and frictionless pulleys'. They are named after the mathematician George Atwood,[15] who invented machines of this type to help illustrate Newton's laws of motion. They have been used to challenge/torture Physics students ever since.[16] We will analyse a three-mass Atwood variant described in Fig. 2.16. We require *four* equations to solve for the four unknowns: the three accelerations $a_{1,2,3}$ of the masses and the string tension T.

First, let's write down Newton's second law for all the masses ($m, \alpha m$ and βm). Assume $\beta > \alpha > 1$. Then,

$$ma_1 = T - mg, \tag{2.371}$$

$$\alpha m a_2 = 2T - \alpha mg, \tag{2.372}$$

$$\beta m a_3 = \beta mg - T. \tag{2.373}$$

Use the inextensibility of the connecting strings, i.e. 'the conservation of string' to generate the fourth equation:

$$2a_2 = a_3 - a_1. \tag{2.374}$$

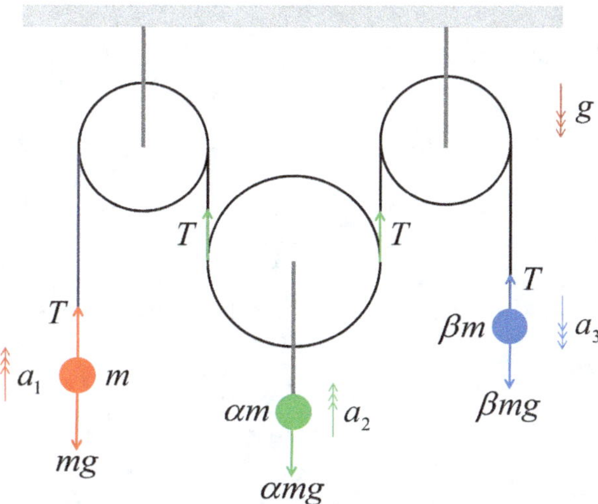

Fig. 2.16. An *Atwood machine* consisting of a mass m connected via a fixed pulley to a freely suspended pulley with mass αm and then, via another fixed pulley, to a mass βm. Assume $\beta > \alpha > 1$, the strings are light and inextensible and the pulleys are light and frictionless. The accelerations are (in terms of gravitational field strength g) $a_3 = \frac{4\beta - 3\alpha + \alpha\beta}{4\beta + \alpha + \alpha\beta} g$, $a_1 = (\beta - 1) g - \beta a_3$ and $a_2 = \frac{1}{2}(a_3 - a_1)$. The string tension is $T = m(a_1 + g)$.

---

[15] George Atwood (1745–1807).

[16] But mathematical torture can be so enjoyable(!), as in the case of Morin's *Introduction to Classical Mechanics* [26].

The two ends of the string connect to the central pulley, so the difference between the *downward* movement of mass βm and the *upward* movement of mass m must balance twice the upward movement of the central pulley (and hence mass αm, which is connected to it via a light rigid vertical rod). Since the string is deemed inextensible, the same relationship must hold for the velocities and accelerations of the three masses.

Eliminating T between the first and third equations,

$$ma_1 + \beta m a_3 = \beta m g - m g. \tag{2.375}$$

$$\therefore a_1 = (\beta - 1)\, g - \beta a_3, \tag{2.376}$$

and between the second and third equations,

$$\alpha m a_2 + 2\beta m a_3 = 2\beta m g - \alpha m g. \tag{2.377}$$

$$\therefore \alpha a_2 + 2\beta a_3 = (2\beta - \alpha)\, g. \tag{2.378}$$

Now, substitute for $a_2 = \frac{1}{2}(a_3 - a_1)$ in $\alpha a_2 + 2\beta a_3 = (2\beta - \alpha)\, g$,

$$\text{i.e. } \tfrac{1}{2}\alpha\,(a_3 - a_1) + 2\beta a_3 = (2\beta - \alpha)\, g, \tag{2.379}$$

and then substitute $a_1 = (\beta - 1)g - \beta a_3$:

$$\tfrac{1}{2}\alpha\,(a_3 - (\beta - 1)\, g + \beta a_3) + 2\beta a_3 = (2\beta - \alpha)\, g \tag{2.380}$$

$$a_3\left(\tfrac{1}{2}\alpha + \tfrac{1}{2}\alpha\beta + 2\beta\right) = g(2\beta - \alpha + \tfrac{1}{2}\alpha\,(\beta - 1)) \tag{2.381}$$

$$a_3\,(\alpha + \alpha\beta + 4\beta) = g(4\beta - 2\alpha + \alpha\beta - \alpha). \tag{2.382}$$

$$\therefore a_3 = \frac{4\beta - 3\alpha + \alpha\beta}{4\beta + \alpha + \alpha\beta}\, g. \tag{2.383}$$

You can now back-substitute to find the other accelerations and T to yield the full solution set:

$$a_3 = \frac{4\beta - 3\alpha + \alpha\beta}{4\beta + \alpha + \alpha\beta}\, g, \tag{2.384}$$

$$a_1 = (\beta - 1)g - \beta a_3, \tag{2.385}$$

$$a_2 = \tfrac{1}{2}(a_3 - a_1), \tag{2.386}$$

$$T = m(a_1 + g). \tag{2.387}$$

For example, with $\alpha = 2$ and $\beta = 3$ (which is Problem 3.30 on p. 76 from Morin [26]):

$$a_3 = \frac{4\,(3) - 3\,(2) + (2)\,(3)}{4(3) + (2) + (2)(3)}\, g = \tfrac{3}{5}g, \tag{2.388}$$

$$a_1 = (3 - 1)\, g - 3a_3 = 2g - \tfrac{9}{5}g = \tfrac{1}{5}g, \tag{2.389}$$

$$a_2 = \tfrac{1}{2}\left(\tfrac{3}{5} - \tfrac{1}{5}\right)g = \tfrac{1}{5}g, \tag{2.390}$$

$$T = \tfrac{6}{5}mg. \tag{2.391}$$

2.12 A Swinging Door

This problem is a classic demonstration of rotational dynamics principles and the combination of (1) Newton's second law, (2) conservation of energy, (3) the use of polar coordinates and (4) torque = moment of inertia $\times$ angular acceleration. This type of problem is an appropriate challenge to complete a basic mechanics course and has the additional interest of analysing the forces acting on a hinge.

Consider a uniform rectangular block, a kind of trapdoor, of mass m. It is $2a$ high and $2b$ deep, as illustrated in Fig. 2.17. The width c doesn't matter, i.e. we treat the trapdoor as a *lamina*, a rigid two-dimensional structure. It is initially placed at a right-angled edge (e.g. atop a door) with one edge (of height $2a$) pointing vertically upwards. It is hinged (frictionlessly) at the edge. The block is initially at rest but

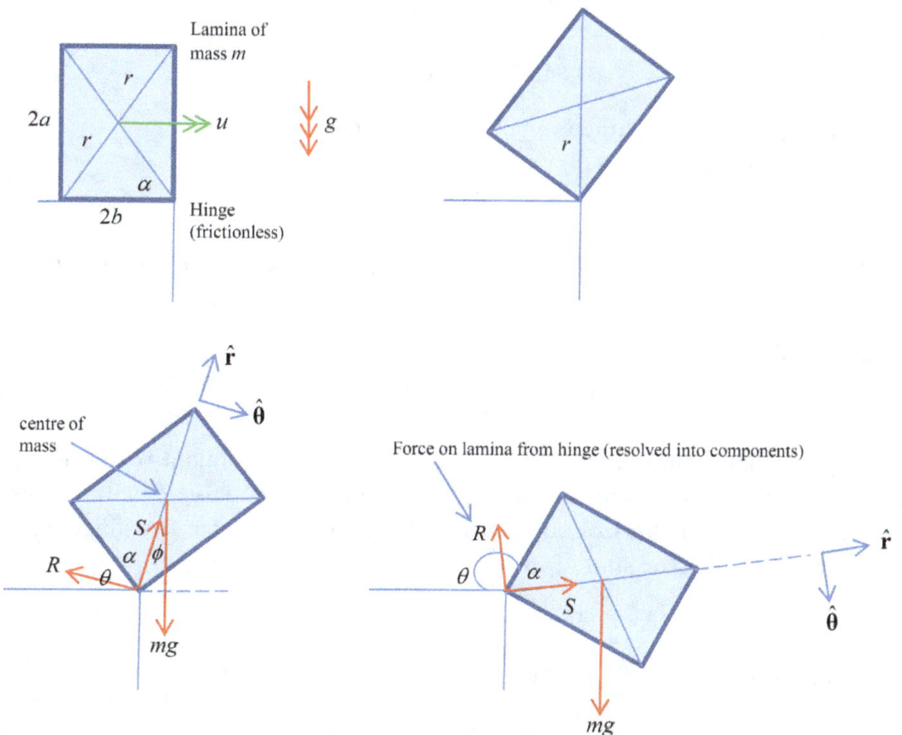

Fig. 2.17. A swinging rectangular trapdoor of height $2a$ and width $2b$ pivots about a (frictionless) hinge. The trapdoor is given an impulse (momentum change), which results in an instantaneous horizontal velocity u. Newton's laws of motion, augmented by principles of rotational dynamics, can be used to model the subsequent motion and compute how the hinge force's components S and R vary with angle θ. $\hat{\mathbf{r}}$ and $\hat{\boldsymbol{\theta}}$ are, respectively, radial and tangential unit vectors for circular motion about the hinge.

then struck with an impulse (i.e. a momentum change) which causes the centre of mass of the block acquiring an instantaneous horizontal velocity u. This is sufficient to cause the trapdoor to rotate about the hinge until it snaps shut in a downward vertical position. What are the equations of motion which describe what happens next, and how does the hinge force vary as the trapdoor rotates?

2.12.1   *Coordinates and geometry*

Define a Cartesian x, y coordinate origin to be at the hinge. The centre of mass of the lamina is initially at $(-b, a)$. Let the back edge of the lamina rise by angle θ from the horizontal. The trapdoor closes when $\theta = \pi$ radians. The angle of the diagonal that passes from the hinge to the top-left corner of the lamina is $\alpha = \tan^{-1} \frac{a}{b}$ from the bottom edge of the lamina. Let the weight mg, which acts vertically downwards, be at angle ϕ from the hinge to top-left corner diagonal. From geometry,

$$\theta + \alpha + \left(\tfrac{1}{2}\pi - \phi\right) = \pi. \tag{2.392}$$

$$\therefore \phi = \theta + \alpha - \tfrac{1}{2}\pi. \tag{2.393}$$

$$\therefore \cos\phi = \sin(\theta + \alpha). \tag{2.394}$$

$$\therefore \sin\phi = -\cos(\theta + \alpha). \tag{2.395}$$

The distance from the hinge to the centre of mass is

$$r = \sqrt{a^2 + b^2}. \tag{2.396}$$

2.12.2   *Applying Newton's second law to the centre of mass of the lamina*

Let hinge force

$$F = \sqrt{R^2 + S^2} \tag{2.397}$$

have perpendicular components R and S. Define the S component to act along the hinge to the top-corner diagonal, i.e. through the centre of mass. S will always act at angle $\theta + \alpha$ from the horizontal. In the radial $\hat{\mathbf{r}}$ direction from the hinge, mass $\times$ acceleration = the vector sum of force, which is

$$-mr\dot{\theta}^2 = S - mg\cos\phi, \tag{2.398}$$

whereas in the tangential $\hat{\boldsymbol{\theta}}$ direction,

$$mr\ddot{\theta} = mg\sin\phi - R. \tag{2.399}$$

2.12.3   *Moment of inertia*

The moment of inertia is *about the hinge*, not the centre of mass. We can use the *parallel axis theorem* to convert the moment of inertia about the centre of a rectangular lamina, which is[17] $\frac{1}{12}m\{(2a)^2 + (2b)^2\} = \frac{1}{3}m(a^2 + b^2)$ for dimensions $2a \times 2b$, to what we need. The correction is to add mr^2, where r is the distance between the centre of mass and the hinge:

$$I = \tfrac{1}{3}m\left(a^2 + b^2\right) + m(a^2 + b^2) \tag{2.400}$$

$$= \tfrac{4}{3}mr^2. \tag{2.401}$$

2.12.4 *Conservation of energy*

Assuming the initial GPE + KE of the system remains constant, i.e. we can ignore work done against air resistance and hinge friction,

$$mga + \tfrac{1}{2}mu^2 = mgr\cos\phi + \tfrac{1}{2}I\dot\theta^2, \tag{2.402}$$

i.e. $mgr\cos\phi$ is the GPE of the centre of mass of the lamina, and all KE is rotation $\frac{1}{2}I\dot\theta^2$ about the hinge. Note that the initial KE must exceed the maximum gain in GPE in order for the lamina to rotate over the hinge. This means

$$\tfrac{1}{2}mu^2 > mg(r - a). \tag{2.403}$$

$$\therefore u > \sqrt{2g(r - a)}. \tag{2.404}$$

2.12.5 *Torque about hinge = moment of inertia × angular acceleration*

Note that the hinge forces both act through the hinge and therefore exert no torque about an axis passing through the hinge. The only force which provides a turning moment is therefore the weight of the lamina. Hence,

$$mgr\sin\phi = I\ddot\theta. \tag{2.405}$$

2.12.6 *Solving for the dynamics of the hinge problem*

We can now combine all the equations above. We express everything in terms of $\phi = \theta + \alpha - \frac{1}{2}\pi$ and hence elevation angle θ. We can then solve for $t(\theta)$ using a numeric integration method. From conservation of energy, $mga + \frac{1}{2}mu^2 = mgr\cos\phi + \frac{1}{2}I\dot\theta^2$, and noting $I = \frac{4}{3}mr^2$,

$$\frac{2mg\left(a - r\cos\phi\right) + mu^2}{\frac{4}{3}mr^2} = \dot\theta^2. \tag{2.406}$$

$$\therefore \frac{3}{2}\frac{g}{r}\left(\frac{a}{r} - \cos\phi\right) + \frac{3}{4}\frac{u^2}{r^2} = \dot\theta^2. \tag{2.407}$$

This means

$$\frac{d\theta}{dt} = \sqrt{\frac{3g}{2r}}\sqrt{\frac{a}{r} - \cos\phi + \frac{1}{2}\frac{u^2}{rg}}. \tag{2.408}$$

---

[17]For example, see Quadling [29, p. 228].

$$\therefore t(\theta) = \sqrt{\frac{2r}{3g}} \int_0^\theta \frac{1}{\sqrt{\frac{a}{r} - \sin(\theta' + \alpha) + \frac{u^2}{2rg}}} \, d\theta'. \tag{2.409}$$

The integral expression for $t(\theta)$ can be evaluated by first fitting cubic splines to the integrand (over a range of $0 \leq \theta \leq \pi$ in a few thousand steps) and then integrating these in a cumulative piecewise fashion. This can readily be achieved using a computer program written in MATLAB, Python or equivalent.

The formula $\frac{3}{2}\frac{g}{r}\left(\frac{a}{r} - \cos\phi\right) + \frac{3}{4}\frac{u^2}{r^2} = \dot\theta^2$ can be used to determine the S component of the hinge force:

$$S = -mr\dot\theta^2 + mg\cos\phi \tag{2.410}$$

$$S = -m\frac{3}{2}g\left(\frac{a}{r} - \cos\phi\right) - \frac{\frac{3}{4}mu^2}{r} + mg\cos\phi \tag{2.411}$$

$$S = \frac{5}{2}mg\cos\phi - \frac{3}{2}mg\frac{a}{r} - \frac{\frac{3}{4}mu^2}{r}. \tag{2.412}$$

$$\therefore S = \frac{1}{2}mg\left(5\sin(\theta + \alpha) - \frac{3a}{r} - \frac{3}{2}\frac{u^2}{rg}\right). \tag{2.413}$$

The torque formula $mgr\sin\phi = I\ddot\theta$ can then be used to find the R component of the hinge force:

$$R = mg\sin\phi - mr\ddot\theta \tag{2.414}$$

$$R = mg\sin\phi - mr \times \frac{mgr\sin\phi}{I} \tag{2.415}$$

$$R = mg\sin\phi - mr \times \frac{mgr\sin\phi}{\frac{4}{3}mr^2} \tag{2.416}$$

$$R = mg\sin\phi\left(1 - \frac{3}{4}\right) \tag{2.417}$$

$$R = \frac{1}{4}mg\sin\phi. \tag{2.418}$$

$$\therefore R = -\frac{1}{4}mg\cos(\theta + \alpha). \tag{2.419}$$

In summary:

$$\alpha = \tan^{-1}\frac{a}{b}, \tag{2.420}$$

$$r = \sqrt{a^2 + b^2}, \tag{2.421}$$

$$u > \sqrt{2g(r - a)}, \tag{2.422}$$

$$F = \sqrt{R^2 + S^2}, \tag{2.423}$$

$$R = -\frac{1}{4}mg\cos(\theta + \alpha), \tag{2.424}$$

$$S = \frac{1}{2}mg\left(5\sin(\theta + \alpha) - \frac{3a}{r} - \frac{3}{2}\frac{u^2}{rg}\right), \tag{2.425}$$

$$t(\theta) = \sqrt{\frac{2r}{3g}} \int_0^\theta \frac{1}{\sqrt{\frac{a}{r} - \sin(\theta' + \alpha) + \frac{u^2}{2rg}}} \, d\theta'. \tag{2.426}$$

A common problem is to define the lamina to have dimensions that form a *Pythagorean triple*. Let $a = 4k$ and $b = 3k$, which means $r = 5k$ since $5^2 = 3^2 + 4^2$. In this case,

$$\alpha = \tan^{-1} \tfrac{4}{3}, \tag{2.427}$$

$$I = \tfrac{4}{3}mr^2 = \tfrac{100}{3}mk^2, \tag{2.428}$$

$$S = \tfrac{1}{2}mg \left(5\sin(\theta + \alpha) - \tfrac{12}{5} - \tfrac{3}{10}\frac{u^2}{gk} \right), \tag{2.429}$$

$$R = -\tfrac{1}{4}mg\cos(\theta + \alpha). \tag{2.430}$$

A model of the rotating lamina is illustrated in Fig. 2.18. Subplot (a) shows screenshots of a MATLAB simulation which animates the motion of a lamina with $a = 10$ m, $b = 2$ m and weight $mg = 3{,}924$ N. Coloured arrows show the magnitude and direction of the hinge force F, decomposed into perpendicular components S and R. Note that the direction of F changes from a reaction force (i.e. pointing radially outward from the hinge into the lamina) to a 'pulling force' as the lamina rotates to larger angles. This causes S to change sign relative to the radial $\hat{\mathbf{r}}$ unit vector direction. Subplots (b) to (d) describe the θ variation of the hinge force and also the time variation of θ and energies. The hinge force angle is measured anticlockwise from the horizontal. Interestingly, the angular speed seems to tend towards a constant rather than accelerate. The coloured horizontal bars in (d) plot the total energy (KE + GPE) of the system, which confirm constancy throughout the simulation. The curved coloured lines correspond to KE, whereas the downward curving black line is GPE. The latter doesn't depend on the initial horizontal velocity u, so it tracks the same curve for all the u variants plotted.

2.13 Lagrangian Mechanics, the Brachistochrone and Asymmetric Catenary

2.13.1 *When a Newton's second law recipe doesn't cut the mustard*

Until now, we have followed the following recipe for solving mechanics problems:

(1) The problem is set out in a diagram, with all variables defined and assumptions made. The Universe temporarily shrinks to this model world.

(2) A coordinate system is defined that matches the symmetry of the problem (e.g. polars for circular motion and Cartesians for planar motion).

(3) If there are N unknowns, we need to find N independent equations, which can be solved in terms of other fixed input parameters such as the angle of a slope, the masses, and the strength of gravity.

(4) The equations are generated from: (i) geometrical relationships particular to the problem (e.g. the isosceles triangle in the drawbridge problem, or the circular constraints of the weeble base or stacked cylinders) and (ii) the laws of physics. The laws of physics include: (a) conservation of energy, (b) conservation of momentum (for 'before and after' collision situations), (c) Newton's second law

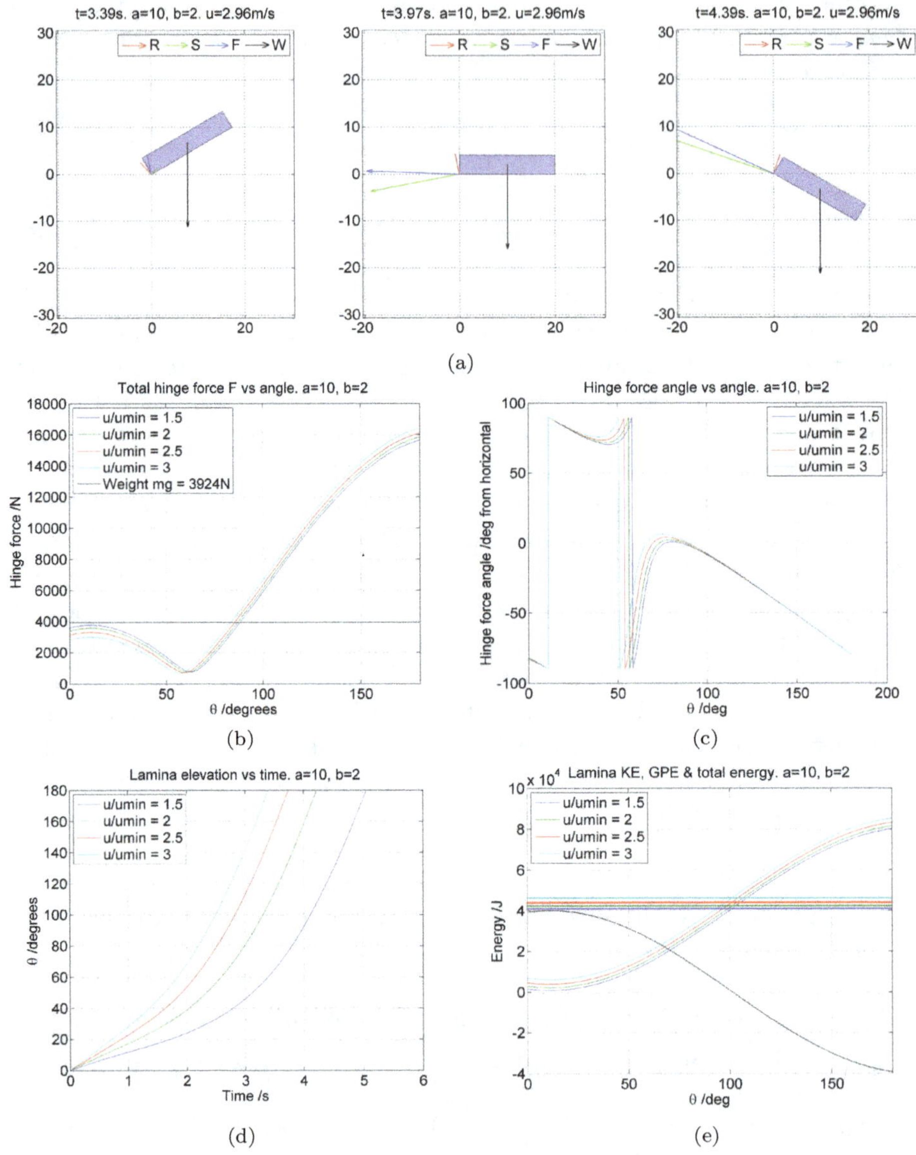

Fig. 2.18. A model of rotating lamina about a fixed, frictionless hinge. $u_{\min} = \sqrt{2g(r-a)}$ is the minimum initial velocity for the lamina to rotate over the hinge, which requires a gain in GPE. The lamina dimensions are a, b and $r = \sqrt{a^2 + b^2}$. Subplot (a) shows screenshots of a MATLAB simulation which animates the motion of a lamina with $a = 10$ m, $b = 2$ m and weight $mg = 3924$ N. Coloured arrows show the magnitude and direction of the hinge force F, decomposed into perpendicular components S and R. Note that the direction of F changes from a reaction force (i.e. pointing radially outward from the hinge into the lamina) to a 'pulling force' as the lamina rotates to larger angles. This causes S to change sign relative to the radial $\hat{\mathbf{r}}$ unit vector direction. Subplots (b) to (e) describe the θ variation of the hinge force and also the time variation of θ and energies. The hinge force angle is measured anticlockwise from the horizontal. The coloured horizontal bars in (e) plot the total energy (KE + GPE) of the system, which confirm constancy throughout the simulation. The curved coloured lines correspond to KE, whereas the downward curving black line is GPE. The latter doesn't depend on the initial horizontal velocity u, so it tracks the same curve for all the u variants plotted.

(i.e. mass $\times$ acceleration = vector sum of force), (d) torque = moment of inertia $\times$ angular acceleration.

A tricky problem can arise when we have systems comprising multiple interacting bodies. For systems such as the Atwood machines and stacked balls, we can calculate string tension, the 'conservation of (assumed light and inextensible) string', or use Newton's third law of equal and opposite forces between two bodies in contact. However, in the general case, it may prove difficult to decompose a system into its component parts and therefore be in a position to write down a Newton's second law expression immediately following inspection of a diagram. At this point, the introduction of the powerful tools of *Lagrangian mechanics* is appropriate. Rather than writing down vector equations, one begins by computing the *total kinetic energy* T and *total potential energy* V of a system. The recipe of Lagrangian mechanics is to apply the *Euler–Lagrange equation* (see Appendix A) to the *Lagrangian* $L = T - V$, which is a *scalar* quantity. If a system has a set of variables $\{x_1, x_2, \ldots, x_N\}$, then the system obeys the equation

$$\frac{d}{dt}\left(\frac{\partial L}{\partial \dot{x}_i}\right) = \frac{\partial L}{\partial x_i}. \tag{2.431}$$

In other words, we obtain one differential equation for each variable x_i. This results in a systematic algebraic recipe, as illustrated by the 'two sliding blocks problem' in the next subsection. The *Euler–Lagrange equation* can also be used to solve more general problems involving the minimisation (or maximisation) of an integral quantity, potentially involving constraints too. The classic problems of the *brachistochrone* (i.e. the minimum time curve for a sliding bead) and the (catenary) shape of a weighted chain shall conclude this mechanics chapter.

2.13.2 *A surprisingly difficult inclined plane problem*

Consider a rectangular block of mass m sliding frictionlessly from the top of a planar wedge of elevation θ and mass M, as illustrated in Fig. 2.19. The *wedge also slides frictionlessly* on a horizontal surface. This is what makes the problem much harder than those associated with a (fixed) inclined plane. h is the vertical displacement of the base of the upper end of the block from the top of the wedge. x_1 is the horizontal displacement of the wedge (to the left), and x_2 is the horizontal displacement of the block (to the right). Assuming that the block has not reached the horizontal surface, after t seconds, the horizontal separation is

$$\Delta x = x_1 + x_2. \tag{2.432}$$

$$\therefore h = (x_1 + x_2)\tan\theta. \tag{2.433}$$

The *Lagrangian* of the system is (noting potential energy $V = -mgh$)

$$L = T - V \tag{2.434}$$

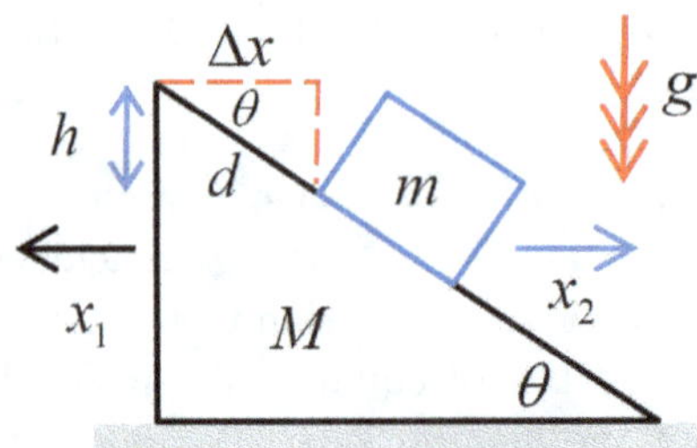

Fig. 2.19. A rectangular block of mass m slides down the diagonal face of a triangular block of mass M. Both the inter-block surface and the surface between the triangular block and horizontal the are assumed to be frictionless. The problem is difficult to analyse using a classic mass $\times$ acceleration = vector sum of force approach since both blocks will accelerate. Instead a *Lagrangian* method is suggested, which generates the equations of motion for each block from the difference between total kinetic energy T and potential energy V.

$$= \underbrace{\tfrac{1}{2}M\dot{x}_1^2 + \tfrac{1}{2}m\dot{x}_2^2 + \tfrac{1}{2}m\dot{h}^2}_{\text{Total kinetic energy } T} + \; mgh \tag{2.435}$$

$$= \tfrac{1}{2}M\dot{x}_1^2 + \tfrac{1}{2}m\dot{x}_2^2 + \tfrac{1}{2}m(\dot{x}_1 + \dot{x}_2)^2 \tan^2\theta + mg(x_1 + x_2)\tan\theta. \tag{2.436}$$

We can apply the Euler–Lagrange equation (see Appendix A) separately for the *wedge* (i.e. involving x_1):

$$\frac{d}{dt}\left(\frac{\partial L}{\partial \dot{x}_1}\right) = \frac{\partial L}{\partial x_1} \tag{2.437}$$

$$\frac{d}{dt}\left(M\dot{x}_1 + m(\dot{x}_1 + \dot{x}_2)\tan^2\theta\right) = mg\tan\theta \tag{2.438}$$

$$M\ddot{x}_1 + m(\ddot{x}_1 + \ddot{x}_2)\tan^2\theta = mg\tan\theta, \tag{2.439}$$

and the *block* (i.e. involving x_2):

$$\frac{d}{dt}\left(\frac{\partial L}{\partial \dot{x}_2}\right) = \frac{\partial L}{\partial x_2} \tag{2.440}$$

$$\frac{d}{dt}\left(m\dot{x}_2 + m(\dot{x}_1 + \dot{x}_2)\tan^2\theta\right) = mg\tan\theta \tag{2.441}$$

$$m\ddot{x}_2 + m(\ddot{x}_1 + \ddot{x}_2)\tan^2\theta. = mg\tan\theta. \tag{2.442}$$

Subtracting the wedge and block equations,

$$M\ddot{x}_1 - m\ddot{x}_2 = 0 \tag{2.443}$$

$$\frac{d}{dt}\left(M\dot{x}_1 - m\dot{x}_2\right) = 0, \tag{2.444}$$

which means

$$M\dot{x}_1 - m\dot{x}_2 = \text{constant}. \tag{2.445}$$

Therefore,

$$\ddot{x}_2 = \frac{M}{m}\ddot{x}_1. \tag{2.446}$$

In other words, a statement of conservation of linear momentum in the horizontal direction. This is of course expected since the only *external* force acting on the system is gravity, which acts downwards.

Substituting $\ddot{x}_2 = \frac{M}{m}\ddot{x}_1$ into $M\ddot{x}_1 + m(\ddot{x}_1 + \ddot{x}_2)\tan^2\theta = mg\tan\theta$,

$$M\ddot{x}_1 + m(\ddot{x}_1 + \ddot{x}_2)\tan^2\theta = mg\tan\theta \tag{2.447}$$

$$M\ddot{x}_1 + m(\ddot{x}_1 + \frac{M}{m}\ddot{x}_1)\tan^2\theta = mg\tan\theta \tag{2.448}$$

$$\ddot{x}_1\left(M + (m + M)\tan^2\theta\right) = mg\tan\theta \tag{2.449}$$

$$\ddot{x}_1 = \frac{mg\tan\theta}{M + (m + M)\tan^2\theta} \tag{2.450}$$

$$\ddot{x}_1 = \frac{mg\tan\theta}{M\left(1 + \tan^2\theta\right) + m\tan^2\theta} \tag{2.451}$$

$$\ddot{x}_1 = \frac{mg\frac{\sin\theta}{\cos\theta}}{\frac{M}{\cos^2\theta} + m\frac{\sin^2\theta}{\cos^2\theta}}. \tag{2.452}$$

$$\therefore \ddot{x}_1 = \frac{g\sin\theta\cos\theta}{\frac{M}{m} + \sin^2\theta}. \tag{2.453}$$

Hence, since $\ddot{x}_2 = \frac{M}{m}\ddot{x}_1$,

$$\ddot{x}_2 = \frac{g\sin\theta\cos\theta}{1 + \frac{m}{M}\sin^2\theta}, \tag{2.454}$$

i.e. constant acceleration motion for both block and wedge, but not a particularly obvious relationship with block slope angle θ and mass ratio $\frac{m}{M}$. If the system starts from rest,

$$\dot{x}_1 = \frac{g\sin\theta\cos\theta}{\frac{M}{m} + \sin^2\theta}t \tag{2.455}$$

$$x_1 = \frac{1}{2}\frac{g\sin\theta\cos\theta}{\frac{M}{m} + \sin^2\theta}t^2 \tag{2.456}$$

$$\dot{x}_2 = \frac{g\sin\theta\cos\theta}{1 + \frac{m}{M}\sin^2\theta}t \tag{2.457}$$

$$x_2 = \frac{1}{2}\frac{g\sin\theta\cos\theta}{1 + \frac{m}{M}\sin^2\theta}t^2. \tag{2.458}$$

The limiting case when $M \gg m$ is

$$\dot{x}_1 \to 0,\ x_1 \to 0, \tag{2.459}$$

$$\dot{x}_2 \to gt\sin\theta\cos\theta, \tag{2.460}$$

$$x_2 \to \tfrac{1}{2}gt^2\sin\theta\cos\theta, \tag{2.461}$$

which is simply the block of mass m sliding down a fixed gradient slope without friction. The wedge doesn't move in this limit. The distance travelled down the slope in this case is

$$d = \frac{x_2}{\cos\theta} = \tfrac{1}{2}gt^2\sin\theta, \tag{2.462}$$

which makes sense since the component of the weight acting down the slope is $mg\sin\theta$.

2.13.3   *The brachistochrone*

If a bead is allowed to slide frictionlessly along a wire, what shape should the wire be such that the bead takes the *least time* to slide between two points? The time taken τ for the bead to slide between two points (x_a,y_a) and (x_b,y_b) along the curve $y(x)$ is

$$\tau = \int_{x_a}^{x_b} \frac{\sqrt{dx^2+dy^2}}{v}, \tag{2.463}$$

where $y \le y_a$. Conservation of energy can be used to determine the bead speed v. Since there is no friction, the kinetic energy gain of the bead will equate to the loss of gravitational potential energy:

$$\tfrac{1}{2}mv^2 + mgy = mgy_a. \tag{2.464}$$

$$\therefore v = \sqrt{2g(y_a - y)}. \tag{2.465}$$

Hence, the sliding time is

$$\tau = \frac{1}{\sqrt{2g}} \int_{x_a}^{x_b} \sqrt{\frac{1+\left(\frac{dy}{dx}\right)^2}{y_a - y}}\,dx. \tag{2.466}$$

To find $y(x)$ that minimises τ is just the sort of problem that the *Euler–Lagrange* equation can solve. If we define

$$\dot{y} = \frac{dy}{dx}, \tag{2.467}$$

$$L = \sqrt{\frac{1+\dot{y}^2}{y_a - y}}, \tag{2.468}$$

then in order to minimise

$$\tau = \frac{1}{\sqrt{2g}} \int_{x_a}^{x_b} L(y,\dot{y})dx, \tag{2.469}$$

we need to solve

$$\frac{d}{dx}\left(\frac{\partial L}{\partial \dot{y}}\right) = \frac{\partial L}{\partial y}. \tag{2.470}$$

Note that since $\frac{\partial L}{\partial x} = 0$, we can shortcut the process by using the *Beltrami identity* (see Appendix A):

$$L - \dot{y}\frac{\partial L}{\partial \dot{y}} = \frac{1}{k}, \tag{2.471}$$

where k is a constant. As you will see in the following arguments, setting the constant to be $\frac{1}{k}$ makes for simpler algebra:

$$\sqrt{\frac{1 + \dot{y}^2}{y_a - y}} - \dot{y}\frac{1}{2}\frac{\left(1 + \dot{y}^2\right)^{-\frac{1}{2}}}{\sqrt{y_a - y}}(2\dot{y}) = \frac{1}{k} \tag{2.472}$$

$$\sqrt{1 + \dot{y}^2} - \dot{y}^2\left(1 + \dot{y}^2\right)^{-\frac{1}{2}} = \frac{1}{k}\sqrt{y_a - y} \tag{2.473}$$

$$\frac{1 + \dot{y}^2 - \dot{y}^2}{\sqrt{1 + \dot{y}^2}} = \frac{1}{k}\sqrt{y_a - y} \tag{2.474}$$

$$(y_a - y)\left(1 + \dot{y}^2\right) = k^2. \tag{2.475}$$

A curve which satisfies this differential equation is a *cycloid*, which can be defined parametrically as

$$y = y_a - \tfrac{1}{2}k^2(1 - \cos\theta), \tag{2.476}$$

$$x = x_a + \tfrac{1}{2}k^2(\theta - \sin\theta), \tag{2.477}$$

with θ in radians. To show that the cycloid equations satisfy $(y_a - y)\left(1 + \dot{y}^2\right) = k^2$, let's use the chain rule to determine $\dot{y}(\theta)$:

$$\dot{y} = \frac{dy}{dx} = \frac{dy}{d\theta}\frac{d\theta}{dx} = \frac{\frac{dy}{d\theta}}{\frac{dx}{d\theta}} \tag{2.478}$$

$$\dot{y} = \frac{-\tfrac{1}{2}k^2\sin\theta}{\tfrac{1}{2}k^2\left(1 - \cos\theta\right)} = \frac{-\sin\theta}{\left(1 - \cos\theta\right)}. \tag{2.479}$$

Hence,

$$(y_a - y)\left(1 + \dot{y}^2\right) = \tfrac{1}{2}k^2\left(1 - \cos\theta\right)\left(1 + \frac{\sin^2\theta}{\left(1 - \cos\theta\right)^2}\right) \tag{2.480}$$

$$= \tfrac{1}{2}k^2\left(1 - \cos\theta\right)\frac{\left(1 - \cos\theta\right)^2 + \sin^2\theta}{\left(1 - \cos\theta\right)^2} \tag{2.481}$$

$$= \tfrac{1}{2}k^2\frac{1 - 2\cos\theta + \cos^2\theta + \sin^2\theta}{1 - \cos\theta} \tag{2.482}$$

$$= \tfrac{1}{2}k^2\frac{2 - 2\cos\theta}{1 - \cos\theta} \tag{2.483}$$

$$= k^2. \tag{2.484}$$

Interestingly, the brachistochrone curve is independent of not only the bead mass m but also the strength of gravity g. From the cycloid equations, $\theta = 0$ corresponds to the initial coordinates (x_a, y_a). So, what is the range of θ? Let's define this to be $0 \le \theta \le \theta_b$, where θ_b satisfies the following pair of equations:

$$y_b = y_a - \tfrac{1}{2}k^2(1 - \cos\theta_b), \tag{2.485}$$

$$x_b = x_a + \tfrac{1}{2}k^2(\theta_b - \sin\theta_b). \tag{2.486}$$

Therefore,

$$\tfrac{1}{2}k^2 = \frac{y_a - y_b}{1 - \cos\theta_b} = \frac{x_b - x_a}{\theta_b - \sin\theta_b}. \tag{2.487}$$

Define

$$\alpha = \frac{y_a - y_b}{x_b - x_a}. \tag{2.488}$$

Therefore, $f(\theta_b) = 0$, where

$$f(\theta) = \frac{1}{\alpha}(1 - \cos\theta) + \sin\theta - \theta. \tag{2.489}$$

The solution for θ_b can be determined numerically using the *Newton–Raphson method*

$$\theta_{n+1} = \theta_n - \frac{f(\theta_n)}{\left.\frac{df}{d\theta}\right|_{\theta_n}} \tag{2.490}$$

as long as the initial guess value of θ_b does not lead to divergence when $f'(\theta_n) = 0$. The derivative of $f(\theta) = \frac{1}{\alpha}(1 - \cos\theta) + \sin\theta - \theta$, evaluated at $\theta = \theta_n$, is

$$\left.\frac{df}{d\theta}\right|_{\theta_n} = \frac{1}{\alpha}\sin\theta_n + \cos\theta_n - 1. \tag{2.491}$$

So, the Newton–Raphson iteration is explicitly in terms of θ_n:

$$\theta_{n+1} = \theta_n - \frac{\frac{1}{\alpha}(1 - \cos\theta_n) + \sin\theta_n - \theta_n}{\frac{1}{\alpha}\sin\theta_n + \cos\theta_n - 1}. \tag{2.492}$$

As illustrated in Fig. 2.20, $x_a = 1$, $x_b = 5$, $y_a = 3$ and $y_b = 1$ yields, using a starting guess of $\theta_1 = 4$, an asymptotic value of $\theta_b = 3.5084$ and hence

$$\tfrac{1}{2}k^2 = \frac{y_a - y_b}{1 - \cos\theta_b} = 1.0344, \tag{2.493}$$

i.e. $k = \sqrt{2 \times 1.0344} = 1.4383$. The brachistochrone curve between $(1, 3)$ and $(5, 1)$ is therefore

$$y = y_a - 1.0344(1 - \cos\theta), \tag{2.494}$$

$$x = x_a + 1.0344(\theta - \sin\theta), \tag{2.495}$$

$$0 \le \theta \le 3.5084. \tag{2.496}$$

What about the actual time τ for the bead to traverse the curve? Recall that the sliding time is

$$\tau = \frac{1}{\sqrt{2g}} \int_{x_a}^{x_b} \sqrt{\frac{1 + \dot{y}^2}{y_a - y}} \, dx. \qquad (2.497)$$

Now, $(y_a - y)\left(1 + \dot{y}^2\right) = k^2$ so $1 + \dot{y}^2 = \frac{k^2}{y_a - y}$, which means

$$\tau = \frac{1}{\sqrt{2g}} \int_{x_a}^{x_b} \sqrt{\left(\frac{k}{y_a - y}\right)^2} \, dx \qquad (2.498)$$

$$= \frac{1}{\sqrt{2g}} \int_{x_a}^{x_b} \frac{k}{y_a - y} \, dx \qquad (2.499)$$

$$= \frac{1}{\sqrt{2g}} \int_{x_a}^{x_b} \frac{k}{\frac{1}{2}k^2 \left(1 - \cos\theta\right)} \, dx \qquad (2.500)$$

$$= \frac{k}{\sqrt{2g}} \int_0^{\theta_b} \frac{1}{\frac{1}{2}k^2 \left(1 - \cos\theta\right)} \, \frac{1}{2}k^2 \left(1 - \cos\theta\right) d\theta, \qquad (2.501)$$

with the last few steps noting $y = y_a - \frac{1}{2}k^2 \left(1 - \cos\theta\right)$, $x = x_a + \frac{1}{2}k^2 \left(\theta - \sin\theta\right)$ and hence

$$dx = \tfrac{1}{2}k^2 \left(1 - \cos\theta\right) d\theta. \qquad (2.502)$$

Therefore,

$$\tau = \frac{k}{\sqrt{2g}} \int_0^{\theta_b} d\theta = \frac{\theta_b k}{\sqrt{2g}}. \qquad (2.503)$$

So, for our example in Fig. 2.20, if performed on Earth with $g = 9.81$ ms^{-2}, with all distances in metres,

$$\tau = \frac{\theta_b k}{\sqrt{2g}} = \frac{3.5084 \times \sqrt{2 \times 1.0344}}{\sqrt{2 \times 9.81}} \approx 1.14\text{s}. \qquad (2.504)$$

2.13.4 *The weighted chain: An asymmetric catenary*

If a uniform, motionless, inextensible chain of length l is connected between two points $(-a, H)$ and (a, h), separated by distance $d < l$, what curve does the chain take if its gravitational potential energy is to be minimised? The total GPE V of the chain, which is of uniform mass per unit length ρ, is

$$V = \rho g \int_{x=-a}^{a} y \, ds = \rho g \int_{x=-a}^{a} y \sqrt{dx^2 + dy^2} = \rho g \int_{x=-a}^{a} y \sqrt{1 + \dot{y}^2} \, dx, \qquad (2.505)$$

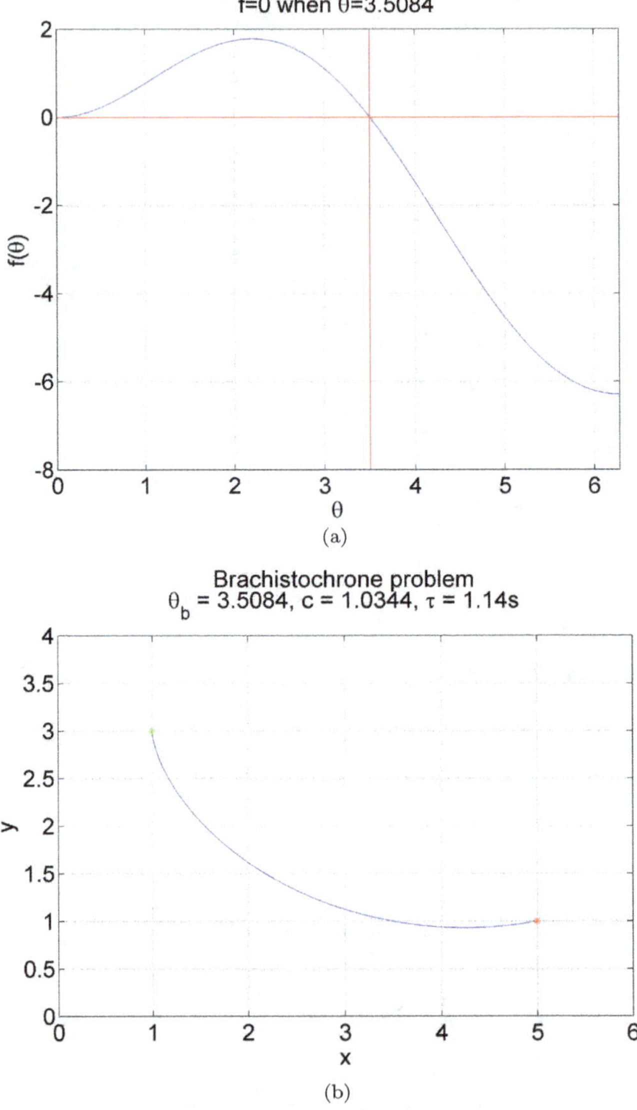

Fig. 2.20. The *brachistochrone* is a curve between two (x, y) coordinates that corresponds to the *shortest time* that a bead sliding frictionlessly would take if it were to slide along the curve. The brachistochrone is a *cycloid* of parametric form $x = x_a + \frac{1}{2}k^2 (\theta - \sin \theta)$, $y = y_a - \frac{1}{2}k^2 (1 - \cos \theta)$ between coordinates (x_a, y_a) and (x_b, y_b), where $x_b > x_a$ and $y_a > y_b$. Parametric angles are within the range $0 \leq \theta \leq \theta_b$, where θ_b is the termination of the iteration: $\theta_{n+1} = \theta_n - \dfrac{\frac{1}{\alpha}(1-\cos \theta_n)+\sin \theta_n -\theta_n}{\frac{1}{\alpha} \sin \theta_n +\cos \theta_n -1}$, where $\alpha = \frac{y_a - y_b}{x_b - x_a}$, i.e. a *Newton–Raphson* iteration $\theta_{n+1} = \theta_n - \dfrac{f(\theta_n)}{\frac{df}{d\theta}\big|_{\theta_n}}$, where $f(\theta) = \frac{1}{\alpha} (1 - \cos \theta) + \sin \theta - \theta = 0$, which is plotted in (a). Note that knowledge of θ_b means you can compute $k = \sqrt{\frac{2(y_a - y_b)}{1 - \cos \theta_b}}$. If released from rest, the bead will take time $\tau = \frac{\theta_b k}{\sqrt{2g}}$ to reach (x_b, y_b) from (x_a, y_a). In plot (b), which is the brachistochrone curve between (1,3) and (5,1) (metres), the cycloid parameter $c = \frac{1}{2}k^2 = 1.0344$, meaning $k = 1.4383$. In this scenario, $\theta_b = 3.5084$.

where s is the distance along the chain and $\dot{y} = \frac{dy}{dx}$. Unlike the brachistochrone, the curve of the chain is also constrained by its length

$$l = \int_{x=-a}^{a} \sqrt{dx^2 + dy^2} = \int_{x=-a}^{a} \sqrt{1 + \dot{y}^2}dx. \qquad (2.506)$$

We can use the Euler–Lagrange equation to solve for $y(x)$ that minimises V, subject to the fixed chain length l, by incorporating the *integrand of the constraint* into the L function using an addition via a *Lagrange multiplier* [15] λ. That is, we aim to minimise

$$V + \lambda l = \int_{x=-a}^{a} L\left(y, \dot{y}\right) dx, \qquad (2.507)$$

where

$$L = \rho g y \sqrt{1 + \dot{y}^2} + \lambda \sqrt{1 + \dot{y}^2} \qquad (2.508)$$

$$= \left(\rho g y + \lambda\right) \sqrt{1 + \dot{y}^2}. \qquad (2.509)$$

The minimum of $V + \lambda l$ is the solution to the Euler–Lagrange equation

$$\frac{d}{dx}\left(\frac{\partial L}{\partial \dot{y}}\right) = \frac{\partial L}{\partial y}. \qquad (2.510)$$

Note that since $\frac{\partial L}{\partial x} = 0$, we can, as for the brachistochrone, shortcut the process by using the *Beltrami identity* (see Appendix A):

$$L - \dot{y}\frac{\partial L}{\partial \dot{y}} = k, \qquad (2.511)$$

where k is a constant. Substituting for our expression for L,

$$\left(\rho g y + \lambda\right)\sqrt{1 + \dot{y}^2} - \dot{y}(\rho g y + \lambda)\left(\tfrac{1}{2}\right)\left(1 + \dot{y}^2\right)^{-\frac{1}{2}}(2\dot{y}) = k \qquad (2.512)$$

$$\sqrt{1 + \dot{y}^2} - \frac{\dot{y}^2}{\sqrt{1 + \dot{y}^2}} = \frac{k}{(\rho g y + \lambda)} \qquad (2.513)$$

$$\frac{1 + \dot{y}^2 - \dot{y}^2}{\sqrt{1 + \dot{y}^2}} = \frac{k}{(\rho g y + \lambda)} \qquad (2.514)$$

$$1 + \dot{y}^2 = \tfrac{1}{k^2}\left(\rho g y + \lambda\right)^2. \qquad (2.515)$$

Now, choose a variable substitution z such that

$$\tfrac{1}{k}\left(\rho g y + \lambda\right) = \cosh z. \qquad (2.516)$$

$$\therefore \tfrac{\rho g}{k}\dot{y} = \sinh z \frac{dz}{dx}. \qquad (2.517)$$

$$\therefore \dot{y} = \frac{k}{\rho g} \sinh z \frac{dz}{dx}. \qquad (2.518)$$

This means, we can express $1 + \dot{y}^2 = \frac{1}{k^2}(\rho g y + \lambda)^2$ in terms of our substitution variable z:

$$1 + \left(\frac{k}{\rho g}\right)^2 \sinh^2 z \left(\frac{dz}{dx}\right)^2 = \cosh^2 z \tag{2.519}$$

$$\left(\frac{k}{\rho g}\right)^2 \sinh^2 z \left(\frac{dz}{dx}\right)^2 = \cosh^2 z - 1 = \sinh^2 z. \tag{2.520}$$

$$\therefore \frac{dz}{dx} = \frac{\rho g}{k}. \tag{2.521}$$

$$\therefore z = \frac{\rho g}{k}(x + c). \tag{2.522}$$

Now, by rearrangement of $\frac{1}{k}(\rho g y + \lambda) = \cosh z$,

$$y = \frac{k}{\rho g}\cosh z - \frac{\lambda}{\rho g}. \tag{2.523}$$

Therefore, the chain curve is a *catenary* (a cosh curve):

$$y = \frac{k}{\rho g}\cosh\left(\frac{\rho g}{k}(x + c)\right) - \frac{\lambda}{\rho g}. \tag{2.524}$$

Our final task is now to find the constants k, λ, c given our chain length l and starting points $(-a, H)$ and end points (a, h). The length of the chain is given by

$$l = \int_{x=-a}^{a} \sqrt{1 + \left(\frac{dy}{dx}\right)^2}\, dx. \tag{2.525}$$

From our catenary curve $y = \frac{k}{\rho g}\cosh\left(\frac{\rho g}{k}(x + c)\right) - \frac{\lambda}{\rho g}$.

$$\therefore \frac{dy}{dx} = \sinh\left(\frac{\rho g}{k}(x + c)\right) \tag{2.526}$$

$$\Rightarrow 1 + \left(\frac{dy}{dx}\right)^2 = 1 + \sinh^2\left(\frac{\rho g}{k}(x + c)\right) = \cosh^2\left(\frac{\rho g}{k}(x + c)\right). \tag{2.527}$$

We can therefore evaluate the integral expression for l analytically:

$$l = \int_{x=-a}^{a} \cosh\left(\frac{\rho g}{k}(x + c)\right) dx \tag{2.528}$$

$$= \frac{k}{\rho g}\left\{\sinh\left(\frac{\rho g}{k}(a + c)\right) - \sinh\left(\frac{\rho g}{k}(-a + c)\right)\right\}. \tag{2.529}$$

This difference of sinh functions implies that the following hyperbolic identity might prove useful:

$$\sinh 2A - \sinh 2B = 2\cosh(A+B)\sinh(A-B). \tag{2.530}$$

Let

$$A = \frac{\rho g}{2k}(a+c), \tag{2.531}$$

$$B = \frac{\rho g}{2k}(-a+c). \tag{2.532}$$

Therefore,

$$A + B = \frac{\rho g c}{k}, \tag{2.533}$$

$$A - B = \frac{\rho g a}{k}, \tag{2.534}$$

which means

$$l = \frac{2k}{\rho g}\cosh\left(\frac{\rho g c}{k}\right)\sinh\left(\frac{\rho g a}{k}\right). \tag{2.535}$$

To match the start $(-a, H)$ and end points (a, h) of $y = \frac{k}{\rho g}\cosh\left(\frac{\rho g}{k}(x+c)\right)-\frac{\lambda}{\rho g}$,

$$H = \frac{k}{\rho g}\cosh\left(\frac{\rho g}{k}(-a+c)\right) - \frac{\lambda}{\rho g}, \tag{2.536}$$

$$h = \frac{k}{\rho g}\cosh\left(\frac{\rho g}{k}(a+c)\right) - \frac{\lambda}{\rho g}. \tag{2.537}$$

Therefore,

$$H - h = \frac{k}{\rho g}\left\{\cosh\left(\frac{\rho g}{k}(-a+c)\right) - \cosh\left(\frac{\rho g}{k}(a+c)\right)\right\}. \tag{2.538}$$

Using the hyperbolic identity $\cosh 2A - \cosh 2B = 2\sinh(A+B)\sinh(A-B)$, where $A = \frac{\rho g}{2k}(a+c)$ and $B = \frac{\rho g}{2k}(-a+c)$,

$$H - h = -\frac{2k}{\rho g}\sinh\left(\frac{\rho g c}{k}\right)\sinh\left(\frac{\rho g a}{k}\right). \tag{2.539}$$

Hence,

$$\frac{H - h}{l} = -\tanh\left(\frac{\rho g c}{k}\right). \tag{2.540}$$

Therefore, noting the identity $\frac{1}{\cosh^2 x} = 1 - \tanh^2 x$,

$$\cosh\left(\frac{\rho g c}{k}\right) = \frac{1}{\sqrt{1 - \left(\frac{H-h}{l}\right)^2}}. \tag{2.541}$$

Therefore,

$$l = \frac{2k}{\rho g} \frac{1}{\sqrt{1 - \left(\frac{H-h}{l}\right)^2}} \sinh\left(\frac{\rho g a}{k}\right), \tag{2.542}$$

which eliminates the variable c. This is an equation of the form $f(k) = \alpha = l\sqrt{1 - \left(\frac{H-h}{l}\right)^2}$, which can be, albeit numerically, solved for k:

$$f(k) = \alpha, \tag{2.543}$$

$$f(k) = \frac{2k}{\rho g} \sinh\left(\frac{\rho g a}{k}\right). \tag{2.544}$$

The derivative of $f(k)$ with respect to k is

$$\frac{df}{dk} = \frac{2}{\rho g} \sinh\left(\frac{\rho g a}{k}\right) + \frac{2k}{\rho g} \cosh\left(\frac{\rho g a}{k}\right)\left(-\frac{\rho g a}{k^2}\right) \tag{2.545}$$

$$= \frac{2}{\rho g} \sinh\left(\frac{\rho g a}{k}\right) - \frac{2a}{k} \cosh\left(\frac{\rho g a}{k}\right). \tag{2.546}$$

Hence, a Newton–Raphson scheme for solving $f(k) - \alpha = 0$ would follow the iteration

$$k_{n+1} = k_n - \frac{\frac{2k_n}{\rho g} \sinh\left(\frac{\rho g a}{k_n}\right) - l\sqrt{1 - \left(\frac{H-h}{l}\right)^2}}{\frac{2}{\rho g} \sinh\left(\frac{\rho g a}{k_n}\right) - \frac{2a}{k_n} \cosh\left(\frac{\rho g a}{k_n}\right)}. \tag{2.547}$$

Now, when $k = \rho g a$,

$$\frac{df}{dk} = \frac{2}{\rho g} \sinh(1) - \frac{2}{\rho g} \cosh(1) \tag{2.548}$$

$$= \frac{1}{\rho g}\left(e - e^{-1} - (e + e^{-1})\right) \tag{2.549}$$

$$= -\frac{2}{\rho g e}. \tag{2.550}$$

The fact that $f(k)$ has a negative derivative for positive k therefore means

$$k_1 = \rho g a \tag{2.551}$$

is probably a good starting point for the iteration.

Once we have determined k, this can be substituted to find λ and c:

$$c = -\frac{k}{\rho g} \tanh^{-1}\left(\frac{H-h}{l}\right), \tag{2.552}$$

$$\lambda = k \cosh\left(\frac{\rho g}{k}(a + c)\right) - h\rho g. \tag{2.553}$$

Note that $\tanh^{-1}(-x) = -\tanh^{-1} x$.

In summary:

Inputs: a, h, H, l, ρ, g:

$$y(x) = \frac{k}{\rho g} \cosh\left(\frac{\rho g}{k}(x + c)\right) - \frac{\lambda}{\rho g},$$

$$c = -\frac{k}{\rho g} \tanh^{-1}\left(\frac{H - h}{l}\right), \tag{2.554}$$

$$\lambda = k \cosh\left(\frac{\rho g}{k}(a + c)\right) - h\rho g, \tag{2.555}$$

where k is determined by the termination of the iteration

$$k_1 = \rho g a, \tag{2.556}$$

$$k_{n+1} = k_n - \frac{\frac{2k_n}{\rho g}\sinh\left(\frac{\rho g a}{k_n}\right) - l\sqrt{1 - \left(\frac{H-h}{l}\right)^2}}{\frac{2}{\rho g}\sinh\left(\frac{\rho g a}{k_n}\right) - \frac{2a}{k_n}\cosh\left(\frac{\rho g a}{k_n}\right)}. \tag{2.557}$$

An example asymmetric catenary (and indeed a Christmas 'cat-enary' too) is illustrated in Fig. 2.21. In this case, a chain of length $l = 10$ m, with mass per unit length $\rho = 1$ kgm^{-1}, is hung between the points (in metres) $(-2, 5)$ and $(2, 1)$, i.e. $a = 2$, $h = 1$ and $H = 5$. The strength of gravity is $g = 9.81$ Nkg^{-1}. These inputs yield constants of $k = 8.1402$, $c = -0.3515$ and $\lambda = 20.423$.

2.14   A Selection of Mechanics Books

Mechanics is such a core topic of physics, engineering and applied mathematics that it will come as no surprise that there are a bounty of excellent textbooks. The following list represents my personal library.

2.14.1   *Pre-university*

The A-Level physics course I teach is (as of 2023) based on the OCR-A textbook by Chadha [3]. It is very clear and beautifully illustrated, and there are many similar books for other UK examination boards, such as AQA. However, you might find the core texts rather thin on mathematical details, so I would personally recommend all the OCR A-Level mathematics mechanics books (M1, M2, M3&4) by Quadling [29]. At sixth form, I gained my confidence in physics by way of mathematics, and one of the key suggestions in this book is that you might find this approach useful too. Revision and example books that are directly associated with the course textbooks are also often a useful purchase, as they contain a wealth of very specific examples and 'facts for immediate recall' that are essential for examination success and mastery of the basics. For enrichment and preparation for university interviews, *The Language of Physics* by Cullerne and Machacek [4] offers an excellent and mathematically rigorous exposition of key physics concepts, and Povey's *Perplexing*

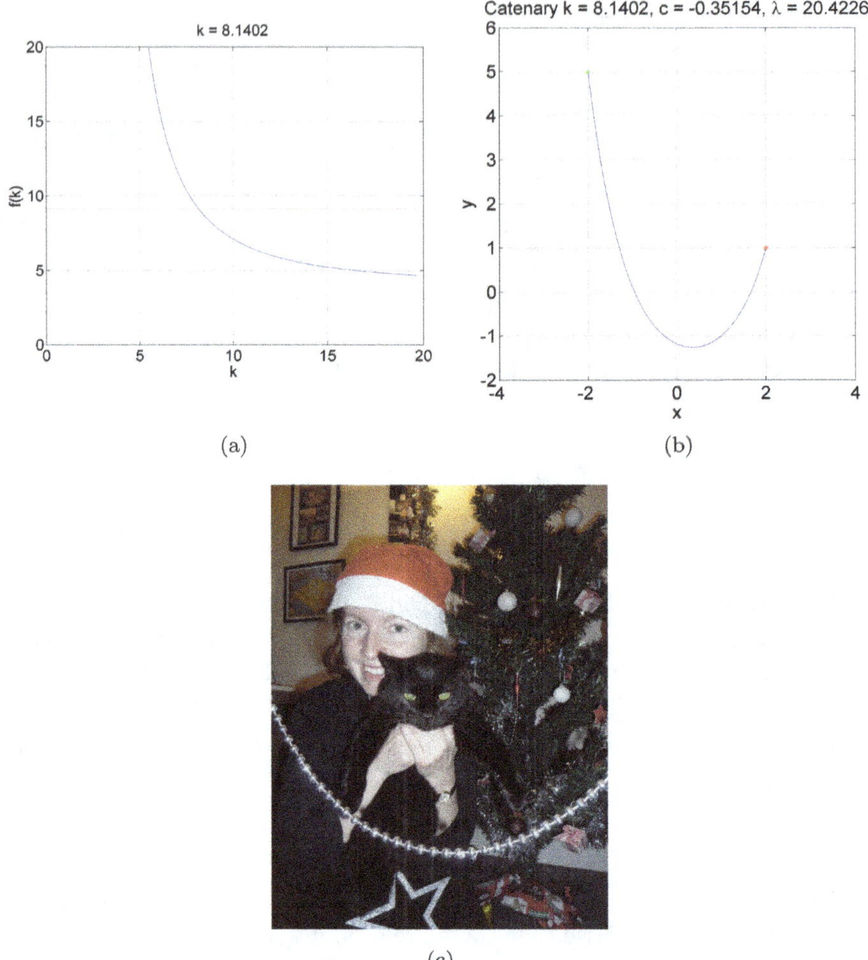

(c)

Fig. 2.21. Subplot (a) is a graphical solution for k to the equation $f(k) = \frac{2k}{\rho g} \sinh\left(\frac{\rho g a}{k}\right) = l\sqrt{1 - \left(\frac{H-h}{l}\right)^2}$, where l is the length of an inextensible chain of mass per unit length ρ hung between $(-a, H)$ and (a, h). The strength of gravity acting on the chain is g Nkg^{-1}. In this example, $l = 10$ m, with mass per unit length $\rho = 1$ kgm^{-1} hung between points (in metres) $(-2, 5)$ and $(2, 1)$, i.e. $a = 2$, $h = 1$ and $H = 5$. The strength of gravity is $g = 9.81$ Nkg^{-1}. From k, one can determine the *catenary* curve of the weighted chain, as illustrated in subplot (b): $y(x) = \frac{k}{\rho g} \cosh\left(\frac{\rho g}{k}(x + c)\right) - \frac{\lambda}{\rho g}$ where $c = -\frac{k}{\rho g} \tanh^{-1}\left(\frac{H-h}{l}\right)$ and $\lambda = k \cosh\left(\frac{\rho g}{k}(a + c)\right) - h\rho g$. Subplot (c) is a *Christmas 'cat-enary'* (!)

Problems [27] provides a fabulous range of beautiful and amusing conundrums to try out and hone your skills. Physics-based quantitative arguments are at the heart of MacKay's incredibly important book, *Sustainable Energy – Without the Hot Air* [25], which I encourage everyone to read, especially those with scientific training. There are many, many books of physics problems. A recommended pair are *Practice in Physics* by Akrill, Bennet and Millar [1] and *3,000 Solved Problems in Physics* from the Schaum's Outline series [32]. *Mastering Pre-University Physics* [23] is also an excellent printed companion to the amazing online *Isaac Physics* educational platform (https://isaacphysics.org). And for the 'gold standard' of pre-university physics problem-solving, try out Round 1 past papers from the British Physics Olympiad (https://www.bpho.org.uk).

2.14.2 *University level*

The textbook I wish I had used as an undergraduate is *An Introduction to Mechanics* by Kleppner and Kolenkow [21]. It is simply marvellous in terms of clarity, illustrations and exposition. A similarly outstanding book is Morin's *Introduction to Classical Mechanics* [26], which is based on hundreds of beautifully compact problems. But beware, a few tantalising lines in Morin may result in lost days of work, frustration and, eventually, satisfaction! Sometimes, what you really need is a step-by-step recipe, and the how-to book that I and all my undergraduate contemporaries used, perhaps above all else, was Rees' *Physics by Example* [30]. For more advanced topics, such as Lagrangian mechanics, Goldstein, Poole and Safko's *Classical Mechanics* [13] is arguably the canonical tome, with Hand and Finch's *Analytical Mechanics* [15] providing a more singular focus on the application of the calculus of variations. And every physics equation you have ever encountered, and many you might be curious about, are presented with outstanding clarity in Woan's *The Cambridge Handbook of Physics Formulas* [36]. This should ideally live on or within reach of your desk. Finally, Longair's *Theoretical Concepts in Physics* [22] should be on every physicist's bookshelf. I was lucky enough to attend a lecture course delivered with fizzing enthusiasm by the man himself back in 1998, and the idea of physics pedagogy placed in the context of historical storytelling is a pattern that I would recommend to all students and teachers.

Chapter 3

Thermodynamics

In this chapter, we follow a very common pedagogical path, from the smooth continuous *classical* theory of heat to *random walks, kinetic theory* and *Boltzmann's*[1] *exponential statistics*. We begin with an analysis of *heat engines* (including *Carnot, Otto* and *Diesel* examples) based on energy changes associated with a (combustible) *ideal gas*. We won't (yet) consider the discrete nature of the molecules that comprise the gas and assert that pressure, volume, temperature and entropy are characteristic quantities of a fluidic macroscopic continuum within the cylinders of an engine. After a model of cooling based on *conduction*, we investigate the consequences of a *kinetic theory* of a huge number of discrete molecules in random motion. We will construct a computational recipe for a *molecular random walk* and define a *Knudsen number* (Kn), which will help us assess the validity of the assumptions of the ideal gas model. The Knudsen number is the *ratio of the mean free path between molecular collisions to the diameter of the molecules*. When Kn is high, the ideal gas assumption of elastic (i.e. 'not sticky') collisions between infinitesimally small particles is appropriate. We will then develop Boltzmann statistics and properly justify the idea that temperature (in kelvin) is proportional to the mean kinetic energy of molecules in random motion. Finally, we shall employ the Boltzmann factor $e^{-\frac{E}{k_B T}}$ to determine the *Maxwell–Boltzmann distribution of molecular speeds*.

3.1 Heat Engines

The fundamental purpose of an engine is the efficient transformation of energy (for example, thermal energy) into useful mechanical work. The operation, and theoretical efficiency $\eta =$ work done/heat input of combustion driven piston engines (e.g. diesel, petrol or, hopefully in the near future, hydrogen fuelled) can be analysed by considering changes to the pressure p, absolute temperature T (in kelvin) and volume V of the gaseous components during a single combustion cycle in the cylinder of an engine. The analysis proceeds by quantifying the energy changes in the gas

---

[1]Ludwig Boltzmann (1844–1906).

as a result of heat applied to and work done by the gas, combined with the *ideal gas equation*, which relates its physical properties p, V, T. Before we consider the different models of heat engines (we'll look at Carnot, Otto and Diesel variants), let us determine the relationships between pressure, volume, temperature, heat added, work done and entropy changes associated with processes involving ideal gases at constant-temperature (*isothermal*), constant-volume (*isochoric*), constant-pressure (*isobaric*) or constant-entropy (*isentropic* or *adiabatic*).

3.1.1　*Ideal gases*

An ideal gas assumes a large number of point particles colliding elastically. It neglects any short-range intermolecular forces resulting from repulsion or attraction due to molecular charges, as well as the fact that molecules have a finite volume. This means a real gas is not infinitely compressible, whereas an ideal gas has no such limits. However, for the range of pressures and temperatures we shall consider in engines, the mean free path between molecular collisions will be much larger than molecular dimensions. Therefore, we can use the ideal gas approximation.

For n moles of ideal gas

$$pV = nRT, \tag{3.1}$$

where $R = 8.314$ Jmol^{-1}K^{-1} is the *molar gas constant*. Let's define the internal energy U of an ideal gas to be directly proportional to its absolute temperature T. (In the section on *Boltzmann's exponential statistics* in this chapter, we formalise the concept that *temperature is directly proportional to the mean kinetic energy of molecules in an ideal gas,* but for the moment, let's just run with the idea.) U must also scale with the number of moles n, and since pV has dimensions of $\frac{\text{force}}{\text{area}} \times \text{volume} = \text{force} \times \text{distance}$ (i.e. work done and therefore energy), it makes dimensional sense for U to be proportional to nRT. In fact,

$$U = \tfrac{1}{2}\alpha nRT, \tag{3.2}$$

where α is the *number of degrees of freedom* of the gas. For molecules that can only translate in three dimensions, $\alpha = 3$. Larger values of α can result if the molecule can stretch, rotate, etc., but these modes are only significant contributors to U at higher temperatures.[2] Why the factor $\frac{1}{2}\alpha$? Well, we could set this to be an arbitrary dimensionless constant and then compare theoretical predictions about *specific heat capacity* with experimental results (see the section on the Mayer relationship and γ later on in this chapter). However, this empirical approach will be theoretically augmented very nicely when we calculate an expression for the mean kinetic energy of molecules using the *Maxwell–Boltzmann speed distribution*, which shall be the culmination of this chapter.

---

[2]What temperatures, you might ask? Well, again, we shall have to wait until the last section of this chapter for the *Boltzmann distribution*, which will enable us to assign a *probability* for a molecule to be in a particular energy state. This will depend on the Boltzmann factor $e^{-\varepsilon/k_B T}$, which is functionally dependent on the dimensionless ratio of energy ε and absolute temperature T multiplied by *Boltzmann's constant* $k_B = 1.38 \times 10^{-23}$ JK^{-1}.

Note that combining $U = \frac{1}{2}\alpha nRT$ with the ideal gas equation yields

$$U = \tfrac{1}{2}\alpha pV, \tag{3.3}$$

which we shall make good use of in the ensuing discussions.

3.1.2 *Heat, work and internal energy*

Consider a cylinder of gas (see Fig. 3.1) of cross-sectional area A and width x, being compressed by $-dx$ via a force F. The negative sign is needed since the gaseous *expansion* dx is negative for compression.[3] The work done *on* the gas *by* the force is, therefore,

$$dW = -Fdx. \tag{3.4}$$

The pressure p acting on the gas is, by definition of pressure,

$$p = \frac{F}{A} \tag{3.5}$$

and the volume change is

$$dV = Adx. \tag{3.6}$$

Therefore, since $F = pA$ and $dx = \frac{dV}{A}$,

$$dW = -Fdx = -pA\frac{dV}{A} \tag{3.7}$$

$$= -pdV. \tag{3.8}$$

If heat dQ is supplied to the gas, then the *first law of thermodynamics* (i.e. energy in a closed system is conserved) means the internal energy change of the gas is

$$dU = dQ + dW = dQ - pdV. \tag{3.9}$$

In a combustion engine, a gas (typically in a cylinder) has heat exchanged with it, which enables the gas (and hence the cylinder piston) to do work. In order to

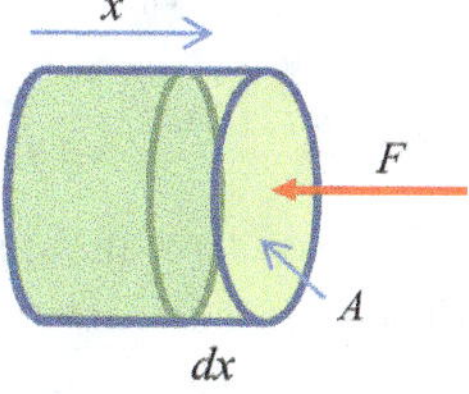

Fig. 3.1. A cylinder of gas of cross-sectional area A is compressed by $-dx$ by the action of a force F. The negative sign is needed since expansion dx is negative for compression. The (positive) amount of work done on the gas by the force is $dW = -Fdx$. This equates to $-pdV$, where p is gas pressure and $-dV$ is the change in the volume of the gas.

---

[3]The ensuing chapter may annoy pure mathematicians in that I have liberally used the derivative notation dx instead of a small (but finite) quantity, δx (and then the differential limit is taken when forming integrals). I've made this choice to keep the rather copious amounts of calculus notation to a manageable level. I hope the implication that dx really means δx as $\delta x \to 0$ is acceptable.

repeat in a sensible mechanical way (e.g. cylinders firing in a petrol engine at many thousands of revolutions per minute), the initial state of the gas must be the same at the end of the cycle as it was when the cycle began. This means we can track the progress of an ideal gas in a p versus V diagram, and the cycle can be represented by a closed loop, as illustrated (for a Carnot cycle) in Fig. 3.3. Around this loop, the total change in internal energy must be zero, i.e. adding up infinitesimal changes in U that correspond to one cycle of a heat engine,

$$\oint dU = 0. \tag{3.10}$$

Therefore, if the net heat $\Delta Q = Q_{\text{in}} - Q_{\text{out}}$ is added to the gas over the cycle,

$$0 = \oint dU = \Delta Q - \oint p\,dV, \tag{3.11}$$

which means the total work done *by* the gas over the cycle is $\oint p\,dV$, which equates to the total heat input ΔQ by conservation of energy. Geometrically, the total work done per cycle $\oint p\,dV$ is the *area enclosed* by a loop in the p versus V diagram.

3.1.3 *Constant-temperature (isothermal) processes*

$$dU = 0,\ T = \text{constant},\ \therefore dT = 0. \tag{3.12}$$

The first law of thermodynamics states that $dU = dQ - p\,dV$. So, if $dU = 0$,

$$dQ = p\,dV. \tag{3.13}$$

Using the ideal gas equation, $p = \frac{nRT}{V}$, which means $p \propto \frac{1}{V}$ or $pV = \text{constant}$. This is a statement of *Boyle's law.*[4] Using $dQ = p\,dV$,

$$dQ = \frac{nRT}{V}dV, \tag{3.14}$$

$$\Delta Q = nRT \int_{V_0}^{V} \frac{1}{V'}dV'. \tag{3.15}$$

$$\therefore \Delta Q = nRT \ln\left(\frac{V}{V_0}\right). \tag{3.16}$$

Since $dU = 0$ for an isothermal process, this means $\Delta Q + \Delta W = 0$. Hence, work done *on* a gas in an isothermal process resulting in a volume change of V_0 to V is

$$\Delta W = -nRT \ln\left(\frac{V}{V_0}\right) \tag{3.17}$$

and the work done *by* a gas is minus this, i.e. $nRT \ln\left(\frac{V}{V_0}\right)$. So, a gas that expands does work on the surroundings, whereas work is required to be done on a gas in order to compress it.

---

[4] Robert Boyle (1627–1691).

3.1.4 *Constant-volume (isochoric) processes*

We define a constant-volume *molar* heat capacity[5] c_V, which enables heat transfer dQ to be calculated from a consequential temperature change dT of n moles of gas:

$$dQ = nc_V dT. \tag{3.18}$$

If c_V is a constant, then the heat supplied to cause a temperature change ΔT is

$$\Delta Q = nc_V \Delta T. \tag{3.19}$$

For a constant-volume process, $dV = 0$; therefore, no work is done on (or indeed by) the gas, as $dW = -pdV$. Hence, by the first law $dU = dQ - pdV$,

$$\therefore dU = dQ. \tag{3.20}$$

Now, internal energy $U = \frac{1}{2}\alpha nRT$ and therefore $dU = \frac{1}{2}\alpha nRdT$. Equating this internal energy change to the heat change $dQ = nc_V dT$,

$$\tfrac{1}{2}\alpha nRdT = nc_V dT. \tag{3.21}$$

$$\therefore c_V = \tfrac{1}{2}\alpha R, \tag{3.22}$$

i.e. we expect the molar heat capacity for a constant-volume process to be half the number of degrees of freedom α multiplied by the molar gas constant.

3.1.5 *Constant-pressure (isobaric) processes*

We define a constant-pressure *molar* heat capacity c_p in a similar fashion to c_V for an isochoric process:

$$dQ = nc_p dT. \tag{3.23}$$

If c_p is a constant, then the heat supplied to cause a temperature change ΔT is

$$\Delta Q = nc_p \Delta T. \tag{3.24}$$

From the ideal gas equation, $pV = nRT$, and noting $p = $ constant for an isobaric process,

$$pdV = nRdT. \tag{3.25}$$

Hence, work done ΔW on the gas can easily be computed from ΔT. Since $dW = -pdV$,

$$dW = -nRdT. \tag{3.26}$$

$$\therefore \Delta W = -nR\Delta T. \tag{3.27}$$

Alternatively, using $dW = -pdV$ directly, since $p = $ constant,

$$\Delta W = -p\Delta V. \tag{3.28}$$

---

[5]The *specific* heat capacity (for a constant-volume process) is c_V/M, where M is the molar mass of an ideal gas. For air, this is about 29 gmol^{-1}. https://www.engineeringtoolbox.com/molecular-mass-air-d_679.html (Accessed 1/1/22).

From the first law, $dU = dQ + dW$. Now, since $U = \frac{1}{2}\alpha nRT$, this means $dU = \frac{1}{2}\alpha nRdT$. Hence, since $dQ = nc_p dT$ and $dW = -nRdT$,

$$\frac{1}{2}\alpha nRdT = nc_p dT - nRdT. \tag{3.29}$$

$$\therefore c_p = \left(1 + \tfrac{1}{2}\alpha\right) R.$$

Comparing this with the (constant-volume) molar heat capacity $c_V = \frac{1}{2}\alpha R$,

$$c_p = c_V + R. \tag{3.30}$$

3.1.6   *The Mayer relationship and γ*

$c_p = c_V + R$ is called the *Mayer relationship*.[6] Let us define a ratio γ of the constant-pressure and constant-volume molar heat capacities:

$$\gamma = \frac{c_p}{c_V} \tag{3.31}$$

$$= \frac{\left(1 + \tfrac{1}{2}\alpha\right) R}{\tfrac{1}{2}\alpha R}. \tag{3.32}$$

$$\therefore \gamma = \frac{2}{\alpha} + 1.$$

Now, if $\gamma = \frac{2}{\alpha} + 1 \Rightarrow \alpha = \frac{2}{\gamma-1}$. Since $c_V = \frac{1}{2}\alpha R$ and $c_p = \gamma c_V$, we can therefore express the constant-volume and constant-pressure molar heat capacities in terms of γ and R:

$$c_V = \frac{R}{\gamma - 1}, \tag{3.33}$$

$$c_p = \frac{\gamma R}{\gamma - 1}. \tag{3.34}$$

3.1.7   *Constant-entropy (isentropic or adiabatic) processes*

An *entropy change* of a body at absolute (i.e. kelvin) temperature T, following the addition of heat dQ, is

$$dS = \frac{dQ}{T}. \tag{3.35}$$

If a process occurs at constant-entropy, this means $dS = 0$ and therefore $dQ = 0$. So a constant-entropy process is one in which no heat is transferred. From the first

---

[6]Experimentally, it is very hard to maintain a constant gas volume as heat is added, so constant-volume heat capacity is difficult to measure directly. However, constant-pressure heat capacity is much easier to calculate, as one can allow a volume to change in order to maintain equilibrium with the ambient pressure. The *Mayer relationship* $c_p = c_V + R$ is therefore very useful for working out the constant-volume heat capacity from the constant-pressure heat capacity.

law, $dU = dQ - pdV$, which implies

$$dU = -pdV. \tag{3.36}$$

Now, internal energy $U = \frac{1}{2}\alpha nRT$ can be written using the ideal gas equation $pV = nRT$ as

$$U = \frac{1}{2}\alpha pV. \tag{3.37}$$

Therefore, noting the product rule result $d(pV) = pdV + Vdp$,

$$dU = \frac{1}{2}\alpha pdV + \frac{1}{2}\alpha Vdp. \tag{3.38}$$

Hence, by equating $dU = -pdV$ with $\frac{1}{2}\alpha pdV + \frac{1}{2}\alpha Vdp$, we can derive a relationship between p and V for an *adiabatic* (or *isentropic*) process:

$$-pdV = \frac{1}{2}\alpha pdV + \frac{1}{2}\alpha Vdp \tag{3.39}$$

$$-\frac{2}{\alpha}pdV = pdV + Vdp \tag{3.40}$$

$$-pdV\left(\underbrace{\frac{2}{\alpha} + 1}_{\gamma}\right) = Vdp \tag{3.41}$$

$$-\gamma p\frac{dV}{V} = \frac{dp}{p} \tag{3.42}$$

$$-\gamma \int_{V_0}^{V} \frac{dV'}{V'} = \int_{p_0}^{p} \frac{dp'}{p'} \tag{3.43}$$

$$-\gamma \ln\left(\frac{V}{V_0}\right) = \ln\left(\frac{p}{p_0}\right) \tag{3.44}$$

$$\ln\left(\left(\frac{V}{V_0}\right)^{-\gamma}\right) = \ln\left(\frac{p}{p_0}\right). \tag{3.45}$$

$$\therefore p_0 V_0^{\gamma} = pV^{\gamma}. \tag{3.46}$$

The work done $\Delta W = -\int pdV$ *on* an ideal gas during an isentropic (i.e. constant-entropy) change can therefore be computed using $p = \frac{p_0 V_0^{\gamma}}{V^{\gamma}}$, for a volume change V_0 to V:

$$\Delta W = -\int pdV \tag{3.47}$$

$$= -p_0 V_0^{\gamma} \int_{V_0}^{V} (V')^{-\gamma} dV' \tag{3.48}$$

$$= -p_0 V_0^\gamma \left[\frac{1}{-\gamma + 1} \left(V' \right)^{-\gamma+1} \right]_{V_0}^{V}. \tag{3.49}$$

$$\therefore \Delta W = -\frac{p_0 V_0^\gamma}{1 - \gamma} \left(V^{1-\gamma} - V_0^{1-\gamma} \right). \tag{3.50}$$

3.1.8 *Maximum efficiency of a heat engine*

It is possible to bound the efficiency of any heat engine (see Fig. 3.2) using (i) the law of conservation of energy (i.e. the *first law of thermodynamics*) and (ii) the *second law of thermodynamics*.

The second law of thermodynamics states that the *total entropy of a closed system will either remain the same or increase as time progresses*. Regardless of the specific details, a heat engine is essentially a flow of energy from a hot reservoir to a colder one. By *reservoir*, we mean an ensemble of molecules that are so numerous that their average temperature will not change when the heat we associate with our engine is taken from or added to it. The difference in heat taken from the hot reservoir Q_{in} and the heat transferred to the cold reservoir Q_{out} is the maximum possible work done W *by* the engine.[7] This must be true to satisfy energy conservation, i.e.

$$Q_{\text{in}} = Q_{\text{out}} + W. \tag{3.51}$$

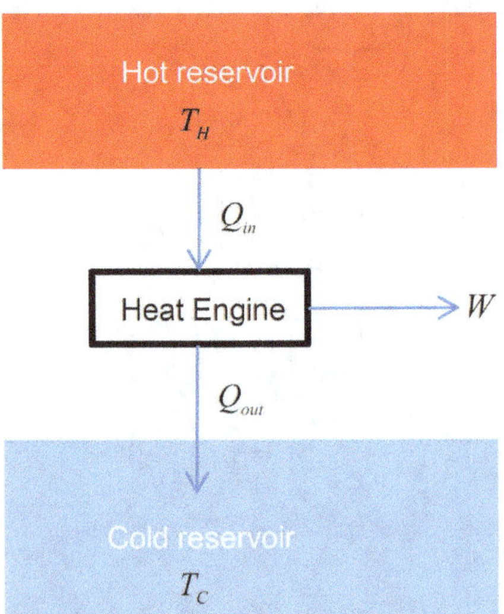

Fig. 3.2. Energy flow diagram for a heat engine. Heat Q_{in} is drawn from a reservoir of temperature T_H. This powers the heat engine which does work W and dumps heat Q_{out} to a colder-temperature reservoir of temperature T_C. One can show from the first and second laws of thermodynamics that the efficiency $\eta = \frac{W}{Q_{\text{in}}}$ has a maximum value of $1 - \frac{T_C}{T_H}$.

---

[7]We will be careful to perpetuate this definition of work done W *by* a gas from now on. In the previous sections developing the thermodynamic relationships for isothermal, isobaric, isochoric and isentropic processes, ΔW was defined as the work done *on* a gas.

The second law of thermodynamics states that for every change, there can *never be an overall decrease in entropy.* We may be able to reduce the entropy of a particular system (e.g. a chemical reaction or, indeed, the tidiness of a study), but only if the surroundings gain in entropy by the same amount or more. For our idealised heat engine and hot-and-cold reservoir system, this means the loss of entropy of the hot reservoir ΔS_H must at least be compensated for by the gain in entropy ΔS_C of the cold reservoir. If the hot reservoir has an (absolute) temperature of T_H and the cold reservoir has a temperature of T_C, the respective entropy changes, following the extraction of heat Q_{in} from the hot reservoir and the dumping of heat Q_{out} to the cold reservoir, are

$$\Delta S_H = -\frac{Q_{\text{in}}}{T_H}, \tag{3.52}$$

$$\Delta S_C = \frac{Q_{\text{out}}}{T_C}. \tag{3.53}$$

The total entropy change in the hot-and-cold reservoir system is

$$\Delta S_{\text{total}} = \Delta S_H + \Delta S_C = -\frac{Q_{\text{in}}}{T_H} + \frac{Q_{\text{out}}}{T_C}.$$

Invoking the second law,

$$\Delta S_{\text{total}} \geq 0. \tag{3.54}$$

$$\therefore -\frac{Q_{\text{in}}}{T_H} + \frac{Q_{\text{out}}}{T_C} \geq 0. \tag{3.55}$$

From the first law $Q_{\text{out}} = Q_{\text{in}} - W$, we can substitute for Q_{out} in $-\frac{Q_{\text{in}}}{T_H} + \frac{Q_{\text{out}}}{T_C} \geq 0$.

$$\therefore -\frac{Q_{\text{in}}}{T_H} + \frac{Q_{\text{in}} - W}{T_C} \geq 0. \tag{3.56}$$

$$\therefore Q_{\text{in}} \left(\frac{1}{T_C} - \frac{1}{T_H} \right) \geq \frac{W}{T_C} \tag{3.57}$$

$$\Rightarrow 1 - \frac{T_C}{T_H} \geq \frac{W}{Q_{\text{in}}}.$$

Let us define the *efficiency of a heat engine*:

$$\eta = \frac{W}{Q_{\text{in}}}. \tag{3.58}$$

Therefore, since $\frac{W}{Q_{\text{in}}} \leq 1 - \frac{T_C}{T_H}$,

$$\eta \leq 1 - \frac{T_C}{T_H}. \tag{3.59}$$

The inequality $\eta \leq 1 - \frac{T_C}{T_H}$ implies that the maximum efficiency allowed by thermodynamics (i.e. satisfying both the first and second laws) is

$$\eta_{\text{max}} = 1 - \frac{T_C}{T_H}. \tag{3.60}$$

We demonstrate in the following section that a *Carnot engine* has an efficiency of $\eta = 1 - \frac{T_C}{T_H}$ and, therefore, is as efficient as any engine can possibly be.

Note that a *reversible* heat engine is one which is assumed to operate at thermodynamic equilibrium at all times; the ideal gas equation, along with its associated relationships for different processes, holds; and there are no losses due to friction etc. In other words, the differential form of the first law holds at all times, i.e. changes in internal energy dU are fully accounted for by heat change dQ and work done $dW = -pdV$. This means that there is no net internal energy change, and indeed entropy change, over the complete cycle. Therefore, the 'ideal' cycle could be run in reverse without breaking the second law since a net zero entropy change is permitted. Real engines will, of course, depart from this ideal, and the idealised models that are discussed in the following should be regarded very much as a first approximation.

3.1.9　*Carnot engine*

A *Carnot*[8] *cycle*, an idealised model of the most efficient heat engine, has four distinct stages, as illustrated in Fig. 3.3. We assume that n moles of gas are used and that there is no leakage during the process. The four stages are as follows:

(1) *Isothermal expansion* of a gas from volume V_1 to V_2: This is achieved via the exchange of heat Q_{in} from a hot reservoir[9] of constant-temperature T_H.

(2) *Isentropic expansion* from volume V_2 to V_3: In this stage, no more heat is added, and the work done by the gas is accounted for by a change in internal energy.

(3) *Isothermal compression*: Heat Q_{out} is transferred from the gas to a cold reservoir at a temperature of T_C. This causes the gas to compress in volume, from V_3 to V_4.

(4) *Isentropic compression*: In this stage, the gas returns to the original volume V_1 and temperature T_H.

To model the Carnot cycle, the inputs are T_H, T_C, n, V_1 and V_2, plus the constants $\gamma = \frac{c_p}{c_V}$ and R. The goal is, therefore, to derive all the other parameters in terms of these. $r = \frac{V_2}{V_1}$ is the (isothermal) *expansion ratio*. To determine the p, V cycle coordinates, we can apply the ideal gas equation, with $pV = $ constant for the isothermal stages and $pV^\gamma = $ constant for the isentropic stages. However, before we do this, consider the entropy definition $dS = \frac{dQ}{T}$, which means we can write the first law as

$$dU = TdS - pdV. \tag{3.61}$$

---

[8]Nicolas Léonard Sadi Carnot (1796–1832).

[9]A *reservoir* is a suitably large 'thermal mass' such that extracting heat from it does not change its temperature. Consider an analogous volumetric example of a water reservoir, such as a deep lake. Taking a litre out won't change its water level very much!

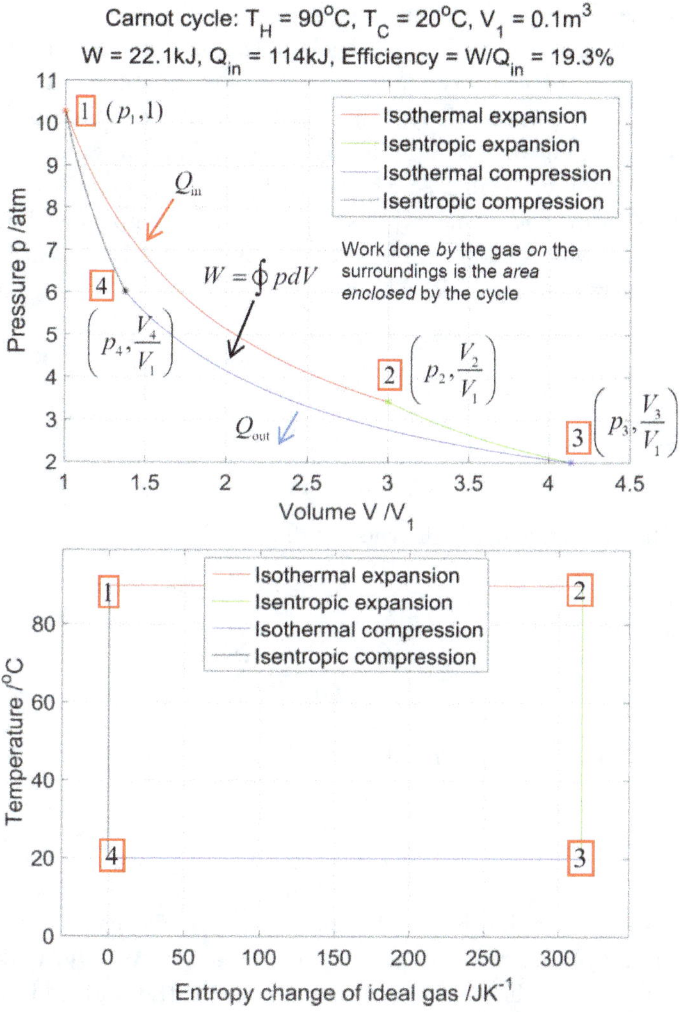

Fig. 3.3. The *Carnot cycle* is a *heat engine* that is maximally efficient. An ideal gas is expanded isothermally at temperature T_H via the absorption of heat Q_{in}. It is then further expanded isentropically (i.e. at constant-entropy, with no heat exchanged) before being compressed via the exchange of heat Q_{out} with a cold reservoir at temperature T_C. Further isentropic compression occurs to complete the cycle. The work done per cycle is $W = \oint p\,dV$, i.e. the area enclosed by the cycle in the p, V space. p is gas pressure and V is gas volume. The efficiency $\eta = \frac{W}{Q_{in}} = 1 - \frac{T_C}{T_H}$, which is the limit imposed by the laws of thermodynamics. The lower trace plots the Carnot cycle in T, S space, where T is temperature and S is entropy. This is a much more straightforward representation since the isothermal stages are horizontal lines and the isentropic stages are vertical lines. The area, that of a rectangle, also enables a more straightforward calculation of work done W. In this example, the ideal gas has a mass of 1.00 kg and a molar mass of 28.966 gmol^{-1}.

Hence, over a cycle,

$$0 = \oint dU = \oint T\,dS - \oint p\,dV. \tag{3.62}$$

$$\therefore \oint T\,dS = \oint p\,dV. \tag{3.63}$$

So, the work done $W = \oint p\,dV$ by the gas is equal to the area enclosed in the T, S space as well as the area enclosed by the cycle in the p, V space. For a Carnot cycle, the pair of isentropic stages (between 2 and 3 and between 4 and 1 in Fig. 3.3) are vertical lines (i.e. of constant-entropy) in T, S space, whereas the isothermal stages (between 1 and 2 and between 3 and 4 in Fig. 3.3) are horizontal lines in T, S space. Therefore, the Carnot cycle in T, S space must be a *rectangle*, which means the entropy change ΔS for isothermal expansion is equal in magnitude (but opposite in sign) to isothermal compression. The work done by the gas $W = \oint p\,dV = \oint T\,dS$ is the area enclosed by the rectangle, which is

$$W = (T_H - T_C)\,\Delta S. \tag{3.64}$$

The heat inputs and outputs in the isothermal stages are

$$Q_{\text{in}} = T_H \Delta S, \tag{3.65}$$

$$Q_{\text{out}} = T_C \Delta S, \tag{3.66}$$

which means the efficiency of the Carnot engine is

$$\eta = \frac{W}{Q_{\text{in}}} \tag{3.67}$$

$$= \frac{(T_H - T_C)\,\Delta S}{T_H \Delta S}. \tag{3.68}$$

$$\therefore \eta = 1 - \frac{T_C}{T_H}. \tag{3.69}$$

Since every heat engine must obey the inequality

$$\eta \leq 1 - \frac{T_C}{T_H}. \tag{3.70}$$

This means a Carnot engine is as efficient as any engine can possibly be. Now, let us solve for the p,V curve that is plotted in Fig. 3.3. We will make use of the relationships derived in the earlier sections for (i) isothermal and (ii) isentropic processes.

3.1.9.1 *Stage 1: Isothermal expansion*

In a programming language, such as MATLAB, we could define a linearly spaced vector of V using the code `V = linspace(V1,V2,1000)`, which would create 1,000 equally spaced V values between V_1 and V_2. A corresponding vector of p values could then be determined from V using the ideal gas equation:

$$p = \frac{nRT_H}{V}.$$

The limits of p are:

$$p_1 = \frac{nRT_H}{V_1}, \tag{3.71}$$

$$p_2 = \frac{nRT_H}{V_2}. \tag{3.72}$$

In this isothermal stage, the heat supplied is

$$Q_{\text{in}} = nRT_H \ln\left(\frac{V_2}{V_1}\right),$$ (3.73)

which means the entropy change

$$\Delta S = \frac{Q_{\text{in}}}{T_H} = nR\ln\left(\frac{V_2}{V_1}\right).$$ (3.74)

3.1.9.2 *Stage 2: Isentropic expansion*

In this stage, the pressure changes from p_2 to p_3 and the volume changes from V_2 to V_3. Using the $p(V)$ relationship for an isentropic expansion $p = \frac{p_2 V_2^\gamma}{V^\gamma}$,

$$p_3 = \frac{p_2 V_2^\gamma}{V_3^\gamma}.$$ (3.75)

Now, this stage intersects with the isothermal compression, which is at temperature T_C. Therefore, using the ideal gas equation,

$$p_3 = \frac{nRT_C}{V_3}.$$ (3.76)

Since $p_2 = \frac{nRT_H}{V_2}$ and hence $p_3 = \frac{p_2 V_2^\gamma}{V_3^\gamma} = \frac{nRT_H}{V_2}\frac{V_2^\gamma}{V_3^\gamma}$, we can find V_3 from V_2 and the ratio of the hot and cold reservoir temperatures:

$$p_3 = \frac{nRT_C}{V_3} = \frac{nRT_H}{V_2}\frac{V_2^\gamma}{V_3^\gamma}.$$

$$\therefore \frac{nRT_C}{V_3} = \frac{nRT_H}{V_2}\frac{V_2^\gamma}{V_3^\gamma}.$$ (3.77)

$$\therefore V_3^{\gamma-1} = \frac{T_H}{T_C}V_2^{\gamma-1}.$$ (3.78)

$$\therefore V_3 = \left(\frac{T_H}{T_C}\right)^{\frac{1}{\gamma-1}} V_2.$$ (3.79)

Now that we know V_3, we can determine `V = linspace(V2,V3,1000)` and then compute a vector of p values from

$$p = p_2\left(\frac{V_2}{V}\right)^\gamma = \frac{nRT_H}{V_2}\left(\frac{V_2}{V}\right)^\gamma.$$ (3.80)

3.1.9.3 *Stage 3: Isothermal compression*

In this stage, the pressure changes from p_3 to p_4 and the volume changes from V_3 to V_4. From the ideal gas equation,

$$p_4 V_4 = nRT_C. \tag{3.81}$$

Now, stage 4 (isentropic compression) completes the cycle with (p_1, V_1). Since $pV^\gamma = $ constant in this stage,

$$p_4 V_4^\gamma = p_1 V_1^\gamma. \tag{3.82}$$

Therefore,

$$p_4 = p_1 V_1^\gamma V_4^{-\gamma}. \tag{3.83}$$

Hence, by substitution of $p_4 = p_1 V_1^\gamma V_4^{-\gamma}$ into $p_4 V_4 = nRT_C$,

$$p_1 V_1^\gamma V_4^{-\gamma} V_4 = nRT_C \tag{3.84}$$

$$V_4^{1-\gamma} = \frac{nRT_C}{p_1 V_1^\gamma}. \tag{3.85}$$

$$\therefore V_4 = \left(\frac{nRT_C}{p_1 V_1^\gamma} \right)^{\frac{1}{1-\gamma}}. \tag{3.86}$$

Now,

$$p_1 = \frac{nRT_H}{V_1}. \tag{3.87}$$

Hence,

$$V_4 = \left(\frac{T_C}{T_H V_1^{\gamma-1}} \right)^{\frac{1}{1-\gamma}} \tag{3.88}$$

$$V_4 = \left(\frac{T_C}{T_H} \right)^{\frac{1}{1-\gamma}} V_1. \tag{3.89}$$

$$\therefore V_4 = \left(\frac{T_H}{T_C} \right)^{\frac{1}{\gamma-1}} V_1. \tag{3.90}$$

Now that we know V_4, we can determine `V = linspace(V3,V4,1000)` and hence compute pressures from

$$p = \frac{nRT_C}{V}, \tag{3.91}$$

$$p_4 = \frac{nRT_C}{V_4}. \tag{3.92}$$

Also, the heat output in this isothermal stage is

$$Q_{\text{out}} = nRT_C \ln\left(\frac{V_3}{V_4}\right).$$ (3.93)

3.1.9.4 Stage 4: Isentropic compression

`V = linspace(V4,V1,1000)`. We know the starting values p_1 and V_1, so all that is needed is for the $p(V)$ path to be computed from

$$p = \frac{p_1 V_1^{\gamma}}{V^{\gamma}}.$$ (3.94)

3.1.9.5 Carnot cycle summary

Use the following recipe to construct a model in Excel or a programming language, such as MATLAB or Python. The goal is to automatically generate annotated plots similar to those in Fig. 3.3. The inputs are temperatures T_H and T_C (in kelvin), moles of gas n, volumes V_1 and V_2 and the constants $\gamma = \frac{c_p}{c_V}$ and R. We can therefore calculate the (p, V) coordinates:

$$p_1 = \frac{nRT_H}{V_1},$$ (3.95)

$$p_2 = \frac{nRT_H}{V_2},$$ (3.96)

$$V_3 = \left(\frac{T_H}{T_C}\right)^{\frac{1}{\gamma-1}} V_2,$$ (3.97)

$$V_4 = \left(\frac{T_H}{T_C}\right)^{\frac{1}{\gamma-1}} V_1,$$ (3.98)

$$p_3 = \frac{nRT_C}{V_3},$$ (3.99)

$$p_4 = \frac{nRT_C}{V_4}.$$ (3.100)

Note that this means

$$\frac{V_3}{V_4} = \frac{V_2}{V_1}.$$ (3.101)

If we define linearly spaced vectors between V_1 and V_2, V_2 and V_3, V_3 and V_4, and V_4 and V_1, we can compute the four $p(V)$ curves:

Stage 1: *Isothermal expansion from V_1 to V_2*

$$p = \frac{nRT_H}{V}.$$ (3.102)

Stage 2: *Isentropic expansion from V_2 to V_3*

$$p = p_2 \left(\frac{V_2}{V}\right)^{\gamma}.$$ (3.103)

Stage 3: *Isothermal compression from V_3 to V_4*

$$p = \frac{nRT_C}{V}. \tag{3.104}$$

Stage 4: *Isentropic compression from V_4 to V_1*

$$p = p_1 \left(\frac{V_1}{V}\right)^{\gamma}. \tag{3.105}$$

The heat input Q_{in}, heat output Q_{out} and work done W (by the gas) are as follows:

$$Q_{\text{in}} = nRT_H \ln\left(\frac{V_2}{V_1}\right), \tag{3.106}$$

$$Q_{\text{out}} = nRT_C \ln\left(\frac{V_2}{V_1}\right), \tag{3.107}$$

$$W = nR\left(T_H - T_C\right)\ln\left(\frac{V_2}{V_1}\right), \tag{3.108}$$

and the magnitude of the entropy change (in both isothermal stages) is

$$\Delta S = nR\ln\left(\frac{V_2}{V_1}\right). \tag{3.109}$$

The efficiency of the Carnot engine is

$$\eta = \frac{W}{Q_{\text{in}}} = 1 - \frac{T_C}{T_H}, \tag{3.110}$$

which is the maximum permitted by thermodynamics.

3.1.10 *Otto (petrol) engine*

An Otto[10] cycle is the basis of spark-ignition piston engines, which is essentially how a typical petrol-driven engine operates. It has four stages, as illustrated in Fig. 3.4. It is not as efficient as a Carnot engine but is perhaps a more accurate description of a heat engine that can actually be constructed in the real world.[11] Again, we assume that n moles of gas are used and that there is no leakage during the process. The four stages of an Otto cycle are as follows:

(1) Air is drawn into a piston–cylinder arrangement at constant-pressure, i.e. at atmospheric pressure $p_1 = 1$ atm. The initial cylinder volume is V_1 and the temperature T_1 is the ambient air temperature. *Isentropic compression* of the air occurs via the piston from volume V_1 to volume V_2. Define the *compression ratio $r = \frac{V_1}{V_2}$*.

---

[10]Nikolaus Otto (1832–1891).

[11]Although, hopefully, as you read this, fossil fuel-based automotive engines may be less common than when this chapter was first drafted.

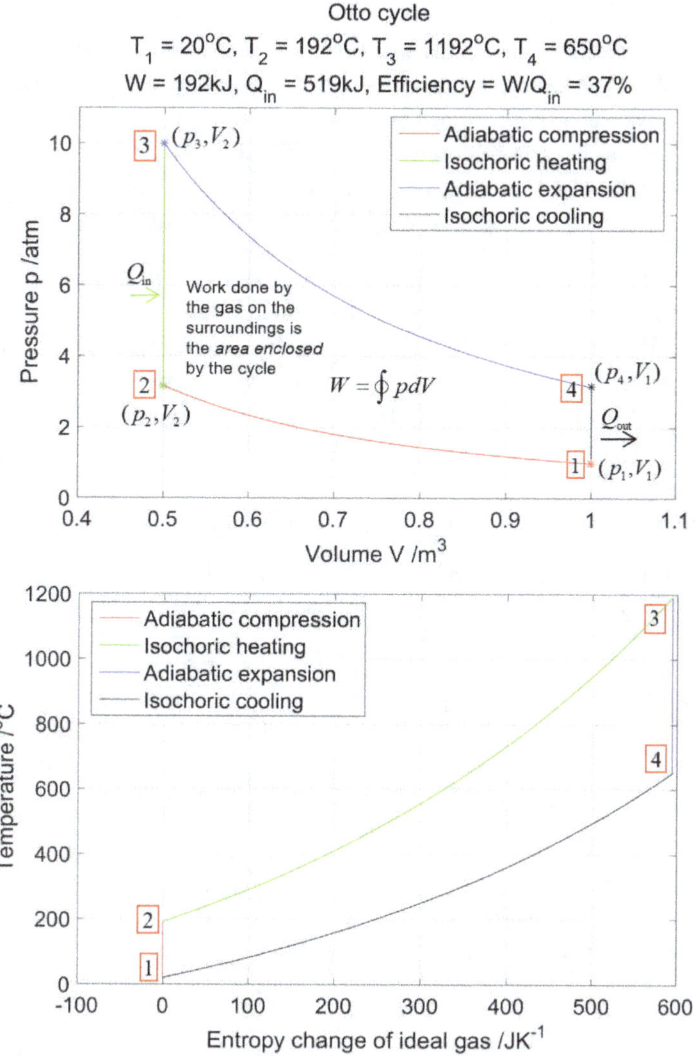

Fig. 3.4. Example of the key processes associated with a spark-ignition petrol engine represented by an *Otto cycle* in p, V and T, S spaces for an ideal gas, which represents a mixture of air and fuel. The gas mixture is drawn into a cylinder at atmospheric pressure and compressed isentropically, i.e. at constant-entropy with no heat exchanged. The spark ignition then liberates heat Q_{in} from the combusted fuel and rapidly increases gas pressure. The gas expands, doing work (isentropically) against a piston. This is the engine's 'power stroke'. At maximum expansion, the gas is exchanged with a fresh mixture via an exhaust system, and the cycle starts again.

(2) *Isochoric (constant-volume) heat transfer* to the gas occurs while the piston is at maximum compression. This is a simple model of a spark igniting the flammable gas, which causes a rapid increase in temperature and pressure but at constant-volume. Assume heat Q_{in} is added to the gas from the combustion of the fuel–air mixture, which causes the pressure to rise from p_2 to p_3. The gas volume remains at V_2.

(3) *Isentropic expansion* of the gas occurs from V_2 to V_1. The change in internal energy of the cooling gas equates to the work done by the piston as the cylinder expands. With no heat transferred, this stage is at constant-entropy.

(4) *Isochoric cooling* (i.e. at constant-volume) takes place at maximum piston expansion. Heat Q_{out} is released. In practice, this occurs via gas exchange. The hot gases are released as an exhaust, and n moles of cool air at temperature T_1 are drawn into the piston, so the cycle can begin again.

To model the Otto cycle, we must determine the equations of the lines and curves in the p, V diagram in a similar fashion to the Carnot cycle and compute Q_{in}, Q_{out} and $W = \oint p\, dV = Q_{\text{in}} - Q_{\text{out}}$ in terms of the input parameters p_1, V_1, V_2, T_1, T_3 and molar mass M. We also need the number of degrees of freedom of molecular motion α, as we need to compute the constant-volume molar heat capacity $c_V = \frac{R}{\gamma - 1}$, where $\gamma = \frac{c_p}{c_V} = \frac{2}{\alpha} + 1$.

3.1.10.1   *Stage 1: Isentropic compression from V_1 to V_2*

The number of moles of gas drawn into the cylinder is

$$n = \frac{p_1 V_1}{R T_1}, \tag{3.111}$$

`V = linspace(V1,V2,1000)` and

$$p = \frac{p_1 V_1^{\gamma}}{V^{\gamma}}. \tag{3.112}$$

Hence, since $r = \frac{V_1}{V_2}$,

$$p_2 = p_1 r^{\gamma}. \tag{3.113}$$

From the ideal gas equation,

$$T_2 = \frac{p_2 V_2}{nR}. \tag{3.114}$$

The work done *on* the gas undergoing isentropic compression is

$$\Delta W_{1,2} = \frac{p_1 V_1}{\gamma - 1}(r^{\gamma - 1} - 1). \tag{3.115}$$

3.1.10.2   *Stage 2: Isochoric heating (i.e. spark ignition)*

The gas volume is $V = V_2$, and hence by the ideal gas equation,

$$T_3 = \frac{p_3 V_2}{nR}. \tag{3.116}$$

Rather than p_3 being an input, T_3 might be more appropriate to model the effect of spark ignition. In this case, compute p_3 from

$$p_3 = \frac{nR T_3}{V_2}. \tag{3.117}$$

Heat input in this isochoric stage is

$$Q_{\text{in}} = nc_V(T_3 - T_2). \tag{3.118}$$

The resulting entropy change can therefore be calculated from

$$dS = \frac{dQ}{T} = \frac{nc_V \, dT}{T} \tag{3.119}$$

$$\implies \Delta S_{2,3} = nc_V \int_{T_2}^{T_3} \frac{dT}{T}. \tag{3.120}$$

Hence,

$$\Delta S_{2,3} = nc_V \ln\left(\frac{T_3}{T_2}\right) \tag{3.121}$$

$$= \frac{nR}{\gamma - 1} \ln\left(\frac{\frac{p_3 V_2}{nR}}{\frac{p_2 V_2}{nR}}\right) \tag{3.122}$$

$$= \frac{nR}{\gamma - 1} \ln\left(\frac{p_3}{p_2}\right) \tag{3.123}$$

$$= \frac{nR}{\gamma - 1} \ln\left(\frac{p_3}{p_1\left(\frac{V_1}{V_2}\right)^{\gamma}}\right). \tag{3.124}$$

$$\therefore \Delta S_{2,3} = \frac{nR}{\gamma - 1}\left(\ln\left(\frac{p_3}{p_1}\right) - \gamma \ln r\right).$$

In order to plot entropy change ΔS versus temperature for this stage, define a T vector `T = linspace(T2,T3,1000)`, and hence compute a vector of ΔS from

$$\Delta S = \frac{nR}{\gamma - 1} \ln\left(\frac{T}{T_2}\right),$$

where $T_2 = \frac{p_2 V_2}{nR} = \frac{p_1 r^{\gamma} V_2}{nR}$.

3.1.10.3 *Stage 3: Isentropic expansion from V_2 to V_1*

`V = linspace(V2,V1,1000)` and

$$p = \frac{p_3 V_2^{\gamma}}{V^{\gamma}}. \tag{3.125}$$

Hence, since $r = \frac{V_1}{V_2}$,

$$p_4 = p_3 r^{-\gamma}.$$

From the ideal gas equation,

$$T_4 = \frac{p_4 V_1}{nR}. \tag{3.126}$$

The work done *by* the gas is

$$\Delta W_{3,4} = \frac{p_3 V_2}{\gamma - 1}(1 - r^{1-\gamma}). \qquad (3.127)$$

3.1.10.4   *Stage 4: Isochoric cooling*

Since the gas is at constant volume:

$$V = V_1. \qquad (3.128)$$

The heat output in this isochoric stage is

$$Q_{\text{out}} = n c_V (T_4 - T_1). \qquad (3.129)$$

The resulting entropy change can therefore be calculated from

$$\Delta S_{4,1} = n c_V \ln\left(\frac{T_1}{T_4}\right) \qquad (3.130)$$

$$= \frac{nR}{\gamma - 1} \ln\left(\frac{\frac{p_1 V_1}{nR}}{\frac{p_4 V_1}{nR}}\right) \qquad (3.131)$$

$$= \frac{nR}{\gamma - 1} \ln\left(\frac{p_1}{p_4}\right) \qquad (3.132)$$

$$= \frac{nR}{\gamma - 1} \ln\left(\frac{p_1}{p_3 \left(\frac{V_2}{V_1}\right)^{\gamma}}\right) \qquad (3.133)$$

$$= \frac{nR}{\gamma - 1} \left(\ln\left(\frac{p_1}{p_3}\right) - \gamma \ln\left(\frac{V_2}{V_1}\right)\right). \qquad (3.134)$$

$$\therefore \Delta S_{4,1} = \frac{nR}{\gamma - 1}\left(\gamma \ln r - \ln\left(\frac{p_3}{p_1}\right)\right).$$

Note that $-\gamma \ln(\frac{V_2}{V_1}) = \gamma \ln(\frac{V_1}{V_2}) = \gamma \ln r$.

In order to plot entropy change ΔS versus temperature for this stage, define a T vector `T = linspace(T4,T1,1000)`, and hence compute a vector of ΔS from

$$\Delta S = \Delta S_{2,3} + \frac{nR}{\gamma - 1} \ln\left(\frac{T}{T_4}\right).$$

3.1.10.5   *Otto cycle efficiency*

The heat input Q_{in} is, noting $V_2 = V_3$,

$$Q_{\text{in}} = n c_V (T_3 - T_2) \qquad (3.135)$$

$$= n c_V \left(\frac{p_3 V_2}{nR} - \frac{p_2 V_2}{nR}\right). \qquad (3.136)$$

$$\therefore Q_{\text{in}} = \frac{V_2}{\gamma - 1}(p_3 - p_1 r^{\gamma}). \qquad (3.137)$$

The heat output is, noting $V_1 = V_4$,

$$Q_{\text{out}} = nc_V\,(T_4 - T_1) \tag{3.138}$$

$$= nc_V\left(\frac{p_4 V_1}{nR} - \frac{p_1 V_1}{nR}\right) \tag{3.139}$$

$$= \frac{c_V V_1}{R}\left(p_3\left(\frac{V_2}{V_1}\right)^{\gamma} - p_1\right). \tag{3.140}$$

$$\therefore Q_{\text{out}} = \frac{V_1}{\gamma - 1}(p_3 r^{-\gamma} - p_1), \tag{3.141}$$

where the compression ratio $r = \frac{V_1}{V_2}$.

The work done per cycle $W = \oint p\,dV$ *by* the gas is

$$W = \Delta W_{3,4} - \Delta W_{1,2} = Q_{\text{in}} - Q_{\text{out}}. \tag{3.142}$$

We can therefore now compute the efficiency of the Otto cycle:

$$\eta = \frac{W}{Q_{\text{in}}} = \frac{Q_{\text{in}} - Q_{\text{out}}}{Q_{\text{in}}} = 1 - \frac{Q_{\text{out}}}{Q_{\text{in}}} \tag{3.143}$$

$$\therefore \eta = 1 - \frac{\frac{V_1}{\gamma-1}\left(p_3 r^{-\gamma} - p_1\right)}{\frac{V_2}{\gamma-1}\left(p_3 - p_1 r^{\gamma}\right)} \tag{3.144}$$

$$= 1 - r\frac{p_3 r^{-\gamma} - p_1}{p_3 - p_1 r^{\gamma}} \tag{3.145}$$

$$= 1 - r r^{-\gamma}\frac{p_3 - p_1 r^{\gamma}}{p_3 - p_1 r^{\gamma}}. \tag{3.146}$$

$$\therefore \eta = 1 - \frac{1}{r^{\gamma-1}}. \tag{3.147}$$

Interestingly, the efficiency η of an Otto engine only depends on the compression ratio $r = \frac{V_1}{V_2}$ and the ratio of molar heat capacities $\gamma = \frac{c_p}{c_V}$.

The total entropy change over one cycle is

$$\Delta S = \Delta S_{2,3} + \Delta S_{4,1} \tag{3.148}$$

$$= \frac{nR}{\gamma - 1}\left(\ln\left(\frac{p_3}{p_1}\right) - \gamma \ln r\right) + \frac{nR}{\gamma - 1}\left(\gamma \ln r - \ln\left(\frac{p_3}{p_1}\right)\right) \tag{3.149}$$

$$= 0, \tag{3.150}$$

which is as expected over a closed loop in the S, T space.

3.1.10.6 *Otto cycle summary*

Use the following recipe to construct a model in Excel or a programming language, such as MATLAB or Python. The goal is to automatically generate annotated plots similar to those in Fig. 3.4. The inputs are T_1, T_3, p_1, V_1, V_2 and molar mass M, plus the constants $\gamma = \frac{c_p}{c_V}$ and R.

The number of moles of gas is

$$n = \frac{p_1 V_1}{R T_1},$$

and the compression ratio $r = \frac{V_1}{V_2}$.

Stage 1: *Isentropic compression from V_1 to V_2*
`V = linspace(V1,V2,1000).`

$$p = \left(\frac{V_1}{V}\right)^{\gamma}, \tag{3.151}$$

$$p_2 = p_1 r^{\gamma}, \tag{3.152}$$

$$T_2 = \frac{p_2 V_2}{nR}. \tag{3.153}$$

Stage 2: *Isochoric (spark-ignition) heating*
$V = V_2$. `p = linspace(p2,p3,1000). T = linspace(T2,T3,1000).`

$$T_3 = \frac{p_3 V_2}{nR}, \tag{3.154}$$

$$\Delta S = \frac{nR}{\gamma - 1} \ln\left(\frac{T}{T_2}\right), \tag{3.155}$$

$$\Delta S_{2,3} = \frac{nR}{\gamma - 1} \ln\left(\frac{T_3}{T_2}\right). \tag{3.156}$$

Stage 3: *Isentropic expansion from V_2 to V_1*
`V = linspace(V2,V1,1000).`

$$p = p_3 \left(\frac{V_2}{V}\right)^{\gamma}, \tag{3.157}$$

$$p_4 = p_3 \left(\frac{V_2}{V_1}\right)^{\gamma}, \tag{3.158}$$

$$T_4 = \frac{p_4 V_1}{nR}. \tag{3.159}$$

Stage 4: *Isochoric cooling*
$V = V_1$. `p = linspace(p4,p1,1000). T = linspace(T4,T1,1000).`

$$\Delta S = \Delta S_{2,3} + \frac{nR}{\gamma - 1} \ln\left(\frac{T}{T_4}\right). \tag{3.160}$$

The heat input Q_{in}, heat output Q_{out} and work done W (by the gas) are as follows:

$$Q_{\text{in}} = n c_V \left(T_3 - T_2\right) = \frac{V_2}{\gamma - 1} \left(p_3 - p_1 r^{\gamma}\right), \tag{3.161}$$

$$Q_{\text{out}} = n c_V \left(T_4 - T_1\right) = \frac{V_1}{\gamma - 1} \left(p_3 r^{-\gamma} - p_1\right), \tag{3.162}$$

$$W = Q_{\text{in}} - Q_{\text{out}}. \tag{3.163}$$

The efficiency $\eta = \frac{W}{Q_{\text{in}}}$ of the Otto engine is

$$\eta = \frac{W}{Q_{\text{in}}} = 1 - \frac{1}{r^{\gamma-1}}. \qquad (3.164)$$

3.1.11 *Diesel engine*

The Diesel[12] cycle is the basis of a diesel engine, which is ubiquitous in modern road haulage and shipping. Unlike petrol-driven engines, diesel variants are more suited to heavy machinery. They can be found powering thousands of marine vessels as well as trucks, buses and cars. The first isentropic compression stage of the diesel cycle is similar to the Otto cycle in a petrol engine. However, in a diesel engine, combustion occurs spontaneously following compression rather than being initiated via spark ignition. An example of a Diesel cycle is illustrated in Fig. 3.5. The four stages of the Diesel cycle are as follows:

(1) Air is drawn into piston–cylinder arrangement at constant-pressure, i.e. at atmospheric pressure, $p_1 = 1$ atm. The initial cylinder volume is V_1, and the temperature T_1 is the ambient air temperature. *Isentropic compression* of the air occurs via a piston from volume V_1 to volume V_2. Define the *compression ratio* $r = \frac{V_1}{V_2}$.

(2) *Constant-pressure (isobaric) heat transfer* Q_{in} occurs while the piston is at maximum compression. This process is intended to represent the ignition of the fuel–air mixture and the subsequent rapid gas expansion. The pressure remains at p_2, while the volume increases from V_2 to V_3. This is different from the Otto cycle, which involves constant-volume (isochoric) heating during this stage, resulting from combustion initiated via a spark. In the Diesel cycle, the change in internal energy (and hence temperature) from gaseous compression is assumed to be sufficient to ignite the fuel vapours.

(3) *Isentropic expansion* occurs from volume V_3 to volume V_1, with resulting temperature T_4 and pressure p_4.

(4) As in an Otto cycle, the final stage is a *constant-volume (isochoric)* exchange of gas via an exhaust. n moles are exchanged for those at temperature T_1 and pressure p_1. This results in a loss of heat Q_{out}.

To model the Diesel cycle, we must determine the equations of the lines and curves in the p, V diagram, in a similar fashion to the Otto and Carnot cycles, and compute Q_{in}, Q_{out} and $W = \oint p\,dV = Q_{\text{in}} - Q_{\text{out}}$ in terms of the input parameters p_1, V_1, T_1, V_2 and V_3. We will also need the number of degrees of freedom of molecular motion α, as we need to compute the molar heat capacities $c_V = \frac{R}{\gamma-1}$

---

[12]Rudolf Diesel (1858–1913).

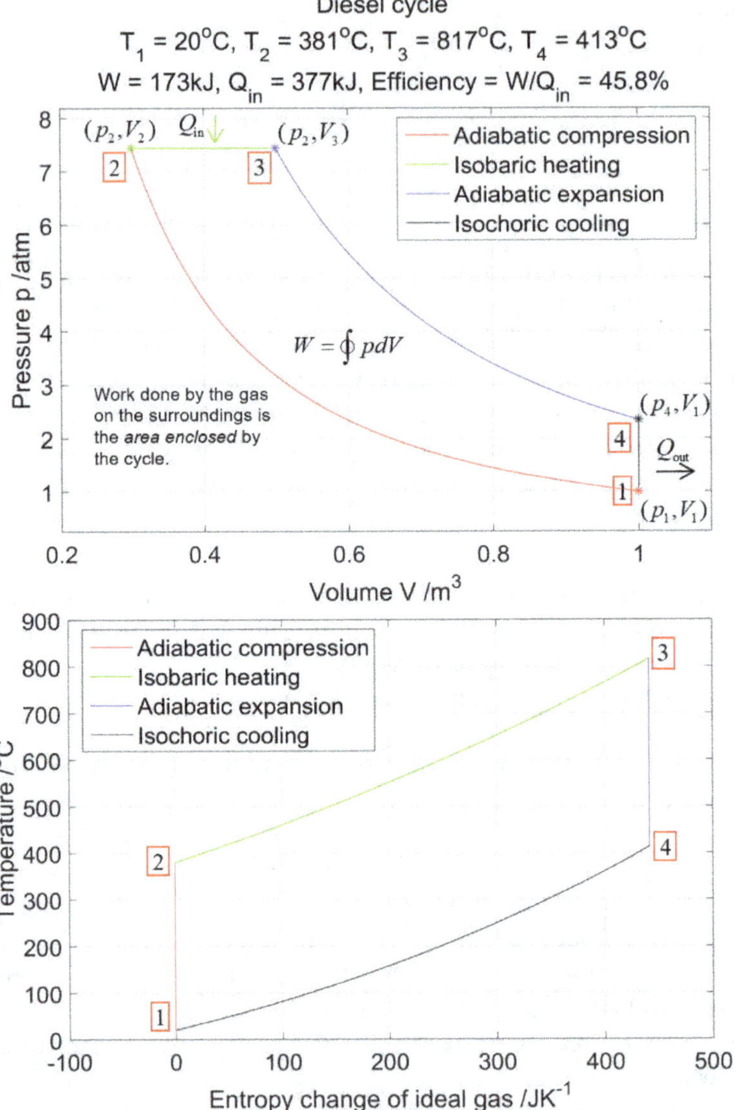

Fig. 3.5. Example of the key processes associated with a diesel engine represented by a *Diesel cycle* in p, V and T, S spaces for an ideal gas, which represents a mixture of air and fuel vapours. The gas mixture is drawn into a cylinder at atmospheric pressure and compressed isentropically, i.e. at constant-entropy with no heat exchanged. The heavier diesel fuel molecules can tolerate greater compression than those of a petrol-based fuel but will eventually spontaneously combust, liberating heat Q_{in} from the fuel. The gas expands at constant-pressure, doing work against a piston. Further expansion (and work) is done isentropically, as the gas loses pressure and temperature. At maximum expansion, the gas is exchanged with a fresh mixture via an exhaust system, and the cycle starts again. A Diesel cycle has the potential to be more efficient than a petrol (Otto) cycle due to the higher compression ratios; therefore, diesel engines are more common in commercial vehicles and ships, where high numbers of cycles per second (i.e. 'engine revs') are typically less of a requirement.

and $c_p = \frac{\gamma R}{\gamma - 1}$, where $\gamma = \frac{c_p}{c_V} = \frac{2}{\alpha} + 1$. Define the compression ratios:

$$r = \frac{V_1}{V_2},\tag{3.165}$$

$$s = \frac{V_3}{V_2}.\tag{3.166}$$

Also, note that

$$\frac{s}{r} = \frac{\frac{V_3}{V_2}}{\frac{V_1}{V_2}} = \frac{V_3}{V_1}.\tag{3.167}$$

3.1.11.1 *Stage 1: Isentropic compression from V_1 to V_2*

The number of moles n of gas drawn into the cylinder is

$$n = \frac{p_1 V_1}{R T_1}.\tag{3.168}$$

Define a vector `V = linspace(V1,V2,1000)`, and since this stage is isentropic,

$$p = \frac{p_1 V_1^{\gamma}}{V^{\gamma}}.\tag{3.169}$$

Hence,

$$p_2 = p_1 r^{\gamma}.\tag{3.170}$$

From the ideal gas equation,

$$T_2 = \frac{p_2 V_2}{nR}.\tag{3.171}$$

For this constant-entropy stage, the work done *on* the gas is

$$\Delta W_{1,2} = \frac{p_1 V_1}{\gamma - 1}\left(r^{\gamma - 1} - 1\right).\tag{3.172}$$

3.1.11.2 *Stage 2: Isobaric expansion during combustion*

`V = linspace(V2,V3,1000)` and

$$p = p_2 = p_1 r^{\gamma},\tag{3.173}$$

$$T_3 = \frac{p_2 V_3}{nR},\tag{3.174}$$

$$Q_{\text{in}} = n c_p \left(T_3 - T_2\right).\tag{3.175}$$

This means

$$Q_{\text{in}} = n\frac{\gamma}{\gamma - 1}R\left(\frac{p_2 V_3}{nR} - \frac{p_2 V_2}{nR}\right).\tag{3.176}$$

$$\therefore Q_{\text{in}} = \frac{\gamma}{\gamma - 1}p_1 r^{\gamma} V_2 (s - 1).\tag{3.177}$$

The work done *by* the gas in this isobaric stage is

$$\Delta W_{2,3} = \int_{V_2}^{V_3} p\,dV = p_2(V_3 - V_2) \tag{3.178}$$

$$= p_2 V_2(s - 1). \tag{3.179}$$

$$\therefore \Delta W_{2,3} = p_1 V_2 r^\gamma (s - 1). \tag{3.180}$$

The (differential) entropy change is

$$dS = \frac{dQ}{T} = \frac{n c_p dT}{T}. \tag{3.181}$$

$$\therefore \Delta S_{2,3} = n c_p \int_{T_2}^{T_3} \frac{dT}{T}. \tag{3.182}$$

Now, using

$$c_p = \frac{\gamma R}{\gamma - 1}, \tag{3.183}$$

we can evaluate the entropy change $\Delta S_{2,3}$ during isobaric expansion in terms of Diesel cycle inputs:

$$\Delta S_{2,3} = nR\frac{\gamma}{\gamma - 1} \ln\left(\frac{T_3}{T_2}\right) \tag{3.184}$$

$$= nR\frac{\gamma}{\gamma - 1} \ln\left(\frac{\frac{p_2 V_3}{nR}}{\frac{p_2 V_2}{nR}}\right). \tag{3.185}$$

$$\therefore \Delta S_{2,3} = \frac{n\gamma R}{\gamma - 1} \ln s. \tag{3.186}$$

In order to plot entropy change ΔS versus temperature for this stage, define a T vector `T = linspace(T2,T3,1000)`, and hence compute a vector of ΔS from

$$\Delta S = \frac{nR\gamma}{\gamma - 1} \ln\left(\frac{T}{T_2}\right).$$

3.1.11.3   *Stage 3: Isentropic expansion from V_3 to V_1*

`V = linspace(V3,V1,1000)` and

$$p = \frac{p_2 V_3^\gamma}{V^\gamma}, \tag{3.187}$$

$$p_4 = p_1 r^\gamma \left(\frac{V_3}{V_1}\right)^\gamma = p_1 s^\gamma. \tag{3.188}$$

From the ideal gas equation,

$$T_4 = \frac{p_4 V_1}{nR}. \tag{3.189}$$

The work done *by* the gas is

$$\Delta W_{3,4} = \frac{p_2 V_3}{\gamma - 1}\left(1 - \left(\frac{V_3}{V_1}\right)^{\gamma-1}\right). \tag{3.190}$$

Hence,

$$\Delta W_{3,4} = \frac{p_1 V_3 r^\gamma}{\gamma - 1}\left(1 - \left(\frac{s}{r}\right)^{\gamma-1}\right). \tag{3.191}$$

3.1.11.4 *Stage 4: Isochoric cooling*

Since the volume of gas is constant:

$$V = V_1. \tag{3.192}$$

Heat transfer:

$$Q_{\text{out}} = nc_V\left(T_4 - T_1\right) \tag{3.193}$$

$$= \frac{nR}{\gamma - 1}\left(\frac{p_4 V_1}{nR} - T_1\right). \tag{3.194}$$

Since $p_4 = p_1 s^\gamma$,

$$Q_{\text{out}} = \frac{nR}{\gamma - 1}\left(\frac{p_1 s^\gamma V_1}{nR} - \frac{p_1 V_1}{nR}\right) \tag{3.195}$$

$$= \frac{p_1 V_1}{\gamma - 1}(s^\gamma - 1). \tag{3.196}$$

Entropy change:

$$\Delta S_{4,1} = nc_V \ln\left(\frac{T_1}{T_4}\right) \tag{3.197}$$

$$= \frac{nR}{\gamma - 1}\ln\left(\frac{\frac{p_1 V_1}{nR}}{\frac{p_4 V_1}{nR}}\right) \tag{3.198}$$

$$= \frac{nR}{\gamma - 1}\ln\left(\frac{p_1}{p_4}\right) \tag{3.199}$$

$$= \frac{nR}{\gamma - 1}\ln\left(\frac{p_1}{p_1 s^\gamma}\right) \tag{3.200}$$

$$= -\frac{n\gamma R}{\gamma - 1}\ln s. \tag{3.201}$$

$\Delta S_{4,1} = -\frac{n\gamma R}{\gamma-1}\ln s$ and $\Delta S_{2,3} = \frac{n\gamma R}{\gamma-1}\ln s$ are a good check of our calculations since over the cycle,

$$\Delta S_{2,3} + \Delta S_{4,1} = 0, \tag{3.202}$$

which we have asserted must be the case for this idealised, reversible, system.

In order to plot entropy change ΔS versus temperature for this stage, define a T vector `T = linspace(T4,T1,1000)`, and hence compute a vector of ΔS from

$$\Delta S = \Delta S_{2,3} + \frac{nR}{\gamma - 1}\ln\left(\frac{T}{T_4}\right).$$

3.1.11.5   *Diesel cycle efficiency*

The work done per cycle is

$$W = \Delta W_{2,3} + \Delta W_{3,4} - \Delta W_{1,2} \tag{3.203}$$

$$= p_1 V_2 r^\gamma (s - 1) + \frac{p_1 V_3 r^\gamma}{\gamma - 1} \left(1 - \left(\frac{s}{r}\right)^{\gamma - 1}\right) - \frac{p_1 V_1}{\gamma - 1}(r^{\gamma - 1} - 1) \tag{3.204}$$

$$= p_1 V_2 r^\gamma (s - 1) + \frac{p_1 V_2}{\gamma - 1}(s r^\gamma - s^\gamma r) - \frac{p_1 V_2}{\gamma - 1}(r^\gamma - r) \tag{3.205}$$

$$= \frac{p_1 V_2}{\gamma - 1}\left((\gamma - 1)(s - 1)r^\gamma + s r^\gamma - s^\gamma r - r^\gamma + r\right) \tag{3.206}$$

$$= \frac{p_1 V_2}{\gamma - 1}\left(r(1 - s^\gamma) + r^\gamma(\gamma - 1)(s - 1) + (s - 1)\right) \tag{3.207}$$

$$= \frac{p_1 V_2}{\gamma - 1}\left(r(1 - s^\gamma) + r^\gamma \gamma (s - 1)\right). \tag{3.208}$$

Efficiency $\eta = \frac{W}{Q_{\text{in}}}$ is therefore

$$\eta = \frac{W}{Q_{\text{in}}} = \frac{Q_{\text{in}} - Q_{\text{out}}}{Q_{\text{in}}} = 1 - \frac{Q_{\text{out}}}{Q_{\text{in}}}. \tag{3.209}$$

$$\therefore \eta = 1 - \frac{\frac{p_1 V_1}{\gamma - 1}(s^\gamma - 1)}{\frac{\gamma}{\gamma - 1}p_1 r^\gamma V_2 (s - 1)}.$$

$$\therefore \eta = 1 - \frac{r(s^\gamma - 1)}{\gamma r^\gamma (s - 1)}.$$

$$\therefore \eta = 1 - \frac{1}{r^{\gamma - 1}}\left(\frac{s^\gamma - 1}{\gamma(s - 1)}\right). \tag{3.210}$$

This is different to the Otto Cycle efficiency $\eta_{otto} = 1 - \frac{1}{r^{\gamma-1}}$ by the extra factor $\frac{s^\gamma - 1}{\gamma(s-1)}$. The Diesel engine is typically *more* efficient than a petrol (Otto) engine since the former works on the basis of self ignition due to high compression. This 'knocking' is undesirable for petrol engines,[13] so a lower r value is required. However, for a *given* r, the ideal Otto cycle is more efficient than the Diesel cycle since $\frac{s^\gamma - 1}{\gamma(s-1)} > 1$. Note since $\gamma = \frac{2}{\alpha} + 1$, this means $\gamma > 1$. Also, $s = \frac{V_3}{V_2}$ and $V_3 > V_2$ since this is an isobaric *expansion*, which implies $s > 1$. As illustrated in Fig. 3.6, $\frac{s^\gamma - 1}{\gamma(s-1)} > 1$ is satisfied for all s. Therefore an (ideal) Diesel engine will always be less efficient than an Otto engine for the *same* r.

3.1.11.6   *Diesel cycle summary*

Use the following recipe to construct a model in Excel or a programming language such as MATLAB or Python. The goal is to automatically generate annotated plots

---

[13]Spontaneous combustion mid-cycle is not desirable for the harmony of piston firings in a petrol engine!

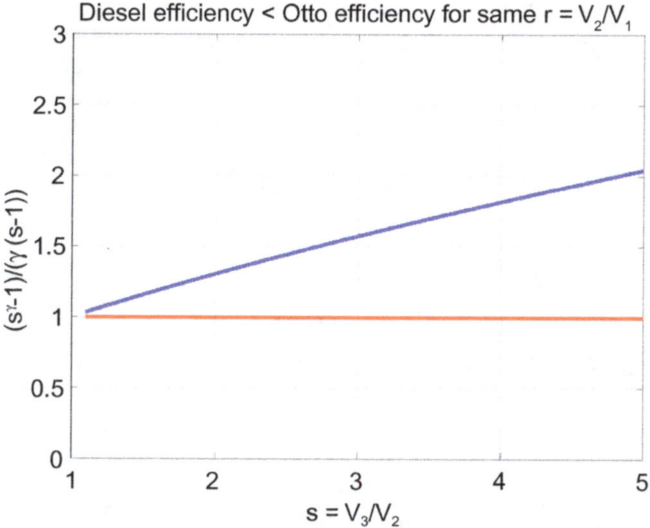

Fig. 3.6. An idealised Diesel engine is always less efficient than an Otto engine, an idealisation of a petrol spark-ignition engine, assuming the two engines have the *same* compression ratio $r = \frac{V_1}{V_2}$. The Diesel engine efficiency is $\eta_D = 1 - \frac{1}{r^{\gamma-1}}\left(\frac{s^\gamma-1}{\gamma(s-1)}\right)$ whereas $\eta_O = 1 - \frac{1}{r^{\gamma-1}}$ for an Otto engine. If $\eta_O > \eta_D$ then $\frac{s^\gamma-1}{\gamma(s-1)} > 1$, which is visually demonstrated by the plot. Efficiency $\eta = \frac{Q_{\text{in}}-Q_{\text{out}}}{Q_{\text{in}}}$. $r = \frac{V_1}{V_2}$ and $s = \frac{V_3}{V_2}$ are the gas volume compression ratios.

similar to those in Fig. 3.5. The input parameters are p_1, V_1, T_1, V_2 and V_3. We also need the number of degrees of freedom of molecular motion α, as we need to compute the molar heat capacities $c_V = \frac{R}{\gamma-1}$ and $c_p = \frac{\gamma R}{\gamma-1}$, where $\gamma = \frac{c_p}{c_V} = \frac{2}{\alpha}+1$. Define the compression ratios:

$$r = \frac{V_1}{V_2}, \tag{3.211}$$

$$s = \frac{V_3}{V_2}. \tag{3.212}$$

Stage 1: *Isentropic compression from V_1 to V_2*
`V = linspace(V1,V2,1000)`.

$$p = \left(\frac{V_1}{V}\right)^\gamma, \tag{3.213}$$

$$p_2 = p_1 r^\gamma, \tag{3.214}$$

$$T_2 = \frac{p_2 V_2}{nR}. \tag{3.215}$$

Stage 2: *Isobaric heating and expansion from V_2 to V_3*
$p = p_2$. `p = linspace(p2,p3,1000).` `T = linspace(T2,T3,1000).`

$$T_3 = \frac{p_3 V_2}{nR}, \tag{3.216}$$

$$\Delta S = \frac{nR\gamma}{\gamma - 1} \ln\left(\frac{T}{T_2}\right), \tag{3.217}$$

$$\Delta S_{2,3} = \frac{nR\gamma}{\gamma - 1} \ln s. \tag{3.218}$$

Stage 3: *Isentropic expansion from V_3 to V_1*
`V = linspace(V2,V1,1000).`

$$p = p_3 \left(\frac{V_3}{V}\right)^\gamma, \tag{3.219}$$

$$p_4 = p_3 \left(\frac{V_3}{V_1}\right)^\gamma, \tag{3.220}$$

$$T_4 = \frac{p_4 V_1}{nR}. \tag{3.221}$$

Stage 4: *Isochoric cooling*
$V = V_1$. `p = linspace(p4,p1,1000).` `T = linspace(T4,T1,1000).`

$$\Delta S = \Delta S_{2,3} + \frac{nR}{\gamma - 1} \ln\left(\frac{T}{T_4}\right). \tag{3.222}$$

The heat input Q_{in}, heat output Q_{out} and work done W (by the gas) are as follows:

$$Q_{\text{in}} = nc_p \left(T_3 - T_2\right) = \frac{\gamma}{\gamma - 1} p_1 r^\gamma V_2 \left(s - 1\right),$$

$$Q_{\text{out}} = nc_V \left(T_4 - T_1\right) = \frac{p_1 V_1}{\gamma - 1} \left(s^\gamma - 1\right),$$

$$W = Q_{\text{in}} - Q_{\text{out}}.$$

The efficiency $\eta = \frac{W}{Q_{\text{in}}}$ of the Diesel cycle is

$$\eta = 1 - \frac{1}{r^{\gamma - 1}} \left(\frac{s^\gamma - 1}{\gamma \left(s - 1\right)}\right). \tag{3.223}$$

3.2   Newtonian Cooling

Consider a vessel containing a fluid of uniform temperature T. The walls of the vessel are of thickness Δx and enclose the fluid. The ambient temperature outside the vessel is T_a, which is less than T and assumed to be constant.[14] If one assumes

---

[14]When investigating Newtonian cooling experimentally, it is recommended that a fan be placed near the fluid vessel. This will disrupt the formation of a boundary layer of hot air around the vessel, which would otherwise invalidate the assumption of a constant-external temperature, T_a.

heat transfer between the fluid and surroundings (e.g. air) is via *conduction* (i.e. not radiative transfer or via physical transfer of fluid particles from the vessel to surroundings, as in evaporative cooling or convection), then we can use *Fourier's law*[15] to determine the heat flux between the vessel and the surroundings, i.e. the *rate of heat transferred per unit area of the vessel is proportional to the temperature gradient between the fluid and the surroundings.*

If Q is the heat transferred from the vessel to its surroundings, k is the *thermal conductivity* of the vessel, Δx is the thickness of the vessel and A is the surface area of the vessel, Fourier's law implies

$$\frac{dQ}{dt} = kA\frac{(T - T_a)}{\Delta x}. \tag{3.224}$$

If the specific heat capacity of the fluid is c and the vessel contains m kg of fluid,

$$dQ = -mcdT. \tag{3.225}$$

If we assume that the heat capacity c is independent of temperature[16] and the initial fluid temperature is T_0,

$$\frac{-mcdT}{dt} = kA\frac{(T - T_a)}{\Delta x} \tag{3.226}$$

$$\int_{T_0}^{T} \frac{dT'}{T' - T_a} = -\frac{kA}{mc\Delta x}\int_0^t dt'. \tag{3.227}$$

$$\ln\left(\frac{T - T_a}{T_0 - T_a}\right) = -\frac{kAt}{mc\Delta x}. \tag{3.228}$$

$$\therefore \ T = T_a + (T_0 - T_a)e^{-\frac{t}{\tau}}, \tag{3.229}$$

where the time constant

$$\tau = \frac{mc\Delta x}{kA}. \tag{3.230}$$

So, we expect an *exponential decay* of temperature difference $\Delta T = T - T_a$. By writing[17] $e = 2^{\frac{1}{\ln 2}}$ and defining $\Delta T_0 = T_0 - T_a$,

$$\Delta T = \Delta T_0 \times 2^{\frac{-t}{\tau \ln 2}}. \tag{3.231}$$

The 'half-life' $t_{1/2}$ is the time for $\Delta T(t)$ to decay from ΔT_0 to $\frac{1}{2}\Delta T_0$:

$$\therefore \frac{1}{2}\Delta T_0 = \Delta T_0 \times 2^{\frac{-t_{1/2}}{\tau \ln 2}}. \tag{3.232}$$

$$\therefore 2^1 = 2^{\frac{t_{1/2}}{\tau \ln 2}}. \tag{3.233}$$

$$\therefore t_{1/2} = \frac{mc\Delta x}{kA}\ln 2. \tag{3.234}$$

---

[15] Joseph Fourier (1768–1830).

[16] This is not a good model at low temperatures; however, at and above room temperature, most liquids and solids have a molar heat capacity of about $3R$, where $R = 8.314$ Jmol^{-1}K^{-1}, the molar gas constant. This is called the *Dulong–Petit law* and is based on the idea of thermal energy in a solid arising from atomic lattice vibrations, which are constrained to have discrete energies by the quantum nature of these oscillators.

[17] $y = 2^{\frac{1}{\ln 2}}$. $\therefore \ln y = \frac{1}{\ln 2}\ln 2 = 1$. Hence, $y = e$.

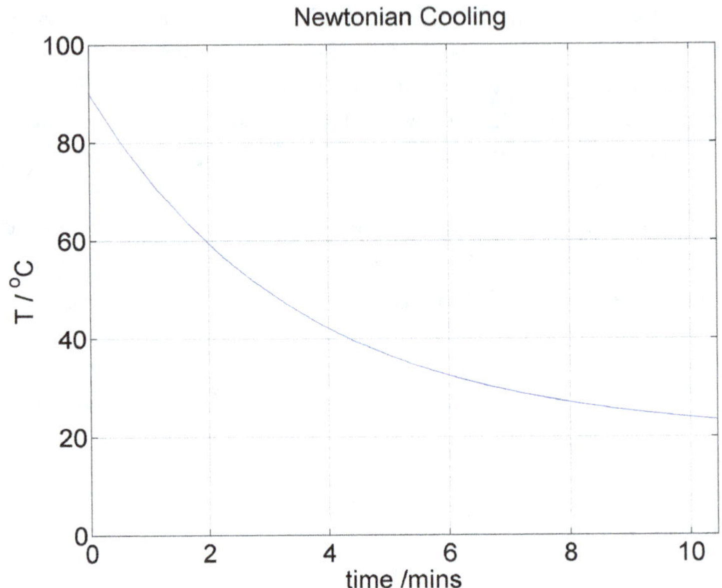

Fig. 3.7. Newton's cooling model for initially hot water in a polystyrene cup. The rate of change of water temperature $\frac{dT}{dt}$ is proportional to the temperature difference between the water temperature T and (fixed) ambient temperature T_a. This implies an exponential decay of water temperature with time, $T(t) = T_a + \frac{T_0 - T_a}{2^{\frac{t}{t_{1/2}}}}$, where $t_{1/2}$ is the time taken for the temperature drop to equalhalf of the initial difference from ambient $T_0 - T_a$. In this case, the initial temperature $T_0 = 90°$ and the ambient temperature $T_a = 20°$. The 'half-life' $t_{1/2} = \frac{mc\Delta x}{kA} \ln 2$ depends on the water mass m, the specific heat capacity of water c, the cup wall area A and the thickness Δx and the thermal conductivity k of the polystyrene. In this example, $\tau = 145$ s, i.e. just under 2.5 min, which should render it moderately useful as a tea cup.

In the example of 'hot water contained in a not-particularly-well-insulated litre-mug' in Fig. 3.7, $k = 2.0$ Wm^{-1}K^{-1}, $c = 4,181$ Jkg^{-1}K^{-1}, $T_0 = 90°, T_a = 20°$, $\Delta x = 0.005$ m, $A = 0.05$ m^2 and $m = 1.00$ kg, which gives time constant $\tau = \frac{mc\Delta x}{kA} = 209$ s and implies a half-life of $t_{1/2} = \ln 2 \times 209$ s $= 145$ s, i.e. just under two-and-a-half minutes.

3.3   Kinetic Theory and Random Walks

3.3.1   *Kinetic theory and molecular transport*

Kinetic theory is essentially a *statistical* theory of the motion of molecules. It can be used to explain the bulk properties of matter since the sizes of molecules (or atoms) are very small compared to human-sized macroscopic objects, such as vials of liquid or bottles of gas. In kinetic theory, we consider *average* physical quantities, such as speed, pressure, and heat flux, since the huge number of molecules involved makes mechanical calculations impractical. Inspired by the observation of *Brownian motion*,[18] we shall assume that molecular motion is a *random process*. This assumption of randomness will enable us to calculate average physical quantities based on our knowledge of their *probability distributions*. Note that a cubic metre of densely packed atoms (e.g. a solid or liquid) will have about 10^{30} atoms since the size of an atom is about 10^{-10} m. A *mole* of a chemical substance (e.g. about 18 g of water) is defined to have *Avogadro's number*[19] of molecules, a truly enormous quantity:

$$N_A = 6.02 \times 10^{23}. \tag{3.235}$$

3.3.2 *Brownian motion, random walks and Knudsen's number*

Brownian motion, initially observed as the chaotic jittering of pollen grains on a microscope slide, is due to the jostling of molecules. In the case of the pollen grains, it is the smaller (invisible) air molecules which are colliding with them (and each other) in a random fashion. How far will a given particle move in a specified time, given that its motion is random? To analyse this in the simplest situation, let us consider motion in one direction in N steps of fixed length l. The caveat is that each step is either forward or backwards, and the direction is determined by chance alone. The model we shall explore is when a forward or backward step is of equal probability, i.e. $P(\text{forward}) = P(\text{backward}) = 0.5$, since there are no other possibilities.

The total displacement is

$$x = l \sum_{i=1}^{N} a_i, \tag{3.236}$$

where

$$a_i = \begin{cases} 1 & r_i \leq 0.5 \\ -1 & r_i > 0.5 \end{cases} \tag{3.237}$$

and r_i is a *uniformly distributed* random number, $0 \leq r_i \leq 1$. Note that most computer languages will have a built-in random number generator of this sort. In MATLAB, `rand(5,10)` will generate a 5×10 matrix of uniformly distributed random numbers, each within the range [0,1].

---

[18]*Brownian motion* is the seemingly chaotic zig-zag dynamics of small particles, such as pollen grains or smoke, when viewed under a microscope. Brownian motion is named after Robert Brown (1773–1858), a Scottish botanist who first described the phenomenon. The key idea is that, rather than the pollen grains being 'alive' in any sentient sense, their motion can be explained by collisions with randomly moving tiny particles that are too small to be seen through a microscope. Explanation of Brownian motion is therefore a *molecular theory*, i.e. a *discrete* theory of matter.

[19]Amedeo Carlo Avogadro (1776–1856).

A sensible measure of the distance travelled is the *root-mean-square*[20] (RMS) displacement:

$$\sqrt{\langle x^2 \rangle} = l \sqrt{\left\langle \left(\sum_{i=1}^{N} a_i \right)^2 \right\rangle} = l \sqrt{\left\langle \sum_{i=1}^{N} a_i^2 \right\rangle + \left\langle \sum_{i=1, i \neq j}^{N} \sum_{j=1}^{N} a_i a_j \right\rangle}. \tag{3.238}$$

Now, since $a_i^2 = 1$, the mean average

$$\left\langle \sum_{i=1}^{N} a_i^2 \right\rangle = N \tag{3.239}$$

and since a_i is a random choice (of equal probability) between -1 and 1,

$$\left\langle \sum_{i=1, i \neq j}^{N} \sum_{j=1}^{N} a_i a_j \right\rangle = 0. \tag{3.240}$$

Therefore,

$$\sqrt{\langle x^2 \rangle} = l \sqrt{\left\langle \sum_{i=1}^{N} a_i^2 \right\rangle + \left\langle \sum_{i=1, i \neq j}^{N} \sum_{j=1}^{N} a_i a_j \right\rangle} = l \sqrt{N + 0} = l \sqrt{N}. \tag{3.241}$$

The prediction that the RMS displacement will vary with the $\sqrt{\text{number of steps}}$ is confirmed in Fig. 3.9. This is a simulation based on tracking $\sqrt{\langle x^2 \rangle}$ from $N = 1$ to 10,000 steps. The step size is $l = 1$ unit. $I = 1{,}000$ independent simulations are run and the RMS of x over the I simulations are computed for each step N. Graphs are plotted of $\sqrt{\langle x^2 \rangle}$ and $l\sqrt{N}$, and you can see that these traces are visually very close to each other. As a natural extension, Fig. 3.10 illustrates a number of 2D random walks. Similar to the 1D example, a fixed step size of l is used,[21] but the angle of each step is uniformly random within a range of 0 to 2π radians.

If the average molecular speed is $v_{\text{rms}} = \sqrt{\langle v^2 \rangle}$, the average distance travelled by a molecule in time t is $v_{\text{rms}}t$, and hence the average number of steps in t seconds is

$$N = \frac{v_{\text{rms}} t}{l}. \tag{3.242}$$

---

[20] The notation $\langle x \rangle$ implies 'take the mean average of quantity x', i.e. if x is an array of numbers $\{1, 2, 3, 4\}$, then $\langle x \rangle = \frac{1}{4}(1 + 2 + 3 + 4) = 2.5$.

[21] The step size could also be a random number based on a probability distribution characterised by l.

Hence, the predicted random walk distance in t seconds is

$$\sqrt{\langle x^2 \rangle} = l\sqrt{N} = \sqrt{lv_{\text{rms}}t}. \tag{3.243}$$

The step size l can be associated with the *mean free path* between molecular collisions. We can define this as the average distance travelled by a molecule in time t, divided by the number of molecules it will likely collide with during that time. This is the number of molecules per unit volume[22] n multiplied by the volume of a 'collision tube' (see Fig. 3.8) with radius d and length $\langle v_{\text{rel}} \rangle t$, where $\langle v_{\text{rel}} \rangle$ is the average *relative velocity*. The tube diameter is set to be small enough to ensure collisions, i.e. there is no point trying to count molecules which are too far apart to collide. The tube radius is, therefore, the *molecular diameter d*, which means a pair of touching molecules constitutes the diameter of our 'collision tube'. The number of molecules in our collision tube is therefore $n\pi d^2 \langle v_{\text{rel}} \rangle t$.

At the scale of this model, we assume molecules to be tiny spheres. Real molecules will, of course, have distinct geometries (e.g. the bent triangle of water, the tetrahedron of methane, and the long hydrocarbon chain of polymers), so what to assign as d will depend on not only the molecule length but also whether we can assume molecules are randomly oriented or not.

The average *relative velocity* between pairs of molecules, labelled i and j, is

$$\langle v_{\text{rel}} \rangle = \sqrt{\left\langle |\mathbf{v}_i - \mathbf{v}_j|^2 \right\rangle} \tag{3.244}$$

$$= \sqrt{\left\langle v_i^2 + v_j^2 - 2\mathbf{v}_i \cdot \mathbf{v}_j \right\rangle} \tag{3.245}$$

$$= \sqrt{2} v_{\text{rms}}, \tag{3.246}$$

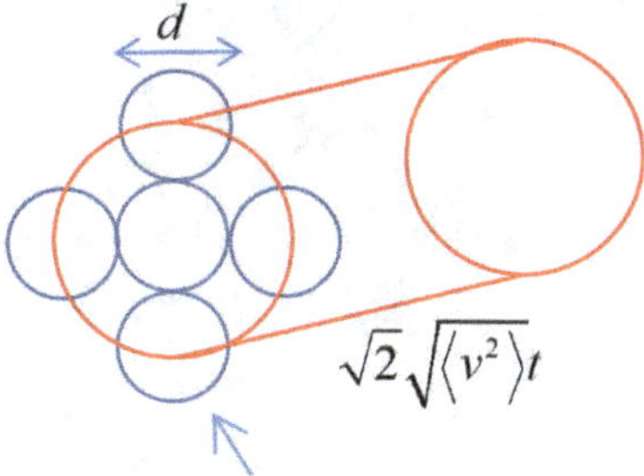

Fig. 3.8. The number of molecular collisions in time t can be modelled by the number of molecules per unit volume n, multiplied by the volume of a 'collision tube' of radius d and length $\sqrt{2}\sqrt{\langle v^2 \rangle}t$, where d is the molecular 'diameter' (we approximate molecules to be spheres) and $\sqrt{\langle v^2 \rangle}$ is the root-mean-square molecular speed.

---

[22] Confusion warning! n is the number of molecules per unit volume, *not* the number of moles!

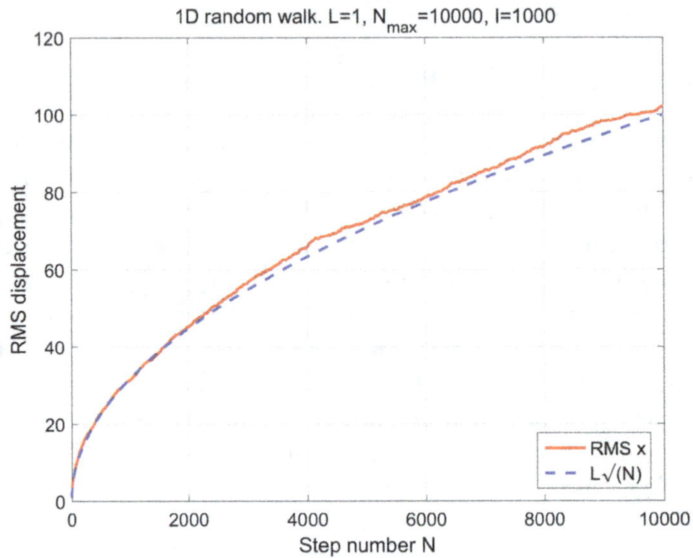

Fig. 3.9. MATLAB simulation based upon tracking $\sqrt{\langle x^2 \rangle}$ for a 1D random walk from $N = 1$ to 10,000 steps. The step size is $l = 1$ unit, and steps of either $+1$ or -1 are determined via a random number generator, with equal probability. $I = 1,000$ independent simulations are run and the RMS of x over the I simulations are computed for each step N. Graphs are plotted of $\sqrt{\langle x^2 \rangle}$ and $l\sqrt{N}$ and you can see that these traces are visually very close to each other.

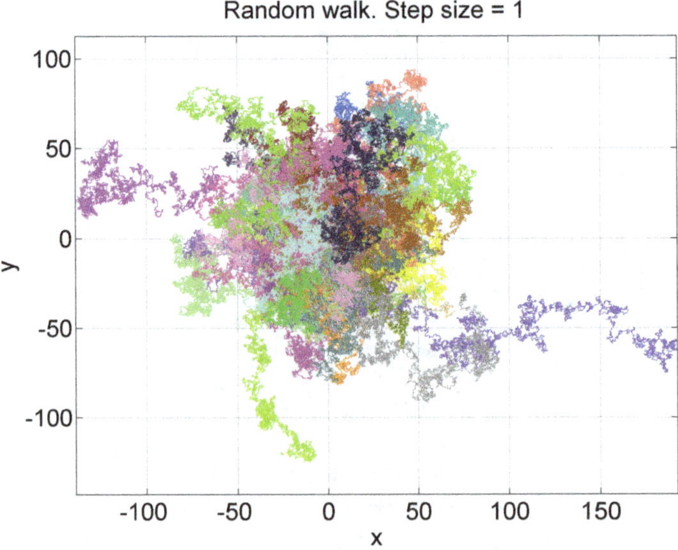

Fig. 3.10. Two-dimensional random walks are easy to generate using a computer programming language. In this example, 42 distinct walks from $(0, 0)$ are overlaid, with each comprising 5,000 steps of fixed size $s = 1$ unit. In the 2D random walk, a random radian angle, θ, is chosen from $0 \leq \theta \leq 2\pi$ for each step, and x, y movements between steps are determined by $\Delta x = \cos\theta$ and $\Delta y = \sin\theta$.

since one would anticipate $\langle \mathbf{v}_i \cdot \mathbf{v}_j \rangle = 0$ for random motion. Therefore, the mean free path is

$$l = \frac{v_{\mathrm{rms}} t}{n \pi d^2 \sqrt{2} v_{\mathrm{rms}} t} \tag{3.247}$$

$$\Rightarrow l = \frac{1}{n \pi d^2 \sqrt{2}}. \tag{3.248}$$

For $\frac{nV}{N_A}$ moles of an ideal gas of pressure p, volume V and temperature T,

$$pV = \frac{nV}{N_A} RT \Longrightarrow n = \frac{N_A p}{RT}. \tag{3.249}$$

Substituting for the number $n = \frac{N_A p}{RT}$ of molecules per m^3 in $l = \frac{1}{n\pi d^2 \sqrt{2}}$,

$$l = \frac{RT}{N_A \pi d^2 \sqrt{2} p} = \mathrm{Kn} \times d. \tag{3.250}$$

Knudsen's number, the ratio $\mathrm{Kn} = \frac{l}{d}$ between mean free path and molecular size, is

$$\mathrm{Kn} = \frac{l}{d} = \frac{RT}{N_A \pi d^3 \sqrt{2} p}. \tag{3.251}$$

The size of the dimensionless constant Kn determines if our statistical mechanics argument is valid for an ideal gas. If $\mathrm{Kn} \ll 1$, then intermolecular collisions are so likely that a *continuum* model is required rather than a kinetic theory model. For $\mathrm{Kn} \ll 1$, an ideal gas model is therefore *not* appropriate. A continuum model is what is used to describe much of *fluid mechanics*,[23] where we consider the fluid to be a continuously varying entity rather than a system of discrete, and randomly moving, molecules. For a typical air molecule on Earth: $d \approx 0.3$ nm, $p \approx 10^5$ Pa and $T \approx 293$ K. This yields $\mathrm{Kn} \approx 333$, which means a mean free path of about $l \approx 10^{-7}$ m. The high Kn means a statistical model is more appropriate for air molecules. In fact, $l \gg d$ is indeed one of the fundamental assumptions of an *ideal gas*: an infinitely compressible 'fluid', where we can ignore the effects of molecular size and electrostatic forces between molecules. Molecules that comprise an ideal gas are modelled as point particles that move in straight lines until they (elastically) collide with each other. This helps justify one of our fundamental assumptions in our earlier heat engine modelling: *in an ideal gas, all internal energy is kinetic energy.* If we can ignore the effect of electric (and magnetic) forces between molecules, then the potential energy associated with these fields can be ignored compared to kinetic energy.

3.4 Boltzmann's Exponential Statistics

3.4.1 *Temperature vs. heat*

The physical model of heat in motion, *thermodynamics*, is in essence a theory of the statistical average movement of a very large number of microscopic particles,

---

[23] A most excellent introduction is Faber's *Fluid Dynamics for Physicists* [5].

which constitute a macroscopic entity (e.g. a litre of liquid water). The basic idea is that *temperature is proportional to the mean kinetic energy of molecules,* and heat is the total amount of energy transferred during some process, e.g. to your foot when you insert it into a warm bath. Temperature is therefore a *microscopic* concept, whereas heat is a *macroscopic* quantity. This explains why passing your hand (briefly!) through a hot flame will typically incur less damage than placing it in a pan of boiling water. Although the average kinetic energy of molecules is much less in the boiling water example, the number of molecules colliding with your hand is significantly higher. The overall amount of energy transferred from the boiling water is higher than from the brief passage through fire, which means more heat is transferred to your hand and hence a greater chance of injury.

3.4.2   *The Boltzmann factor and the Boltzmann distribution*

Consider a large number, N, of particles, which have *discrete* possibilities for their energy.[24] The total amount of energy U is constant, i.e. the system is isolated in terms of energy and, indeed, total particle count. Out of the particles, n_i (with labels $i = 1, 2, 3, \ldots$) have a (fixed) energy of ε_i:

$$U = \sum_i n_i \varepsilon_i,$$

$$\therefore dU = \sum_i \varepsilon_i dn_i = 0. \tag{3.252}$$

Now,

$$N = \sum_i n_i$$

and since N is a constant,

$$dN = \sum_i dn_i = 0.$$

There are $N!$ ways of arranging the particles; however, since n_i in level i all have the same energy, the number of possible energy configurations is

$$W = \frac{N!}{n_1! n_2! \ldots} = \frac{N!}{\prod_i n_i!}. \tag{3.253}$$

Let us assert that the equilibrium (or steady state, or asymptotic) distribution of energy is such that it *maximises* the number of possible energy arrangements. We can make more analytical progress by considering maximising the natural logarithm

---

[24] The Boltzmann analysis, scaled up to large numbers of particles, allows us to comprehend the *classical* continuum, which is the scope of this book. However, it is worth noting that concepts based on the discrete (energies or states) have a natural home in *quantum mechanics*, which will be explored in the next volume of *Science by Simulation*.

of W, rather than W, since by using the logarithm, we can turn products into summations and divisions into subtractions:

$$\ln W = \ln N! - \sum_i \ln n_i! \tag{3.254}$$

For a modest number of moles of gas, N is likely to be very large indeed (i.e. of the order of Avogadro's number $N_A = 6.02 \times 10^{23}$); therefore, we can use *Stirling's approximation*:

$$\ln N! \approx N \ln N - N. \tag{3.255}$$

A quick proof of *Stirling's approximation*:

$$\ln N! = \sum_{n=1}^{N} \ln n.$$

Now, since $N \gg 1$, we can regard $dn = 1$ to be 'differentiably small' compared to N and, in the limit of very large N, $\ln N$ too. Hence, we can approximate the sum $\sum_{n=1}^{N} \ln n$ as equivalent to the area under a graph of n versus $\ln n$. This means we can turn the sum $\sum_{n=1}^{N} \ln n$ into an integral, $\int_1^N \ln n\, dn$:

$$\ln N! = \sum_{n=1}^{N} \ln n \approx \int_1^N \ln n\, dn \tag{3.256}$$

$$= [n \ln n - n]_1^N \tag{3.257}$$

$$= N \ln N - N + 1 \tag{3.258}$$

$$\approx N \ln N - N. \tag{3.259}$$

Now, let's return to $\ln W = \ln N! - \sum_i \ln n_i!$. Assuming that both n_i and N are large numbers, we can apply Stirling's approximation for both $\ln N!$ and $\ln n_i!$:

$$\ln W \approx \underbrace{N \ln N - N}_{\ln N!} - \sum_i \underbrace{(n_i \ln n_i - n_i)}_{\ln n_i!} \tag{3.260}$$

$$= N \ln N - N - \sum_i n_i \ln n_i + N \tag{3.261}$$

$$= N \ln N - \sum_i n_i \ln n_i. \tag{3.262}$$

Differentiating with respect to n_i, which we shall treat in the limit $n_i \gg 1$ as a continuously varying quantity,

$$d(\ln W) \approx -\sum_i \left(n_i \frac{1}{n_i} dn_i + dn_i \ln n_i \right) \tag{3.263}$$

$$\approx -\sum_i \left(1 + \ln n_i\right) dn_i \tag{3.264}$$

$$\approx -\sum_i \ln n_i dn_i. \tag{3.265}$$

We assume $\ln n_i \gg 1$ for the last step.

Now, we seek to maximise $\ln W$, subject to the constraints $dU = 0$ and $dN = 0$. To achieve this, we can use the method of *Lagrange multipliers,* i.e. combining the (zero) differentials and solving for the constants,[25] which are α and β in our case:

$$0 = d(\ln W) + \alpha dU + \beta dN.$$

Using $d(\ln W) \approx -\sum_i \ln n_i dn_i$, $dU = \sum_i \varepsilon_i dn_i$ and $N = \sum_i dn_i$,

$$0 = -\sum_i \ln n_i dn_i + \alpha \sum_i \varepsilon_i dn_i + \beta \sum_i dn_i \tag{3.266}$$

$$0 = \sum_i \left(\alpha\varepsilon_i + \beta - \ln n_i\right) dn_i. \tag{3.267}$$

For this to be true for all i,

$$\alpha\varepsilon_i + \beta - \ln n_i = 0, \tag{3.268}$$

$$\therefore n_i = e^{\alpha\varepsilon_i + \beta} \tag{3.269}$$

which means

$$N = \sum_i n_i = e^{\beta} \sum_i e^{\alpha\varepsilon_i}. \tag{3.270}$$

Hence, the probability of particles with energy ε_i is

$$p(\varepsilon_i) = \frac{n_i}{N} = \frac{e^{-\alpha\varepsilon_i}}{\sum_i e^{-\alpha\varepsilon_i}}. \tag{3.271}$$

This is called the (discrete) *Boltzmann distribution*. Note that, in the last step, we have swapped the sign of α to avoid exponentially increasing probabilities (i.e. α must be positive and therefore $e^{-\alpha\varepsilon_i} \leq 1$, since (kinetic) energies ε_i are positive quantities). We shall show by the conclusion of this section that α is inversely related to absolute (i.e. kelvin-scale) temperature T by the relation

$$\alpha = \frac{1}{k_B T}, \tag{3.272}$$

where *Boltzmann's constant* $k_B = 1.381 \times 10^{-23}$ JK^{-1}.

Figure 3.11 is from a MATLAB simulation performed to demonstrate the emergence of Boltzmann statistics. $N = 10{,}000$ energy quanta of fixed value ε joules are

---

[25] An entire book devoted to the methods of mechanics developed by Joseph Louis Lagrange (1736–1813) is the most elegant *Analytical Mechanics* by Hand and Finch [15]. See p. 56 for a utilisation of the method of *Lagrange multipliers.*

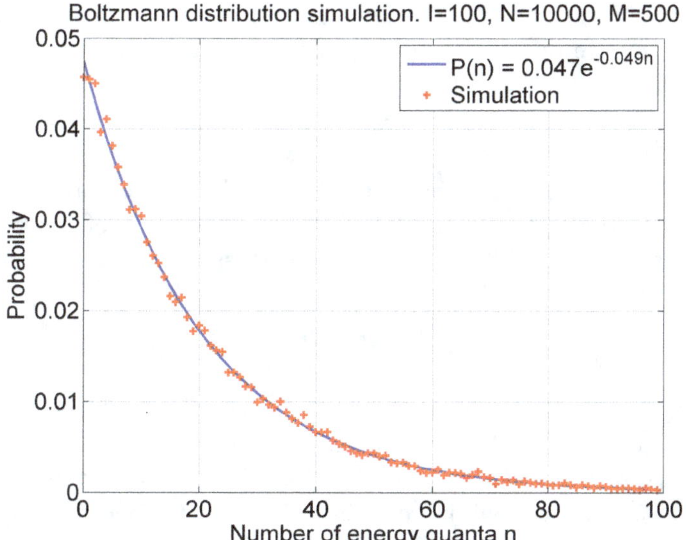

Fig. 3.11. MATLAB simulation performed to demonstrate the emergence of Boltzmann statistics. $N = 10,000$ energy quanta of fixed value ε joules are shared randomly among $M = 500$ molecules, and then the entire process is repeated $I = 100$ times. A frequency table is computed, which corresponds to the number of molecules which have n energy quanta, i.e. energy $\varepsilon_n = n\varepsilon$. Once scaled by total frequency to yield a probability $P(n)$, this curve tends to an exponential decay, characteristic of a Boltzmann distribution, $P(n) \propto e^{-\frac{n\varepsilon}{k_B T}}$. In this simulation, there is no model of temperature T or, indeed, an energy quanta value in J that can be compared to the average thermal energy characterised by $k_B T$. This simulation simply demonstrates that you get an exponential distribution, $P(n) \propto e^{-an}$, where a is a constant, when you randomly share a fixed number of energy quanta among a fixed number of molecules.

shared randomly among $M = 500$ molecules, and then the entire process is repeated $I = 100$ times. A frequency table is computed, which corresponds to the number of molecules that have n energy quanta, i.e. energy $\varepsilon_n = n\varepsilon$. Once scaled by total frequency to yield a probability $P(n)$, this curve tends to an exponential decay, characteristic of a Boltzmann distribution, $P(n) \propto e^{-\frac{n\varepsilon}{k_B T}}$. In this simulation, there is no model of temperature T or, indeed, an energy quanta value in J that can be compared to the average thermal energy characterised by $k_B T$. This simulation simply demonstrates that you get an exponential distribution, $P(n) \propto e^{-an}$, where a is a constant, when you randomly share a fixed number of energy quanta among a fixed number of molecules.

We can make the Boltzmann distribution *continuous* (i.e. for a smoothly varying continuum of possible energy states) by defining $p(\varepsilon)d\varepsilon$ to be the probability of an energy state being between ε and $\varepsilon + d\varepsilon$. Using the Boltzmann distribution idea,

$$p(\varepsilon) \propto e^{-\alpha\varepsilon}. \tag{3.273}$$

So, what is α? To assign an appropriate constant, we shall return to the definition of *temperature being proportional to the mean kinetic energy of molecules*. Consider a molecule of mass m moving at speed v freely in x, y, z, dimensions. If it is an ideal gas, then only the kinetic energy ε is a significant contributor to total energy. If we assume non-relativistic dynamics (i.e. speeds much less than the speed of light), the kinetic energy is given by

$$\varepsilon = \tfrac{1}{2}mv^2. \tag{3.274}$$

Let's start by deriving a probability distribution for a molecular speed of v. Using the *Boltzmann factor* $e^{-\alpha\varepsilon}$, we can write an expression for probability density, i.e. where $p(v)dv$ is the probability of a molecule with speed between v and $v + dv$:

$$p(v)dv = A \times 4\pi v^2 dv \times e^{-\alpha\frac{1}{2}mv^2}. \tag{3.275}$$

The factor $4\pi v^2 dv$ is required since this is the volume of a spherical shell of radius v and thickness dv. As v increases, the number of possible velocities that correspond to speed v, where $v^2 = v_x^2 + v_y^2 + v_z^2$, (i.e. the radius of a sphere in Cartesian v_x, v_y, v_z space) will increase in proportion to this shell volume. Let us define the constant A such that $\int_0^\infty p(v)dv = 1$, i.e. a fundamental requirement for $p(v)$ to be a probability density function.

$$\therefore A = \frac{1}{4\pi \int_0^\infty v^2 e^{-\alpha\frac{1}{2}mv^2}\,dv}. \tag{3.276}$$

Note the following standard integral(s):

$$I_n(a) = \int_0^\infty v^n e^{-av^2}\,dv, \tag{3.277}$$

$$I_0(a) = \frac{1}{2}\sqrt{\frac{\pi}{a}}, \tag{3.278}$$

$$I_1(a) = \frac{1}{2a}, \tag{3.279}$$

$$I_n(a) = \frac{n-1}{2a}I_{n-2}(a). \tag{3.280}$$

Hence, using $a = \tfrac{1}{2}\alpha m$,

$$\int_0^\infty v^2 e^{-\alpha\frac{1}{2}mv^2}\,dv = I_2(\tfrac{1}{2}\alpha m) \tag{3.281}$$

$$= \frac{(2-1)}{\alpha m}I_0\left(\frac{1}{2}\alpha m\right) \tag{3.282}$$

$$= \frac{1}{\alpha m}\frac{1}{2}\sqrt{\frac{\pi}{\frac{1}{2}\alpha m}} \tag{3.283}$$

$$= \frac{1}{\sqrt{2}}\left(\frac{1}{\alpha m}\right)^{\frac{3}{2}}\sqrt{\pi}. \tag{3.284}$$

Therefore,

$$A = \frac{1}{4\pi \frac{1}{\sqrt{2}} \left(\frac{1}{\alpha m}\right)^{\frac{3}{2}} \sqrt{\pi}}, \tag{3.285}$$

$$A = \left(\frac{1}{2\pi}\alpha m\right)^{\frac{3}{2}}. \tag{3.286}$$

Note that we have used $\frac{4}{\sqrt{2}} = 2^{2-\frac{1}{2}} = 2^{\frac{3}{2}}$ to help simplify the final expression.

In summary, the continuous molecular speed distribution is

$$p(v) = 4\pi \left(\frac{1}{2\pi}\alpha m\right)^{\frac{3}{2}} v^2 e^{-\alpha\frac{1}{2}mv^2}. \tag{3.287}$$

In the heat engine section of this chapter, we asserted the internal energy of n moles of ideal gas to be

$$U = \tfrac{1}{2}\alpha n R T, \tag{3.288}$$

where the molar gas constant $R = 8.314$ Jmol^{-1}K^{-1}. If the gas can only move in x, y, z directions, then the number of degrees of freedom $\alpha = 3$. Hence, the average energy *per molecule* is $\langle \varepsilon \rangle = \frac{U}{nN_A}$, which means

$$\langle \varepsilon \rangle = \tfrac{3}{2}k_B T, \tag{3.289}$$

where

$$k_B = \frac{R}{N_A} = 1.381 \times 10^{-23} \text{ JK}^{-1} \tag{3.290}$$

is known as *Boltzmann's constant*. Now, if $p(\varepsilon)d\varepsilon$ is the probability of energy ε between ε and $\varepsilon + d\varepsilon$, the mean energy per molecule is

$$\tfrac{3}{2}k_B T = \langle \varepsilon \rangle = \int_0^\infty \varepsilon p(\varepsilon)d\varepsilon. \tag{3.291}$$

Now, the energy $\varepsilon = \tfrac{1}{2}mv^2$ is a function of one single variable v since in our molecular gas, we assume all the molecules have the same mass m.

$$\therefore p(\varepsilon)d\varepsilon = p(v)dv. \tag{3.292}$$

Hence,

$$\tfrac{3}{2}k_B T = \langle \varepsilon \rangle = \int_0^\infty \underbrace{\tfrac{1}{2}mv^2}_{\varepsilon} \times \underbrace{4\pi \left(\tfrac{1}{2\pi}\alpha m\right)^{\frac{3}{2}} v^2 e^{-\alpha\frac{1}{2}mv^2}}_{p(v)} dv \tag{3.293}$$

$$= 2\pi m \left(\frac{1}{2\pi}\alpha m\right)^{\frac{3}{2}} \int_0^\infty v^4 e^{-\alpha\frac{1}{2}mv^2} dv. \tag{3.294}$$

Using the standard integral $I_n(a) = \frac{n-1}{2a}I_{n-2}(a)$, where $I_n(a) = \int_0^\infty v^n e^{-av^2}dv$ and n is an integer ≥ 2,

$$\int_0^\infty v^4 e^{-\alpha\frac{1}{2}mv^2}dv = I_4\left(\alpha\frac{1}{2}m\right) = \frac{3}{\alpha m}I_2\left(\alpha\frac{1}{2}m\right) = \frac{3}{\alpha m} \times \frac{1}{\sqrt{2}}\left(\frac{1}{\alpha m}\right)^{\frac{3}{2}}\sqrt{\pi}. \tag{3.295}$$

Hence,

$$\tfrac{3}{2}k_BT = \langle \varepsilon \rangle = 2\pi m \left(\frac{1}{2\pi}\alpha m\right)^{\frac{3}{2}} \times \frac{3}{\alpha m \sqrt{2}} \left(\frac{1}{\alpha m}\right)^{\frac{3}{2}} \sqrt{\pi},$$

$$\therefore \ k_BT = 2^{2-\frac{3}{2}-\frac{1}{2}}\pi^{1-\frac{3}{2}+\frac{1}{2}}\alpha^{-1} \tag{3.296}$$

$$\implies k_BT = \alpha^{-1}, \tag{3.297}$$

i.e.

$$\alpha = \frac{1}{k_BT}. \tag{3.298}$$

So, to match our distribution of molecular speeds with the results of classical macroscopic thermodynamics, we must have a *Boltzmann factor*:

$$e^{-\frac{\varepsilon}{k_BT}} \tag{3.299}$$

We can put all these results together to define the *Maxwell*[26]*–Boltzmann distribution* of molecular speeds (see Fig. 3.12):

$$p(v) = 4\pi \left(\frac{m}{2\pi k_BT}\right)^{\frac{3}{2}} v^2 e^{-\frac{\frac{1}{2}mv^2}{k_BT}}. \tag{3.300}$$

Now, since

$$\langle \varepsilon \rangle = \frac{3}{2}k_BT = \frac{1}{2}m\langle v^2\rangle \tag{3.301}$$

and therefore

$$\langle v^2\rangle = \frac{3k_BT}{m} \tag{3.302}$$

the RMS molecular speed is

$$v_{\text{RMS}} = \sqrt{\langle v^2\rangle} = \sqrt{\frac{3k_BT}{m}}. \tag{3.303}$$

For air molecules of molar mass 29.0×10^{-3} kgmol^{-1} at 300 K, $m = \frac{29.0\times10^{-3}}{6.02\times10^{23}}$ kg.

$$\therefore v_{\text{RMS}} = \sqrt{\frac{3 \times 1.38 \times 10^{-23} \times 300}{\frac{29.0\times10^{-3}}{6.02\times10^{23}}}} \approx 508 \ \text{ms}^{-1} \tag{3.304}$$

which is surprisingly fast, particularly given that the speed of sound in air is about 340 ms^{-1}.

---

[26]James Clerk Maxwell (1831–1879).

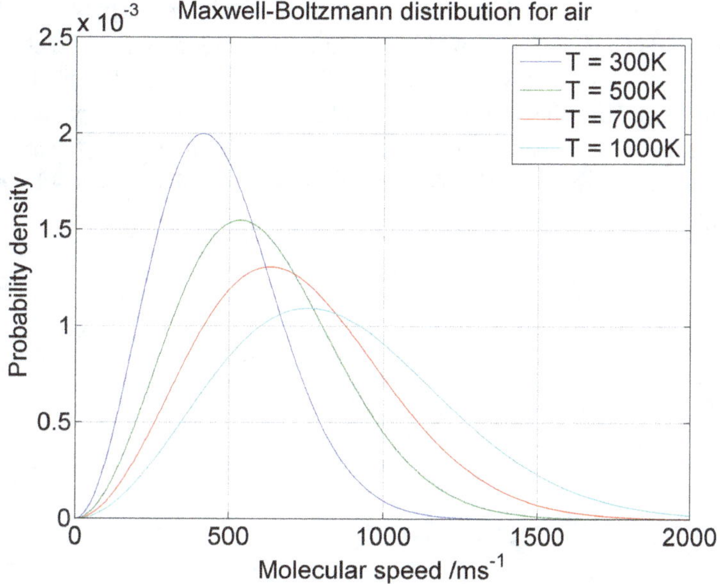

Fig. 3.12. The *Maxwell–Boltzmann* distribution of molecular speeds has probability density function $p(v) = 4\pi \left(\frac{m}{2\pi k_B T} \right)^{\frac{3}{2}} v^2 e^{-\frac{\frac{1}{2}mv^2}{k_B T}}$. The probability of a molecule having speed v in the range from v to $v+dv$ is $p(v)dv$. Molecular mass is m, absolute temperature is T and $k_B = 1.381 \times 10^{-23}$ JK^{-1} is *Boltzmann's constant*, which is the molar gas constant $R = 8.314$ Jmol^{-1}K^{-1} divided by Avogadro's number $N_A = 6.022 \times 10^{23}$. Different coloured traces correspond to different temperatures over the range of 300 K to 1,000 K. As the temperature T increases, the range of possible molecular speeds v increases, as well as their mean average. This means the peak probability density will tend to reduce as T increases.

3.5   A Selection of Thermodynamics Books and Resources

This chapter followed a fairly traditional path of starting with a classical theory of heat, temperature, energy and entropy and then an assembly of models involving ideal gases in heat engines. The focus then moved to the idea of macroscopic thermodynamic quantities being statistical averages of the energies and entropies of vast numbers of randomly moving molecules. Longair's extraordinary book, *Theoretical Concepts in Physics* [22], is highly recommended and follows the same pedagogical journey. It also provides plenty of historical insight into how the science of thermodynamics was developed. My personal standard text in thermodynamics is *Statistical Physics* by Mandl [24], which, in my opinion, is both comprehensive and very clearly written. *Physics by Example* by Rees [30] contains a useful chapter, including several classic problems and their solutions with student-friendly explanations. At a more introductory pre-university level, I would recommend *The Language of Physics* by Cullerne and Machacek [4]. In terms of online resources, there is a huge amount that can be Googled, but I would personally start with R. Nave's amazing *Hyperphysics* (http://hyperphysics.phy-astr.gsu.edu/hbase/hframe.html). Much of the material presented here forms the basis of the notes on my website http://www.eclecticon. info/physics_notes_thermo.htm, which includes downloads of the heat engine, Boltzmann and kinetic theory simulations. There are plenty of experiment videos, too. For problems at the pre-university level, see http://www.eclecticon.info/ physics_questions.htm, *Mastering Pre-University Physics* [23], *Isaac Physics* at https://isaacphysics.org/ and the British Physics Olympiad website https://www. bpho.org.uk/.

Chapter 4

Waves and Ray Optics

4.1 An Anatomy of Waves

A wave is essentially a *disturbance* represented by a quantity $\psi(x, t)$ that propagates spatially at a speed[1] c. The distance from some origin at which the disturbance is described is x, and t is the time as measured from some agreed clock-starting event. One could imagine the spatial extent of a wave at any given time t as if captured via a photograph, e.g. a snapshot of ocean waves about to break upon a beach as taken from a camera mounted on a hovering drone. Alternatively, the time variation at some fixed point could be recorded. In electronics, this is the function of an *oscilloscope*. The variation of voltage across the terminals of the 'scope' input is displayed on a screen with a width which corresponds to a fixed time interval, set by the *timebase* of the oscilloscope. The vertical axis of the oscilloscope corresponds to voltage, with a maximum range set by the *voltage division* setting.

A wave-like disturbance could be all sorts of things—a change in gas pressure, displacement of a string under tension, ground or water movement, or indeed fluctuations in electromagnetic (EM) fields, which constitute radio waves, microwaves, infrared, visible and ultraviolet light, X-rays and gamma rays, i.e. the different wavelengths of the *electromagnetic spectrum*.

Longitudinal waves correspond to disturbances in the same direction as the wave propagation, e.g. the *compressions* and *rarefactions* of air pressure in sound waves or earthquake P-waves. *Transverse* waves correspond to disturbances that are perpendicular to the wave propagation direction, such as electric and magnetic fields in light waves or ground movement in earthquake shear (S) waves. *Surface waves* can occur between the boundaries of different materials. Water waves, earthquake

---

[1]We will make the assumption that *wave speed is a constant* until a boundary is reached, where c changes discontinuously. Models of wave propagation where wave speed varies continuously are beyond the scope of this book—although the short section on *Snell's law* of refraction derived from *Fermat's principle* ('the ray path corresponds to the shortest time') and, indeed, the section on the *calculus of variations* in the chapter on mechanics (and in the appendix) will point the way to how such a problem might be solved. The problem of calculating the path of rays in a 2D medium with a linearly changing wave speed with depth (i.e. applicable to seismic waves) is solved in http://www.eclecticon.info/index_htm_files/Calculus%20-%20Variational%20Calculus.pdf.

Rayleigh[2] (vertical motion) and *Love*[3] (side-to-side, i.e. shear) waves are classic examples. Wave patterns can also originate from *instabilities* in layers of a moving fluid of different densities, velocities and viscosities. The 'mare's tails' (which is called the *Kelvin–Helmholtz instability*) that form on aircraft contrails represent an everyday example.

A wave[4] $\psi(x,t)$ can be described as a *spatial translation* of a disturbance curve $f(x)$ as time progresses. If the wave speed is c and the spatial translation is ct, we can therefore write

$$\psi(x,t) = f(x - ct). \tag{4.1}$$

All waves of the form $\psi(x,t) = f(x - ct)$ obey the *wave equation*

$$\frac{\partial^2 \psi}{\partial x^2} = \frac{1}{c^2} \frac{\partial^2 \psi}{\partial t^2}. \tag{4.2}$$

Note that an alternative form is

$$\left(\frac{\partial \psi}{\partial x}\right)^2 = \frac{1}{c^2} \left(\frac{\partial \psi}{\partial t}\right)^2. \tag{4.3}$$

Periodic waves will repeat spatially with *wavelength* λ. We could measure this distance if we were to take a photograph of a wave at a given time. Alternatively, we could measure $\psi(x,t)$ at a fixed position x and record the time T for the wave to repeat. This is called the *period*. The number of waves per second is the *frequency* $f = \frac{1}{T}$. Hence, since wave speed $c = \frac{\lambda}{T}$,

$$c = f\lambda. \tag{4.4}$$

All periodic waveforms can be constructed from a sum of sine and cosine waves of different amplitudes, frequencies and phases,[5] so we can use the mathematical representation

$$\psi(x,t) = A\cos\left(2\pi\frac{x - ct}{\lambda} - \phi\right) \tag{4.5}$$

to represent the basic components of all waves. A spatially translated cosinusoidal wave matching this description is illustrated in Fig. 4.1. A is the *amplitude* of the wave, the maximum value of $\psi(x,t)$. The phase[6] of the wave is $2\pi\frac{x-ct}{\lambda} - \phi$; therefore, ϕ is a fixed phase shift that may be required to describe waves of the same frequency but are not always 'in-phase'. An array antenna used in radar and high-end audio systems operates by transmitting waves with a controllable phase shift between antenna elements such that the net radiation pattern forms a peak at a desired angle.[7]

---

[2] John William Strutt, 3rd Baron Rayleigh (1842–1919).

[3] Augustus Edward Hough Love (1863–1940).

[4] We will use one-dimensional examples, but the ideas generalise to 2D or 3D.

[5] i.e. a *Fourier series*.

[6] The *phase*, i.e. the input of a sine or cosine function, or 'where you are within a wave'.

[7] See *Science by Simulation – Volume 1: A Mezze of Mathematical Models* (Chapter 9) [9].

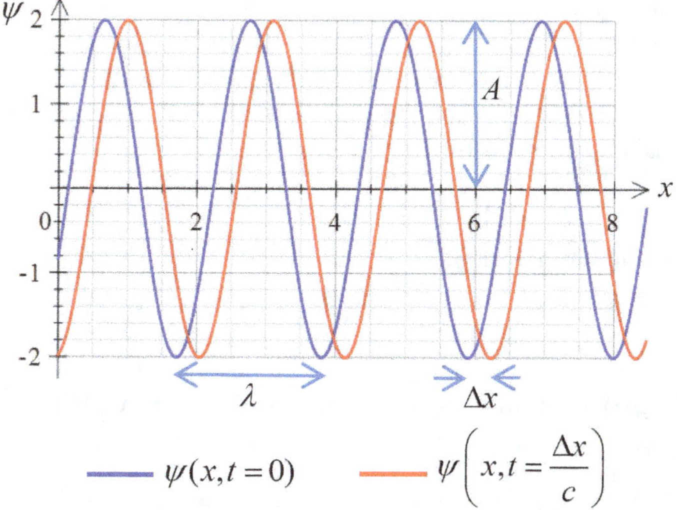

Fig. 4.1. All wave-like disturbances $\psi(x, t)$ can be constructed from a sum of cosinusoids of the form $\psi(x, t) = A\cos(kx - \omega t - \phi)$. x is the displacement of the wave in the propagation direction and t is time. In this figure, $\psi(x, 0)$ and $\psi(x, t = \frac{\Delta x}{c})$ are overlaid to demonstrate that waveforms at different times represent a *spatial translation* (in this case, left to right by Δx along the x direction). We can see this algebraically by noting that $\psi(x, t) = A\cos\left(2\pi\frac{x - ct}{\lambda} - \phi\right)$. The wavelength is λ, amplitude is A, *wavenumber* $k = \frac{2\pi}{\lambda}$ and *angular speed* $\omega = ck$, where c is the wave speed. Note that $c = f\lambda = \frac{\lambda}{T}$, where f is the wave frequency and T is the wave period. ϕ is the initial phase (in radians) when t and x are zero.

The basic waveform component $\psi(x, t) = A\cos\left(2\pi\frac{x - ct}{\lambda} - \phi\right)$ can be simplified by first defining *wavenumber* $k = \frac{2\pi}{\lambda}$ and *angular frequency* $\omega = \frac{2\pi}{T} = 2\pi f$. Using these quantities, we can write $\psi(x, t)$ as

$$\psi(x, t) = A\cos\left(kx - \omega t - \phi\right) \tag{4.6}$$

and relate angular frequency ω, wave speed c and wavenumber k via

$$\omega = ck. \tag{4.7}$$

Note that in many situations, it is convenient to use *complex numbers* to represent wave phenomena. *De Moivre's theorem*[8]

$$e^{i\theta} = \cos\theta + i\sin\theta \tag{4.8}$$

can be employed to express $\psi(x, t)$ as

$$\psi(x, t) = Ae^{i(kx - \omega t - \phi)}. \tag{4.9}$$

Take the real part (or, if you prefer sine, the imaginary part) to get the 'real' wave! This is very useful in a mathematical sense, as exponentials differentiate and integrate to form the same exponentials (but scaled by a constant), whereas sine and cosine functions interchange between each other under calculus operations,[9] and

---

[8] Abraham de Moivre (1667–1754).

[9] Assuming that the phase of a sine or cosine function is expressed in *radians*, not degrees.

their signs change too, i.e. compare $\frac{d}{dx}e^{ikx} = ike^{ikx}$ with $\frac{d}{dx}\cos kx = -k\sin kx$ and $\frac{d}{dx}\sin kx = k\cos kx$. When considering waves moving in two or three dimensions, the *wavevector* $\mathbf{k}$ defines the direction of propagation, as illustrated in Fig. 4.30 on p. 182. The magnitude of the wavevector is the wavenumber $k = \frac{2\pi}{\lambda}$. If $\mathbf{r} = x\hat{\mathbf{x}} + y\hat{\mathbf{y}} + z\hat{\mathbf{z}}$ is a position vector (x, y, z) from some origin,

$$\psi(x, y, z, t) = \psi(\mathbf{r}, t) = A\cos(\mathbf{k} \cdot \mathbf{r} - \omega t - \phi) \tag{4.10}$$

or, in complex number notation,

$$\psi(\mathbf{r}, t) = Ae^{i(\mathbf{k} \cdot \mathbf{r} - \omega t - \phi)}. \tag{4.11}$$

4.2   Standing Waves on Guitar Strings and in Organ Pipes

4.2.1   *Standing waves*

Standing waves can form when the geometry of a wave system constrains *nodes* and *antinodes* to be fixed spatially. A node is a point of zero disturbance, whereas an antinode is a variation of maximum amplitude. Standing waves form from a superposition (i.e. 'signed addition') of incoming and reflected waves, such that the standing wave's mathematical representation $\psi(x, t)$ is a *product* of temporal and spatial oscillations. This explains why the nodes and antinodes are fixed spatially. The entire standing wave still oscillates at frequency f, but the effective amplitude of these vibrations varies in a sinusoidal fashion along the length of the oscillating system. Note that waves on a string will be *inverted* upon reflection off a fixed end. We assume that our standing wave starts with a node, so $A\sin(kx - \omega t)$ will be our representation of a left-to-right incoming wave. We could easily phase shift this to a cosine to start with an antinode instead, since $\cos(x - \frac{\pi}{2}) = \sin x$.

Therefore

$$\psi(x, t) = \underbrace{A\sin(kx - \omega t)}_{\text{incoming wave}} - \underbrace{A\sin(-kx - \omega t)}_{\text{inverted, reflected wave}}. \tag{4.12}$$

Using the trigonometric identity

$$\sin(A \pm B) = \sin A\cos B \pm \cos A\sin B \tag{4.13}$$

and the oddness and evenness of sine and cosine functions

$$\sin(-x) = -\sin x, \tag{4.14}$$

$$\cos(-x) = \cos x, \tag{4.15}$$

we can write

$$A\sin(kx - \omega t) = A\sin kx\cos\omega t - A\cos kx\sin\omega t, \tag{4.16}$$

$$A\sin(-kx - \omega t) = -A\sin kx\cos\omega t - A\cos kx\sin\omega t. \tag{4.17}$$

$$\therefore -A\sin(-kx - \omega t) = A\sin kx\cos\omega t + A\cos kx\sin\omega t. \tag{4.18}$$

Hence,

$$\psi(x,t) = A\sin(kx - \omega t) - A\sin(-kx - \omega t) \tag{4.19}$$

$$= A\sin kx \cos \omega t - A\cos kx \sin \omega t + A\sin kx \cos \omega t + A\cos kx \sin \omega t$$

$$= 2A\sin kx \cos \omega t, \tag{4.20}$$

which is a product of a *purely spatial variation* $\sin kx$ and a *time-only variation* $\cos \omega t$.

In general, propagating waves will lose energy and be attenuated in amplitude, but not in frequency—we treat these 'Fourier components' separately. However, because energy may constantly be fed into a standing wave, it may achieve a *resonance* effect. This is the basis of sound production from most tuned instruments.

4.2.2 *Guitar strings*

4.2.2.1 *Standing waves*

The fixed positions of nodes or antinodes will *quantise* the allowed frequencies of oscillation of standing waves. A tensioned string must have nodes at both ends, and any integer number of antinodes in between. This means that the length L of the string must be an integer number n of *half-wavelengths*.

$$L = \tfrac{1}{2}n\lambda_n. \tag{4.21}$$

Possible standing wave wavelengths λ_n and frequencies $f_n = \frac{c}{\lambda_n}$ are therefore

$$\lambda_n = \tfrac{2}{n}L, \tag{4.22}$$

$$f_n = \tfrac{c}{2L}n, \tag{4.23}$$

i.e. *harmonics* of frequency spaced by $\frac{c}{2L}$ Hz. The wave disturbance for the nth harmonic ($n = 1$ is the *fundamental*, or *first harmonic*; $n = 2$ is the *first overtone*, or *second* harmonic etc.) can be written as

$$\psi(x,t) = 2A\sin\left(\frac{2\pi x}{\lambda_n}\right)\cos\left(2\pi f_n t\right) \tag{4.24}$$

$$= 2A\sin\left(\frac{n\pi x}{L}\right)\cos\left(\frac{n\pi c t}{L}\right). \tag{4.25}$$

A fundamental plus three overtones are illustrated in Fig. 4.2.

4.2.2.2 *Wave speed*

The speed c of waves on a tensioned string is given by

$$c = \sqrt{\frac{T}{\mu}} \tag{4.26}$$

where T is string tension and μ is the mass per unit length. This can be derived by considering an infinitesimally small section of tensioned string of length dx, and hence mass μdx, as illustrated in Fig. 4.3. The vertical displacement of the string

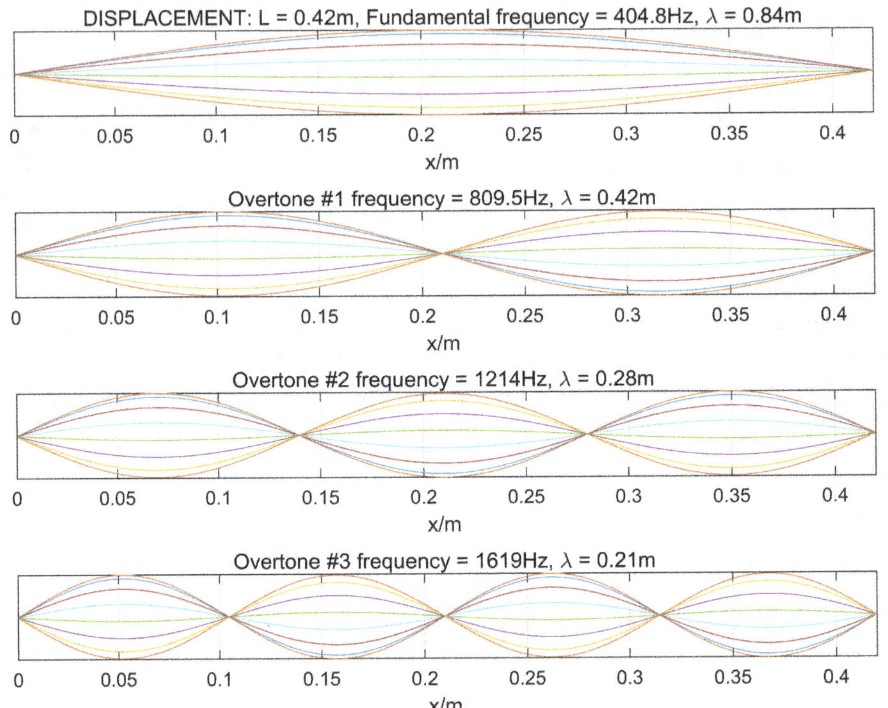

Fig. 4.2. Plots of the horizontal displacement $\psi(x,t)$ of a tensioned string that is clamped at both ends. The coloured curves represent the position of the sting at different times. The oscillatory motion is a *standing wave*, with fixed *nodes* (positions of zero movement) and *antinodes* (positions of maximum movement) along the string. The disturbance $\psi(x,t) = 2A\sin\left(\frac{2\pi x}{\lambda_n}\right)\cos\left(2\pi f_n t\right)$ where $2A$ is the maximum antinode amplitude, wavelength $\lambda_n = \frac{2}{n}L$ and frequency $f_n = \frac{c}{2L}n$. The length of the string is L and c is the wave speed, i.e. for standing waves on a tensioned string, the string length is a whole number of half-wavelengths. For a tensioned string, wave speed $c = \sqrt{\frac{T}{\mu}}$ where T is tension and μ is the mass per unit length. $n+1$ is the number of nodes. $n = 1$ is the fundamental, and $n = 2, 3, 4, \ldots$ are *overtones*, i.e. higher order harmonics. When a guitar string is plucked, the resulting waveform is a superposition of harmonics, but these may not be equally weighted. Some of these (typically higher frequencies) will *attenuate* over time, changing the resulting sound waveform, until the string is plucked again.

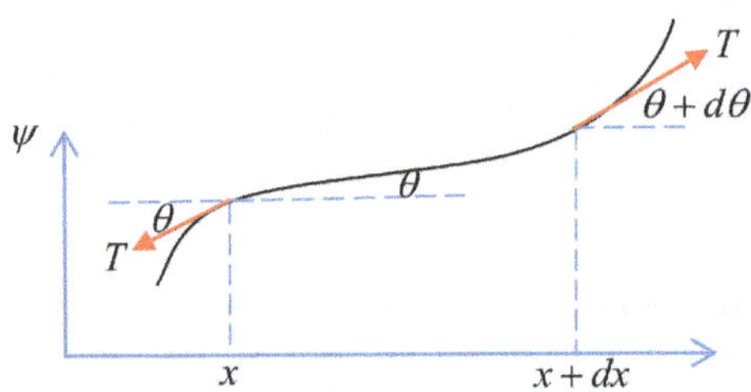

Fig. 4.3. Diagram used to determine the wave speed $c = \sqrt{\frac{T}{\mu}}$ for a tensioned string, where T is the string tension and μ is the mass per unit length. This is achieved by writing down Newton's second law for a small section of string and comparing this (using a small-angle approximation $\theta \ll 1$ radians) to the wave equation.

is ψ. We can use Newton's second law to relate the mass $\times$ acceleration of the string element to the vector sum of forces upon it. We express the acceleration as $\frac{\partial^2 \psi}{\partial t^2}$, with the partial derivative notation since ψ varies with x and t. In the vertical direction,

$$\mu dx \frac{\partial^2 \psi}{\partial t^2} = T \sin(\theta + d\theta) - T \sin \theta. \tag{4.27}$$

If we apply a small angle approximation (i.e. assume ψ is small such that $\theta \ll 1$ radian), then

$$\sin \theta \approx \tan \theta \approx \theta. \tag{4.28}$$

Hence,

$$\frac{\mu}{T} \frac{\partial^2 \psi}{\partial t^2} \approx \frac{\tan(\theta + d\theta) - \tan \theta}{dx}. \tag{4.29}$$

Now, from the geometry of the diagram in Fig. 4.3, $\tan \theta$ is the gradient $\frac{\partial \psi}{\partial x}$ of the string element snapshot. This means $\frac{\tan(\theta + d\theta) - \tan \theta}{dx}$ is the 'gradient of the gradient'.

$$\therefore \frac{\tan(\theta + d\theta) - \tan \theta}{dx} \approx \frac{\partial}{\partial x} \left(\frac{\partial \psi}{\partial x} \right) = \frac{\partial^2 \psi}{\partial x^2}. \tag{4.30}$$

We can therefore express Newton's second law for our string element in the form of the *wave equation* and hence determine an expression for the wave speed.

Newton's second law:

$$\frac{\mu}{T} \frac{\partial^2 \psi}{\partial t^2} \approx \frac{\partial^2 \psi}{\partial x^2}. \tag{4.31}$$

Wave equation:

$$\frac{1}{c^2} \frac{\partial^2 \psi}{\partial t^2} = \frac{\partial^2 \psi}{\partial x^2}. \tag{4.32}$$

By comparing the coefficients of $\frac{\partial^2 \psi}{\partial t^2}$, the wave speed is related to μ and T by $\frac{1}{c^2} = \frac{\mu}{T}$.

$$\therefore c = \sqrt{\frac{T}{\mu}}. \tag{4.33}$$

4.2.2.3 *Guitar string tension calculations*

An important feature of a guitar (or bass, violin, mandolin, cello etc.) is to be able to tune multiple strings such that their fundamental frequency of vibration conforms to a musical note and that the various strings form a progression of notes that allow for strings to be vibrated, singly or together, to produce a desired tone. Guitar strings are typically of a fixed length L between the boundary nodes between the *bridge* and the *nut*. To avoid excessive stresses across the (usually) wooden neck of the instrument, string tensions are ideally quite similar. Therefore, to achieve different fundamental frequencies, each string will have a different mass per unit length.

Let's consider a string to be a cylinder of diameter d and density ρ. The mass of the string over length L is

$$m = \rho\pi \left(\tfrac{1}{2}d\right)^2 L \tag{4.34}$$

and hence the mass per unit length is

$$\mu = \tfrac{1}{4}\rho\pi d^2. \tag{4.35}$$

From the analysis of the previous section on standing waves on strings with nodes at both ends, the fundamental frequency f requires

$$f = \tfrac{c}{2L}. \tag{4.36}$$

Since $c = \sqrt{\tfrac{T}{\mu}}$ and $\mu = \tfrac{1}{4}\rho\pi d^2$,

$$f^2 = \tfrac{1}{4L^2}\frac{T}{\mu} \tag{4.37}$$

$$= \tfrac{1}{4L^2}\frac{T}{\tfrac{1}{4}\rho\pi d^2}. \tag{4.38}$$

Hence, string tension T can be calculated from the rather elegant equation

$$T = \rho\pi \left(fLd\right)^2. \tag{4.39}$$

In Fig. 4.5, the calculations for string tension are performed, which correspond to D'Addario EXL110 electric guitar strings that I like to attach to my own 'axe', a Yamaha Pacifica 212VFM. String density data is from D. Achilles of the University of Illinois (12 December 2000).

The fundamental frequencies associated with musical notes can be modelled using

$$f(n) = 110 \text{ Hz} \times 2^{\frac{1}{12}n}, \tag{4.40}$$

where f is the frequency in Hz and n is the number of semitones above the A_0 note, which is conventionally set at 110 Hz. Frequencies from $n = 0$ to $n = 12$ are evaluated in the table in Fig. 4.4. The formula $f(n) = 110 \text{ Hz} \times 2^{\frac{1}{12}n}$ is a good approximation of the Pythagorean 'harmonious proportions' of antiquity, e.g. a *perfect fifth* interval like A to E is, ideally, a frequency ratio of $3/2$. Using $f(n) = 110 \text{ Hz} \times 2^{\frac{1}{12}n}$, a fifth corresponds to $\frac{f(n+7)}{f(n)} = 2^{7/12} = 1.4983$, which is pretty close!

A	A# or Bb	B	C	C# or Db	D	D# or Eb	E	F	F# or Gb	G	G# or Ab	A (next octave)
n 0	1	2	3	4	5	6	7	8	9	10	11	12
Hz 110	116.5	123.5	130.8	138.6	146.8	155.6	164.8	174.6	185.0	196	207.7	220

Fig. 4.4. The fundamental frequencies associated with musical notes can be modelled using $f(n) = 110 \text{ Hz} \times 2^{\frac{1}{12}n}$, where f is the frequency in Hz and n is the number of semitones above the A_0 note, which is conventionally set at 110 Hz. $f(n) = 110 \text{ Hz} \times 2^{\frac{1}{12}n}$ is a good approximation of the Pythagorean 'harmonious proportions', e.g. a perfect fifth interval like A to E is, ideally, a frequency ratio of $3/2$. Using $f(n) = 110 \text{ Hz} \times 2^{\frac{1}{12}n}$, this corresponds to $\frac{f(n+7)}{f(n)} = 2^{7/12} = 1.4983$.

String tensions can be calculated if one knows the geometry of each string, the material density and the desired note frequency. The following results correspond to D'Addario EXL110 electric guitar strings. (Density data from https://courses.physics.illinois.edu/phys406/Student_Projects/Fall00/DAchilles/Guitar_String_Tension_Experiment.pdf)

Note	n	Frequency /Hz	Diameter (inches)	String length /m	String density /kgm^{-3}	Tension /N
E	19	329.63	0.010 plain	0.75	7690	95.3
B	14	246.94	0.013 plain	0.73	7950	88.5
G	10	196	0.017 plain	0.71	8220	93.2
D	5	146.83	0.026 wound	0.69	6930	97.5
A	0	110	0.036 wound	0.67	6610	94.3
E	-5	82.41	0.046 wound	0.66	6540	83.0

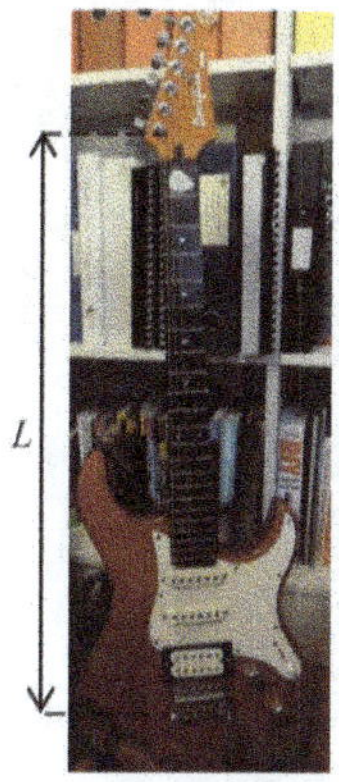

Assume *fundamental* mode of string vibration i.e. two nodes at either end of the string

$$c = f\lambda = \sqrt{\frac{T}{\mu}} \qquad \rho = \frac{\mu L}{L\pi\left(\tfrac{1}{2}d\right)^2} = \frac{4\mu}{\pi d^2}$$

$$\therefore T = 4f^2 L^2 \mu \qquad \therefore \mu = \tfrac{1}{4}\rho\pi d^2$$

$$\therefore T = 4f^2 L^2 \tfrac{1}{4}\rho\pi d^2$$

$$\boxed{T = \pi\rho\left(fLd\right)^2}$$

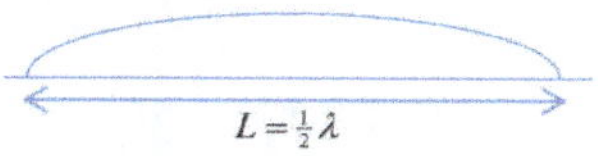

$$L = \tfrac{1}{2}\lambda$$

A wave which has fixed nodes, i.e. does not propagate in the x direction, is called a **standing wave**.

Standing waves can be written as $\psi(x) = \Phi(x)\tau(t)$ i.e. the time dependent part is separated.

For the guitar string fundamental mode: $\boxed{\psi(x,t) = \sin\left(\pi x / L\right)\sin\left(2\pi ft\right)}$

Plucking a real guitar string activates many more *harmonics*, which gives it its distinctive sound.

Fig. 4.5. Calculation of tension T in guitar strings using the formula $T = \rho\pi(fLd)^2$. String density is ρ, diameter is d, and length is L. The string's fundamental frequency f is determined from musical notes using the formula $f(n) = 110 \text{ Hz} \times 2^{\frac{1}{12}n}$.

4.2.2.4 *Guitar vs. recorder spectrographs*

The time variation of harmonics from a plucked guitar string can be visualised using a *spectrograph*. We can create one by recording a sound wave at the audio standard of 44 kHz and then, let's say, every 2048 samples,[10] determine the frequency content via a mathematical operation called a *Discrete Fourier Transform* (DFT).[11] In essence, we describe our waveform recording by a sum of sine and cosine waves of increasing frequency and work out the amplitudes of each. Plotting $\log_{10}(\text{amplitudes}^2)$ is the output of the DFT and a measure of power associated with each frequency, or *Fourier component*. This means every $\frac{2048}{44}$ ms = 46.54 ms, we have a distribution of the fundamentals plus harmonics within a sound recording. Figure 4.6 compares a spectrograph and short-duration variation of the signal (3 and 10 ms, respectively) for a recording of an A note played with a recorder and a guitar.

---

[10] You can make a DFT computationally more efficient by recording a number of samples that is an integer power of two. This can result in a *fast Fourier transform* (FFT). See *Numerical Recipes* [28].

[11] If a signal $x(t)$ is sampled at times $t_n = \frac{n-1}{f_s}$, where f_s is the sample rate, we can define a DFT by $y_k = \sum_{n=1}^{N} w_{n,k} x(t_n)$, where *filter weights* $w_{n,k} = \frac{1}{N}e^{-\frac{2\pi i(n-1)(k-1)}{N}}$. The power spectrum output $|y_k|^2$ can be plotted against frequency f using $f = \frac{k-1}{N}f_s$. Alternatively, compute $y(f) = \frac{1}{N}\sum_{n=1}^{N} x(t_n)e^{-2\pi i(n-1)\frac{f}{f_s}}$ using a finer grid of frequencies within the range $0 \le f \le f_s$, which will *interpolate* the DFT (but not add any extra information).

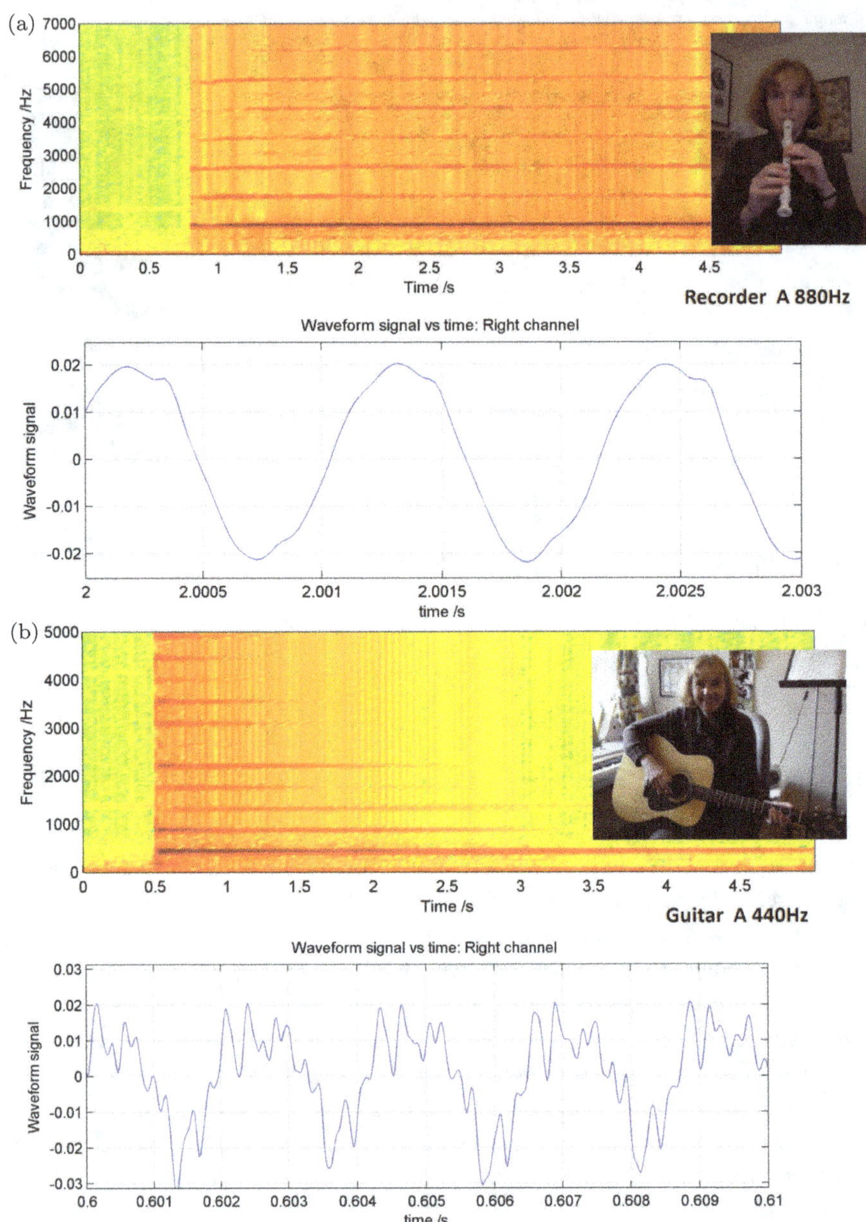

Fig. 4.6. Spectrographs (and a short duration waveform plot vs. time) of 44 kHz samples of sound waves produced from (a) a recorder and (b) a guitar. A single note is played for each instrument. The fundamental frequency for the guitar note is 440 Hz and that for the recorder is 880 Hz. Note how the spectral power is more spread for the guitar, resulting in the jagged wave shape, whereas the fundamental is more dominant for the recorder, resulting in a smoother sinusoidal pattern. Also, note that the higher frequency harmonics diminish for the guitar due to resistive forces of air and friction, whereas continuous blowing sustains the recorder waveform.

The fundamental frequency of the guitar note is 440 Hz and that of the recorder is 880 Hz. The spectrographs reveal a similar set of harmonics of $2, 3, 4, 5, \ldots$ times the fundamental frequency. This works for the recorder too because, although the sound is clearly driven at one end by blowing, there is an air hole near the mouthpiece, so standing waves of pressure must have nodes at both ends of the tube, meaning that the harmonic series has the same $f_n = \frac{c}{2L} n$ pattern as the guitar string. The major difference between the recorder and the guitar is that the higher-frequency harmonics of the guitar string diminish rapidly, leaving a more bass response as time progresses. This is because the string is plucked, and then it vibrates against the air (and the nut and bridge holding the string), which removes energy from the oscillation. Air is continuously blown into the recorder, so the harmonic content is sustained. Another major difference is the waveform itself. The recorder waveform is almost sinusoidal (indicating the fundamental is dominant), whereas the guitar waveform has a jagged pattern of peaks, which implies energy is more spread over higher harmonics. In both cases, the period between wave peaks is the reciprocal of the fundamental frequency ($\frac{1000}{440}$ ms $= 2.28$ ms for the guitar and $\frac{1000}{8800}$ ms $= 1.14$ ms for the recorder), but for the recorder, the fundamental line in the spectrograph is, as predicted by the shape of the waveform, dominant over the harmonics. We can tell this because the colour scale in the spectrograph relates to the spectral power. A deep red means a high power. For the guitar, certainly early on in the recording, more of the higher harmonics are activated and the spectral power is spread more evenly.

4.2.3 *Organ pipes*

An open-ended tube such as a driven organ pipe must have an *antinode of pressure at the driven end* (i.e. a maximum variation above and below atmospheric pressure) and a *node of pressure at the open end* (i.e. at atmospheric pressure). This means that the length of the pipe L is an *odd number of quarter wavelengths*, as illustrated in Fig. 4.7. Note that for pressure waves, an antinode of pressure corresponds to a node of particle displacement, and vice versa. Open-ended tubes also have an *end correction*[12] due to the vibrations of the end of a tube (which cause slightly more air to vibrate). For tubes of radius r, the end correction means the effective length of the tube (i.e. from a standing wave perspective) is increased by about $\delta L \approx \frac{2}{3} r$. We can express this mathematically as

$$L + \tfrac{2}{3} r = \tfrac{2n-1}{4} \lambda_n. \tag{4.41}$$

---

[12]Levine, H., Schwinger, J. (1948). "On the Radiation of Sound from an Unflanged Circular Pipe". *Physical Review.* **73** (4): 383–406. The exact value of the end correction for a tube of radius r turns out to be $\delta L = 0.6133r$. Note that if the tube has two open ends (e.g. a pan pipe or equivalent), you'll have to apply the end correction twice.

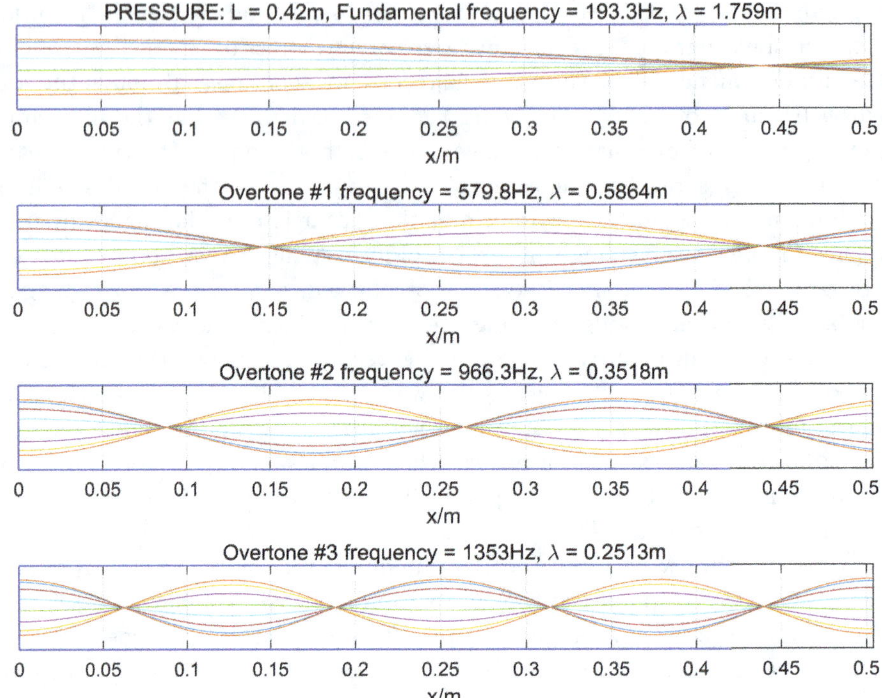

Fig. 4.7. Standing wave of air pressure inside an open-ended tube (right side), driven at the left end, e.g. an organ pipe or a similar wind instrument. The disturbance ψ_n is $\psi_n(x,t) = 2A\cos\left(\frac{2\pi x}{\lambda_n}\right)\cos\left(2\pi f_n t\right)$. The wavelength is $\lambda_n = \frac{4}{2n-1}\left(L + \frac{2}{3}r\right)$, where L is the tube length and 'end correction' $\approx \frac{2}{3}r$, where r is the tube radius. The frequency is $f_n = \frac{c}{4L+\frac{8}{3}r}\left(2n-1\right)$, i.e. the (end corrected) tube length is an odd number of quarter-wavelengths of standing waves characterised by positive integer n.

So, possible standing wave wavelengths λ_n and frequencies f_n are

$$\lambda_n = \frac{4}{2n-1}\left(L + \tfrac{2}{3}r\right), \tag{4.42}$$

$$f_n = \frac{c}{4L+\frac{8}{3}r}\left(2n-1\right), \tag{4.43}$$

i.e. *harmonics* spaced by $\dfrac{c}{2L+\frac{4}{3}r}$ Hz since $2(n+1)-1-(2n-1) = 2n+2-1-2n+1 = 2$.

4.2.4   *Wave energy and power*

In general, the power in a wave is given by

$$P = \tfrac{1}{2}ZA^2\omega^2, \tag{4.44}$$

where Z is the 'wave impedance', i.e. double the amplitude and you quadruple the wave power. Wave impedance is defined as

$$Z = \left|\frac{\text{driving force}}{\text{wave velocity}}\right|. \tag{4.45}$$

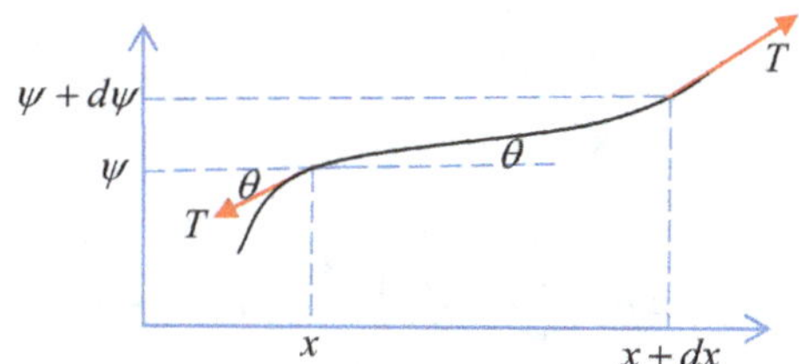

Fig. 4.8. The energy E associated with waves of amplitude $\psi(x,t)$, wavevector k and angular frequency ω on a string of tension T and mass per unit length μ can be shown to be $E = \frac{1}{2}T\left(\frac{\partial\psi}{\partial x}\right)^2 + \frac{1}{2}\mu\left(\frac{\partial\psi}{\partial t}\right)^2$. The time- averaged energy per unit length over period τ is $\overline{E} = \frac{1}{\tau}\int_0^\tau \mu\left(\frac{\partial\psi}{\partial t}\right)^2 dt = \frac{1}{2}\mu A^2\omega^2$, assuming Fourier components of the form $\psi = A\cos(kx-\omega t-\phi)$. This means the input wave power is $P = \frac{1}{2}ZA^2\omega^2$, where *impedance* $Z = \frac{T}{c}$. Wave speed c isc $= \frac{\omega}{k} = \sqrt{\frac{T}{\mu}}$.

Let's explore this relationship for a wave on a tensioned string. If a wave under constant tension T propagates from x to $x + dx$, as illustrated in Fig. 4.8, the work done dW to achieve this is the tension $\times$ the extension of the string if we can assume a constant tension approximation for small extensions. If we can model the disturbed string as a straight line at angle θ, then the extension is the hypotenuse $\sqrt{(d\psi)^2 + (dx)^2}$ minus the original (unstretched) length dx. Hence,

$$dW = T\left(\sqrt{(d\psi)^2 + (dx)^2} - dx\right) \tag{4.46}$$

$$= Tdx\left(\sqrt{1 + \left(\frac{\partial\psi}{\partial x}\right)^2} - 1\right). \tag{4.47}$$

Note that, in the final step, I have written $\frac{\partial\psi}{\partial x}$ instead of $\frac{d\psi}{dx}$ since ψ is a function of displacement x and time t.

If we can assume small gradient disturbances $\frac{\partial\psi}{\partial x} = \tan\theta \ll 1$, we can perform a single-term binomial expansion on $\sqrt{1 + \left(\frac{\partial\psi}{\partial x}\right)^2}$ to yield

$$dW \approx Tdx\left(1 + \frac{1}{2}\left(\frac{\partial\psi}{\partial x}\right)^2 - 1\right). \tag{4.48}$$

$$\therefore \frac{dW}{dx} \approx \frac{1}{2}T\left(\frac{\partial\psi}{\partial x}\right)^2. \tag{4.49}$$

$\frac{dW}{dx}$ is the work done per unit length, which is often described as the 'potential energy per unit length'. Therefore, to determine the total energy per unit length, we must add the kinetic energy (KE) per unit length to this. Since the string oscillations are transverse, the string element's vertical velocity is $\frac{\partial\psi}{\partial t}$, and hence the KE per unit length is $\frac{1}{2}\mu\left(\frac{\partial\psi}{\partial t}\right)^2$. The total energy per unit length is therefore

$$E = \frac{1}{2}T\left(\frac{\partial\psi}{\partial x}\right)^2 + \frac{1}{2}\mu\left(\frac{\partial\psi}{\partial t}\right)^2. \tag{4.50}$$

Now, the wave speed is $c = \sqrt{\frac{T}{\mu}}$; therefore, $T = \mu c^2$. Since we can write the wave equation as $\left(\frac{\partial \psi}{\partial x}\right)^2 = \frac{1}{c^2}\left(\frac{\partial \psi}{\partial t}\right)^2$,

$$E = \tfrac{1}{2}\mu c^2 \frac{1}{c^2}\left(\frac{\partial \psi}{\partial t}\right)^2 + \tfrac{1}{2}\mu\left(\frac{\partial \psi}{\partial t}\right)^2 \tag{4.51}$$

$$= \mu\left(\frac{\partial \psi}{\partial t}\right)^2. \tag{4.52}$$

The time-averaged total energy per unit length is therefore

$$\overline{E} = \frac{1}{\tau}\int_0^\tau \mu\left(\frac{\partial \psi}{\partial t}\right)^2 dt, \tag{4.53}$$

where τ is the period of a cosinusoidal wave which represents a Fourier component of the string vibration. Now, $\psi = A\cos(kx-\omega t-\phi)$ and therefore $\frac{\partial \psi}{\partial t} = A\omega\sin(kx - \omega t - \phi)$. This means

$$\overline{E} = \frac{1}{\tau}\mu A^2\omega^2 \int_0^\tau \sin^2(kx - \omega t - \phi)dt \tag{4.54}$$

$$= \frac{1}{\tau}\mu A^2\omega^2 \tfrac{1}{2}\int_0^\tau \left(1 - \cos\left(2kx - 2\omega t - 2\phi\right)\right)dt \tag{4.55}$$

$$= \frac{1}{\tau}\mu A^2\omega^2 \tfrac{1}{2}\left[t\right]_0^\tau \tag{4.56}$$

$$= \tfrac{1}{2}\mu A^2\omega^2. \tag{4.57}$$

After time dt, an extra cdt of string will be oscillating; therefore, the average input power to the wave is the energy of this string element $\overline{E}cdt$ divided by dt:

$$P = \frac{\overline{E}cdt}{dt}. \tag{4.58}$$

$$\therefore P = \tfrac{1}{2}\mu c A^2\omega^2. \tag{4.59}$$

Referring to our definition of wave impedance $Z = \left|\frac{\text{driving force}}{\text{wave velocity}}\right|$, in the case of our string, this is

$$Z = \frac{T}{c}. \tag{4.60}$$

Since $T = \mu c^2$,

$$Z = \mu c. \tag{4.61}$$

So, we have shown that, for a vibrating string, wave power is expressed by

$$P = \tfrac{1}{2}ZA^2\omega^2. \tag{4.62}$$

4.3 Wave Impedance, Reflection and Transmission

4.3.1 *Further notes on wave impedance*

In the previous section, we defined wave impedance $Z = \left|\frac{\text{driving force}}{\text{wave velocity}}\right|$. If we extend our wave scenario to consider a boundary between two regions with different wave speeds c_1 and c_2, we might anticipate the driving force to be continuous across the boundary, which implies that it takes the same value at either side. If this were not the case, an infinitesimal length of string (or infinitesimally small element of an elastic solid, or a puff of gas), which lies across the boundary, would be subject to an infinite acceleration. This is clearly unphysical. Therefore, the ratio of impedances for the regions either side of a boundary must be inversely related to the ratio of wave speeds:

$$\frac{Z_1}{Z_2} = \frac{c_2}{c_1}. \tag{4.63}$$

For waves on a tensioned string, wave speed $c = \sqrt{\frac{T}{\mu}}$, where string tension is T and mass per unit length is μ. If T is continuous across a boundary of two strings with different mass per unit lengths, then

$$\frac{c_2}{c_1} = \sqrt{\frac{\mu_1}{\mu_2}}. \tag{4.64}$$

In the previous section, we defined wave impedance for a tensioned string to be $Z = \mu c = \mu\sqrt{\frac{T}{\mu}} = \sqrt{\mu T}$.

$$\therefore \frac{Z_1}{Z_2} = \sqrt{\frac{\mu_1}{\mu_2}}, \tag{4.65}$$

which is consistent with our general assertion that $\frac{Z_1}{Z_2} = \frac{c_2}{c_1}$.

4.3.2 *Reflection and transmission coefficients*

We can use the concept of wave impedance to calculate the proportion of wave energy transmitted and reflected at an impedance change and, indeed, the amplitudes of reflected and transmitted waves and whether a wave is inverted or not. As illustrated in Fig. 4.9, let's consider an incident wave moving from left to right in the x direction of the form $\psi_I = A_I e^{i(k_1 x - \omega t)}$, which results in a reflected wave, $\psi_R = A_R e^{i(-k_1 x - \omega t)}$, and a transmitted wave, $\psi_T = A_T e^{i(k_2 x - \omega t)}$, at the impedance boundary Z_1 to Z_2. Note that we have assumed the *wave frequency to be the same either side of the boundary*,[13] so $\omega_1 = \omega_2 = \omega$. Let us also assume that the wave amplitude and its gradient are both continuous across the boundary. This means

$$\psi_I + \psi_R = \psi_T, \tag{4.66}$$

$$\frac{\partial \psi_I}{\partial x} + \frac{\partial \psi_R}{\partial x} = \frac{\partial \psi_T}{\partial x}. \tag{4.67}$$

---

[13]We will address this important point of frequency continuity in the section on *Snell's law of refraction*.

Reflection and transmission of waves on boundaries

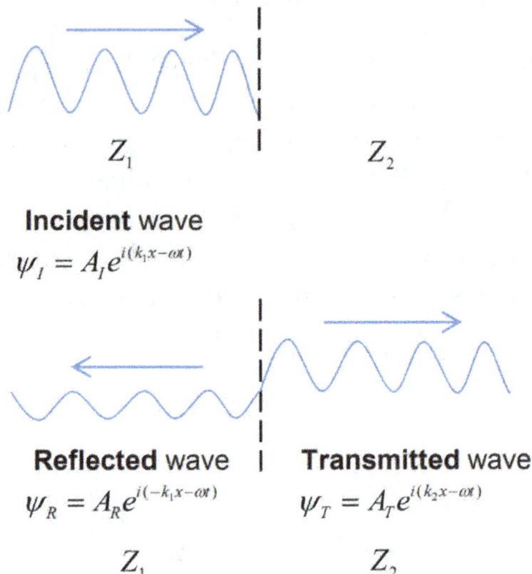

Fig. 4.9. Incident, reflected and transmitted waves upon a boundary of wave impedance. For acoustic waves $Z = \rho c$, i.e. density $\times$ wave speed; for waves on a string $Z = \frac{T}{c}$, i.e. string tension/wave speed. For DC electrical circuits, Z would equal electrical resistance (see Chapter 5). From the continuity of wave amplitude $\psi_I + \psi_R = \psi_T$ and its derivative $\frac{\partial \psi_I}{\partial x} + \frac{\partial \psi_R}{\partial x} = \frac{\partial \psi_T}{\partial x}$, one can derive reflection and transmission coefficients $A_R = \left(\frac{Z_1 - Z_2}{Z_1 + Z_2} \right) A_I$ and $A_T = \frac{2Z_1}{Z_1 + Z_2} A_I$. Power fractions are $\frac{P_R}{P_I} = \left| \frac{Z_1 - Z_2}{Z_1 + Z_2} \right|^2$ and $\frac{P_T}{P_I} = 1 - \left| \frac{Z_1 - Z_2}{Z_1 + Z_2} \right|^2$.

$$\therefore \quad A_I + A_R = A_T. \tag{4.68}$$

$$\therefore \quad k_1 A_I - k_1 A_R = k_2 A_T \tag{4.69}$$

$$\implies \frac{k_1}{k_2} A_I - \frac{k_1}{k_2} A_R = A_T. \tag{4.70}$$

Note that it may seem unintuitive that $\psi_I + \psi_R = \psi_T$ rather than $\psi_I = \psi_R + \psi_T$ since the incident wave *causes* the reflected and transmitted waves. However, at the snapshot in time that we consider, an incident wave and a reflected wave are on the left side of the boundary, whereas a transmitted wave is on the right. Now, from earlier arguments in this chapter,

$$\omega = c_{1,2} k_{1,2}, \tag{4.71}$$

$$\frac{Z_1}{Z_2} = \frac{c_2}{c_1}. \tag{4.72}$$

Hence, since $\omega = c_1 k_1 = c_2 k_2$,

$$\frac{k_1}{k_2} = \frac{c_2}{c_1} = \frac{Z_1}{Z_2}. \tag{4.73}$$

This means $\frac{k_1}{k_2} A_I - \frac{k_1}{k_2} A_R = A_T$ becomes

$$\frac{Z_1}{Z_2} A_I - \frac{Z_1}{Z_2} A_R = A_T. \tag{4.74}$$

Using $A_T = A_I + A_R$,

$$\frac{Z_1}{Z_2} A_I - \frac{Z_1}{Z_2} A_R = A_I + A_R \tag{4.75}$$

$$\left(\frac{Z_1}{Z_2} - 1 \right) A_I = \left(\frac{Z_1}{Z_2} + 1 \right) A_R. \tag{4.76}$$

$$\therefore A_R = \left(\frac{Z_1 - Z_2}{Z_1 + Z_2} \right) A_I. \tag{4.77}$$

This agrees with common experience. If you create a transverse wave with a slinky across the desk of a classroom with a pupil firmly holding one end, the reflected wave will clearly invert. In this case, $Z_2 \gg Z_1$ and hence $A_R \approx -A_I$, i.e. an inversion. The substitution of $A_R = \left(\frac{Z_1 - Z_2}{Z_1 + Z_2} \right) A_I$ into $A_T = A_I + A_R$ yields

$$A_T = \left(\frac{Z_1 - Z_2}{Z_1 + Z_2} + 1 \right) A_I \tag{4.78}$$

$$A_T = \left(\frac{Z_1 - Z_2 + Z_1 + Z_2}{Z_1 + Z_2} \right) A_I. \tag{4.79}$$

$$\therefore A_T = \frac{2Z_1}{Z_1 + Z_2} A_I, \tag{4.80}$$

which is also consistent with our slinky experience. If $Z_2 \gg Z_1$, then $A_T \ll A_I$. Now, since wave power is proportional to wave amplitude, i.e. $P = \frac{1}{2} Z A^2 \omega^2$, we can express the ratio of reflected wave power to incident wave power as

$$\frac{P_R}{P_I} = \left| \frac{Z_1 - Z_2}{Z_1 + Z_2} \right|^2, \tag{4.81}$$

and by conservation of energy, $P_T = P_I - P_R$.

$$\therefore \frac{P_T}{P_I} = 1 - \left| \frac{Z_1 - Z_2}{Z_1 + Z_2} \right|^2. \tag{4.82}$$

Here is an example calculation from ultrasound imaging. Let's consider a sound wave at the interface between soft tissue (region 1) and bone (region 2). By analogy to the case of waves on a tensioned string, where $Z = \mu c$, we define *acoustic wave impedance* as

$$Z = \rho c, \tag{4.83}$$

i.e. *wave impedance = density of material $\times$ wave speed.*

Typical values will be taken as follows: *Soft tissue:* $\rho_1 = 1{,}060$ kgm^{-3}, $c_1 = 1{,}540$ ms^{-1}. *Bone:* $\rho_2 = 1{,}900$ kgm^{-3}, $c_2 = 4{,}000$ ms^{-1}.

Hence,

$$\frac{A_R}{A_I} = \frac{Z_1 - Z_2}{Z_1 + Z_2} = \frac{1,060 \times 1,540 - 1,900 \times 4,000}{1,060 \times 1,540 + 1,900 \times 4,000} = -0.646, \tag{4.84}$$

$$\frac{A_T}{A_I} = \frac{2Z_1}{Z_1 + Z_2} = 0.354, \tag{4.85}$$

$$\frac{P_R}{P_I} = \left| \frac{Z_1 - Z_2}{Z_1 + Z_2} \right|^2 = 0.418, \tag{4.86}$$

$$\frac{P_T}{P_I} = 1 - \left| \frac{Z_1 - Z_2}{Z_1 + Z_2} \right|^2 = 0.582. \tag{4.87}$$

An example CRO trace of such a signal is presented in Fig. 4.10. The voltage associated with the incident ultrasound is 0.7 V, the ultrasonic pulse length (associated with a Gaussian shaped pulse) is $\tau = 0.2\,\mu s$, the CRO timebase is 1 μs and voltage divisions are 0.1 V. The trace illustrates reflections from (i) a soft tissue to bone boundary (first inverted pulse) and (ii) transmission through a 1.23 cm bone, then reflection at a bone to soft tissue boundary, *and then* transmission through a bone to soft tissue boundary back to the ultrasonic transducer. (This is the second pulse with peak at about 8.2 μs.) The Gaussian pulse model adopted for the time variation of intensity I is

$$I = I_{\max} e^{-\frac{1}{2}\left(\frac{t - t_{peak}}{\tau}\right)^2}. \tag{4.88}$$

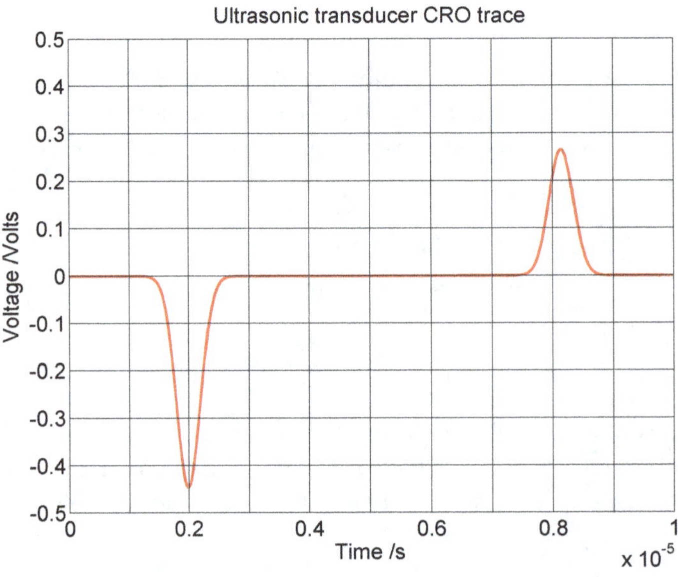

Fig. 4.10. The trace illustrates ultrasound reflections from (i) a soft tissue to bone boundary and (ii) transmission through a 1.23 cm bone then reflection at a bone to soft tissue boundary and finally transmission through a bone to soft tissue boundary. The voltage associated with the incident ultrasound is 0.7 V, the ultrasonic pulse length (associated with a gaussian shaped pulse) is 0.2 μs, the CRO timebase is 1 μs and voltage divisions are 0.1 V.

4.3.3 *Wave speeds and impedances in elastic solids and gases*

It can be shown (e.g. Shearer [33], p. 21) that waves in elastic solids will have a wave speed of the form

$$c = \sqrt{\frac{K}{\rho}}, \tag{4.89}$$

where K is the *coefficient of stiffness* (the 'elastic modulus') and ρ is the density of the solid. In a similar manner to waves on a tensioned string (where $Z = \mu c$), the wave impedance is

$$Z = \rho c, \tag{4.90}$$

which is the formula that we used for acoustic waves in the previous section. The difference is waves on a string involve a mass per unit length as their metric of density, whereas for elastic waves in solids, the density is the more conventional three-dimensional concept of mass per unit volume.

It can be shown[14] that pressure waves in ideal gases have wave speed given by

$$c = \sqrt{\frac{\gamma p}{\rho}}, \tag{4.91}$$

where p is pressure, ρ is gas density and $\gamma = \frac{c_p}{c_V}$ is the ratio of specific heat capacities at, respectively, constant pressure and constant volume. γ depends on the number of degrees of freedom of gas molecule motion and can vary between 1.3 and about 1.7 depending on the molecule and the temperature. (See the Heat Engine section of the previous chapter.)

Using the sea-level density of air as $\rho = 1.29$ kgm^{-3}, 1 atm of pressure $p = 1.01 \times 10^5$ Pa and $\gamma = 1.4$, we can calculate the speed of sound in air to be

$$c = \sqrt{\frac{1.4 \times 1.01 \times 10^5}{1.29}} \text{ ms}^{-1} \tag{4.92}$$

$$= 330 \text{ ms}^{-1}. \tag{4.93}$$

Recall from the earlier chapter on Thermodynamics that $\gamma = \frac{c_p}{c_V} = \frac{2}{\alpha} + 1$, where α is the number of degrees of freedom of molecular motion. Air is, to a sensible approximation, a diatomic gas given its volumetric composition[15] of N$_2$ (78%) and O$_2$ (21%). Therefore, we might expect (at least) two degrees of freedom from either stretching of the molecular bond or molecular rotation, in addition to the x, y, z translation of the molecular centre of mass. This means $\alpha = 5$ and $\gamma = \frac{2}{5} + 1 = 1.4$. One could imagine a smaller γ as air gets hotter, as more vibrational and rotational modes have a more significant Boltzmann factor $e^{-\frac{\text{mode energy}}{k_B T}}$, i.e. α increases.

---

[14] http://hyperphysics.phy-astr.gsu.edu / hbase/Sound/souspe3.html#:~:text=Sound%20Speed%20in%20Gases&text = For%20the%20specific%20example%20of,air%20molecules%20of%2029%20 amu. (Accessed 7 June 2023).

[15] https://www.engineeringtoolbox.com/air-composition-d_212.html (Accessed 7 June 23).

Sound waves in air travel at about 340 m/s. Sound waves in water travel about 1,500 m/s, while sound waves in rocks can travel at speeds between 5,000 m/s and 10,000 m/s. By comparison, the speed of light in a vacuum is

$$c = 2.998 \times 10^8 \, \text{ms}^{-1}. \tag{4.94}$$

4.4 Ray Optics: Refraction, Reflection, Lenses and Mirrors

Understanding the nature of light has been a motivator for the development of much of physics. Indeed, visible light (and its larger and smaller wavelength variants) is, in most cases the only viable means of gaining information about the Cosmos beyond our closest planetary neighbours.[16] Light is an EM wave, a disturbance in the electric and magnetic (vector) fields sourced from, respectively, charges and moving charges. A charge q in an electric field $\mathbf{E}$ will feel a force $q\mathbf{E}$. An additional force $q\mathbf{v} \times \mathbf{B}$ will be experienced if the charge is moving at velocity $\mathbf{v}$ in a magnetic field of flux density $\mathbf{B}$. Unlike a sound or water wave, EM waves can travel through empty space. No 'medium' is required—*it itself moves.* In a vacuum, light travels at a speed of $c = 2.998 \times 10^8$ ms^{-1}, which appears to be the speed limit for the conveyance of information in the Universe.[17] Like other forms of waves, light will travel in straight lines unless the wave speed changes. If light passes through water or glass, the presence of charges in the atomic structure of these substances will impede its progress, making the path more tortuous and longer. The effective speed of light c' will therefore reduce by a factor of n, i.e. $c' = \frac{c}{n}$, where n is the *refractive index*. For a vacuum, $n = 1$; for air, $n \approx 1.00$; for water, $n \approx 1.34$; and for glass, $n \approx 1.50$.

Light can travel through any path between two points, but the most *probable*[18] is the path which *takes the least time.* This is known as *Fermat's principle.*[19] The result is that light *refracts* (bends) at a boundary of wave speeds (e.g. between glass and air), or it *bends continuously if the wave speed also varies.* Fermat's principle of least time also explains the *law of reflection,* i.e. the *angle of incidence from the normal to a surface = the angle of reflection from this normal.*

4.4.1 *The law of reflection*

Consider a beam of light reflecting off a surface. What path does the light take? Feynman[20] proposed that it takes *all possible paths,* although the probability of

---

[16]Although, in recent years, the possibilities of *gravitational wave astronomy* are being realised too, which will add to our field-disturbance remote sensing toolbox. This enables us to gain a quantitative understanding of structures far beyond the viable range of direct physical contact.

[17]The idea that the speed of light is a fundamental limit is one of the axiomatic principles of the theory of *special relativity.* This will be discussed in the next volume of *Science by Simulation.*

[18]This idea is explored in the theory of *quantum electrodynamics* (QED) [7].

[19]Pierre de Fermat (1607–1665).

[20]Richard Feynman (1918–1988).

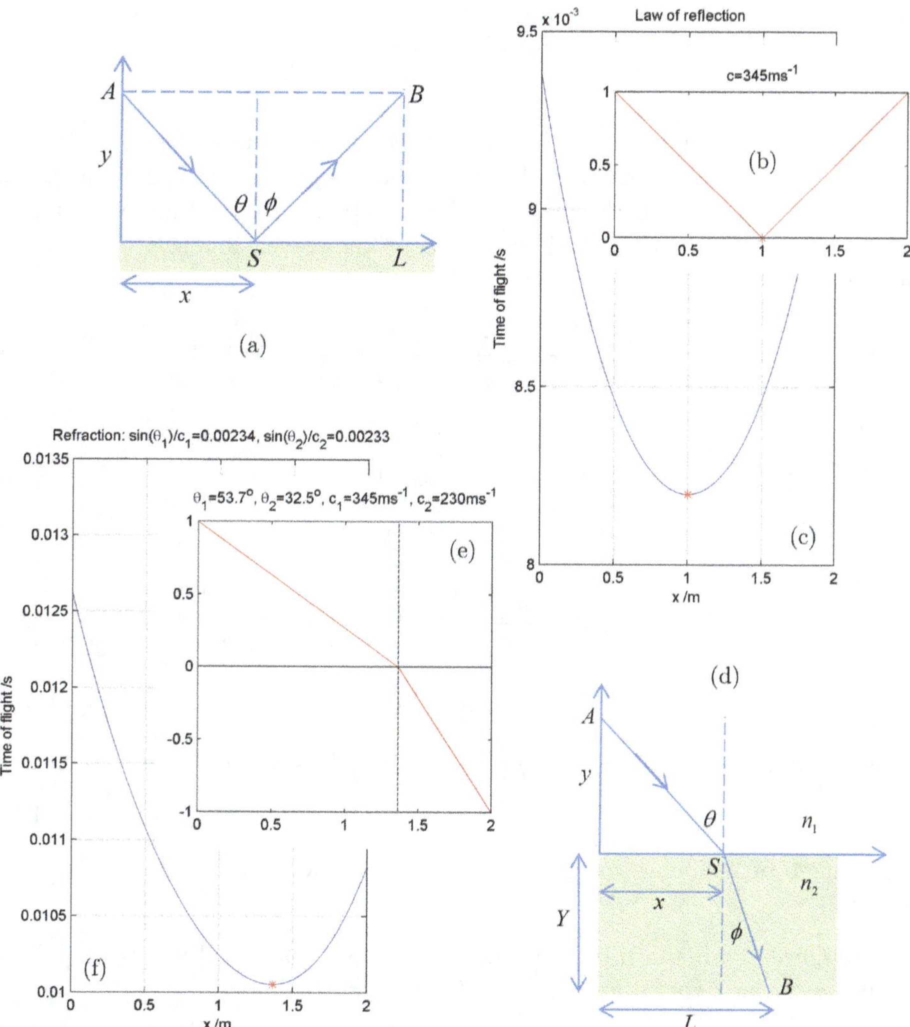

Fig. 4.11. The law of reflection, illustrated in (a), (b) and (c) can be derived from *Fermat's principle*. This states that the ray path from A to B via a reflection from a surface at S will take the *least time*. The time A to S to B is plotted vs. distance x and has a minima at $x = \frac{1}{2}L$, where L is the horizontal distance between A and B. This means the angle of incidence θ from the normal at S equals the angle of reflection ϕ. (d), (e) and (f) apply Fermat's principle to a path ASB, where AS is within a medium with wave speed c_1 and SB is within a medium with wave speed c_2. The minimum travel time from A to B via S is when $\frac{\sin\theta}{c_1} = \frac{\sin\phi}{c_2}$, where θ is the angle of incidence to the normal at S and ϕ is the angle of refraction. The relationship between θ and ϕ is known as *Snell's law of refraction*.

certain paths is more likely. Fermat's principle is more prescriptive—it states the path is the one which takes the *least time*. Let the beam of light traverse a path ASB, reflecting off the surface at S, as shown in Fig. 4.11(a). The refractive index n above the surface is assumed to be constant, i.e. the medium is isotropic (which means light travels at the same speed in any direction). The travel time for path

ASB is

$$t = \frac{\sqrt{x^2 + y^2}}{c/n} + \frac{\sqrt{(L - x)^2 + y^2}}{c/n}. \tag{4.95}$$

Since points A and B are at the same height y from the surface, variations in travel time can be characterised by changes in a single parameter, x, and an example graph of $t(x)$ is plotted in Fig. 4.11(c). We can, therefore, determine the path of least travel time by finding a stationary value of the travel time as a function of x. The gradient of $t(x)$ is

$$\frac{\partial t}{\partial x} = \frac{n}{c} \left(\frac{\frac{1}{2}2x}{\sqrt{x^2 + y^2}} + \frac{\frac{1}{2}2(L - x)(-1)}{\sqrt{(L - x)^2 + y^2}} \right). \tag{4.96}$$

From geometry,

$$x = y \tan\theta, \tag{4.97}$$

$$L - x = y \tan\phi \tag{4.98}$$

$$\Rightarrow \frac{\partial t}{\partial x} = \frac{n}{c} \left(\frac{y \tan\theta}{y\sqrt{\tan^2\theta + 1}} - \frac{y \tan\phi}{y\sqrt{\tan^2\phi + 1}} \right). \tag{4.99}$$

$$\therefore \frac{\partial t}{\partial x} = \frac{n}{c} (\sin\theta - \sin\phi). \tag{4.100}$$

The travel time is minimised when $\frac{\partial t}{\partial x} = 0$, which is when

$$\theta = \phi, \tag{4.101}$$

i.e. the angle of incidence θ equals the angle of reflection ϕ. This is the *law of reflection*.

Snow Queen: "Mirror, mirror on the wall, what is the simplest law of all?"
Mirror: "I think that is pretty self-evident your Highness."

4.4.2   *Snell's law of refraction*

Now, consider a similar ray path situation ASB, but with a ray *refracted* due to its passing through an interface between media of different refractive indices n_1 and n_2 (i.e. where the speed of light changes discontinuously). With reference to Fig. 4.11(d), the travel time ASB is now

$$t = \frac{\sqrt{x^2 + y^2}}{c/n_1} + \frac{\sqrt{(L - x)^2 + Y^2}}{c/n_2}. \tag{4.102}$$

$$\therefore \frac{\partial t}{\partial x} = \frac{1}{c} \left(\frac{\frac{1}{2}2xn_1}{\sqrt{x^2 + y^2}} + \frac{\frac{1}{2}2(L - x)(-1)n_2}{\sqrt{(L - x)^2 + Y^2}} \right). \tag{4.103}$$

Again, noting the geometrical relationships

$$x = y \tan\theta, \tag{4.104}$$

$$L - x = Y \tan \phi \tag{4.105}$$

$$\implies \frac{\partial t}{\partial x} = \frac{1}{c} \left(\frac{y n_1 \tan \theta}{y \sqrt{\tan^2 \theta + 1}} - \frac{Y n_2 \tan \phi}{Y \sqrt{\tan^2 \phi + 1}} \right). \tag{4.106}$$

$$\therefore \; \frac{\partial t}{\partial x} = \frac{1}{c} \left(n_1 \sin \theta - n_2 \sin \phi \right). \tag{4.107}$$

The travel time is minimised when $\frac{\partial t}{\partial x} = 0$, which is when

$$n_1 \sin \theta = n_2 \sin \phi. \tag{4.108}$$

This is *Snell's law of refraction*. Note that we have asserted the refracted ray is that of light, i.e. with speeds $c_{1,2} = \frac{c}{n_{1,2}}$. Since the derivation for Snell's law applies to all disturbances, we can write a more general expression (i.e. that we could apply to acoustic waves for example):

$$\frac{\sin \theta}{c_1} = \frac{\sin \phi}{c_2}, \tag{4.109}$$

i.e. '*the sine of angle from normal to surface/wave speed is a constant on both sides of a wave-speed boundary*'.

The number of waves passing per second (i.e. the frequency f) must be a *conserved quantity* on both sides of the boundary, otherwise there must be some form of energy, and also *information*, discontinuity. Since the speed of waves is the frequency multiplied by the wavelength,

$$f = \frac{c_1}{\lambda_1} = \frac{c_2}{\lambda_2}, \tag{4.110}$$

which means

$$\frac{\lambda_2}{\lambda_1} = \frac{c_2}{c_1}, \tag{4.111}$$

and for the special case of light with associated refractive indices n_1 and n_2,

$$\frac{\lambda_2}{\lambda_1} = \frac{n_1}{n_2}. \tag{4.112}$$

Alternatively, we can simply state that '*wavelength is proportional to wave speed*'. So, a 500 nm light ray in air (where $n_1 \approx 1$) will become about 333 nm in glass with a refractive index of $n_2 = 1.5$.

An interesting special case is when $c_2 > c_1$. Snell's law implies that, in this situation, $\phi > \theta$, i.e. 'rays bend away from the normal'. The maximum value of $\phi = 90°$, i.e. $\sin \phi = 1$. At this *critical angle* of incidence θ_c,

$$\frac{\sin \phi}{c_2} = \frac{1}{c_2} = \frac{\sin \theta_c}{c_1}. \tag{4.113}$$

$$\therefore \theta_c = \sin^{-1} \left(\frac{c_1}{c_2} \right). \tag{4.114}$$

Rays will always be reflected and refracted at a boundary when $\theta < \theta_c$. However, for $\theta \geq \theta_c$, *only reflection occurs*. The lack of transmitted power through a boundary for $\theta \geq \theta_c$ is the key principle why optical fibres are an effective mechanism for long-distance communication. Light pulses can bounce along the interior of a fibre for hundreds of kilometres with minimal attenuation. For a glass–air interface, where $c_1 = \frac{3.00 \times 10^8 \text{ ms}^{-1}}{1.5}$ and $c_2 = \frac{3.00 \times 10^8 \text{ ms}^{-1}}{1.0}$, the critical angle is $\theta_c = \sin^{-1}\left(\frac{1}{1.5}\right) = 41.8°$.

At a boundary, the total power of reflected and transmitted rays must equate to the incident power. The exact balance depends on the angle of incidence, the ratio of refractive indices and the polarisation of the light. (See Section 5.12 on the *Fresnel equations*). Recall that (as seen earlier in this chapter) wave power is proportional to $A^2 f^2$, where A is the amplitude of the EM wave.

4.4.3   *Geometric optics*

Light might be a thin laser beam, but it might also be a broad illumination from an extended source, such as the Sun. In both extremes, we can think of the direction that light propagates. We call this the *wavevector*, or more simply, a *ray*. The branch of physics known as *geometric optics* is, in essence, the drawing of diagrams that represent the directions light rays take from source to observer. When multiple rays can be drawn, the intersection of these different paths will predict where an image[21] of a light source will form. The consideration of rays (rather than their wave properties) from extended light sources explain the phenomenon of eclipses, as well as shadows more generally, and indeed a huge range of optical effects. Hecht [17] provides a veritable cornucopia[22] of beautifully illustrated examples. Geometric optics is a powerful and, appropriately for the subject matter, highly visual approach to scientific problem-solving. In modelling the behaviour of rays in simple systems such as lenses and mirrors, our 'ray rules' methodology typically involves considering the intersection of an undeviated ray and one which starts horizontally and is refracted or reflected by a known amount. We'll explore this for a thin converging lens and for concave and convex spherical mirrors.

4.4.4   *A thin lens*

A *lens* is an optical device which focuses light via refraction. An idealised thin lens (i.e. where the physical width has a negligible effect on the ray paths) will converge rays parallel to the lens surface normal (i.e. the 'optic axis') through a single point. The distance of this point from the lens is called the *focus*. Lenses are typically ground into spherical shapes, so a ray which intersects the thin lens on its optical axis will not be refracted; this is because the angle of incidence from the normal to the lens surface will be zero since all radials of a sphere are also normals. Our two 'ray rules' allow one to determine *where* rays emanating from a source ('the object') will converge ('the image').

---

[21]Or, indeed, *virtual image*, the *apparent source* of diverging light rays.

[22]I would like to think a *cornucopia* is particularly appropriate for edge diffraction(!) See Section 4.5.5.

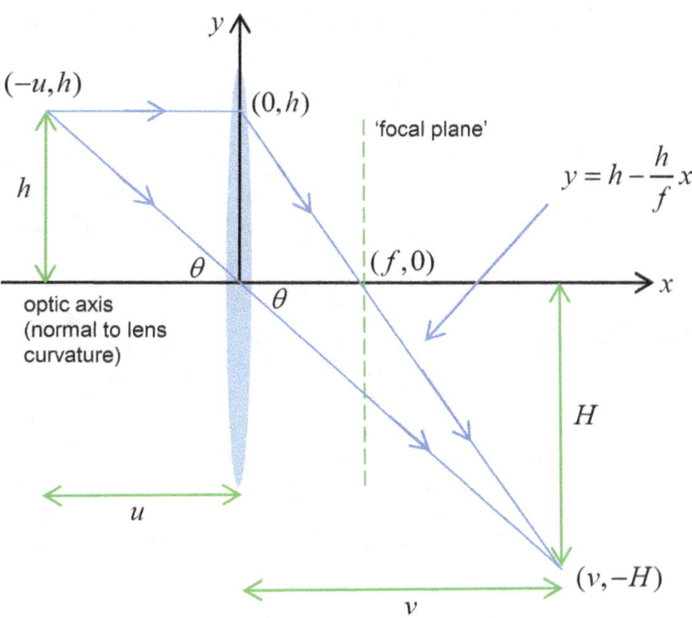

Fig. 4.12. Geometry associated with a model of an idealised thin lens. The assumption is that horizontal light rays (i.e. normal to the plane of the lens) will be refracted in such a way that they pass through a focal point $(f, 0)$, whereas rays incident upon the centre of the lens will pass through undeviated. These assumptions can be used to derive the thin lens formula $\frac{1}{f} = \frac{1}{v} + \frac{1}{u}$, which relates the focal length of the lens to the horizontal extent of object (i.e. the source of light) at $(-u, h)$ and image (where light converges) at $(v, -H)$. The magnification formula $H = \frac{v}{u} h = \frac{f}{u-f} h$ relates image height H to object height h.

Consider the centre of a thin lens to be at $(0, 0)$, i.e. the origin of a Cartesian coordinate system. The ensuing calculation refers to Fig. 4.12. The object location is $(-u, h)$. The image location $(v, -H)$ is the intersection of two known ray paths from the object through the lens. Let's write down equations for the straight-line sections that form part of the two ray paths.

Path I: through the centre of the lens:

$$y = -\frac{h}{u} x. \tag{4.115}$$

Path II: perpendicular to lens and then through focus at $(f, 0)$:

$$y = h - \frac{h}{f} x. \tag{4.116}$$

These ray paths intersect at $(v, -H)$. Substituting for $y = -\frac{h}{u} x$ into $y = h - \frac{h}{f} x$,

$$-\frac{h}{u} v = h - \frac{h}{f} v. \tag{4.117}$$

$$\therefore \frac{1}{f} = \frac{1}{v} + \frac{1}{u}. \tag{4.118}$$

This is known as the *thin lens equation*. Rearranging to make v the subject,

$$v = \left(\frac{1}{f} - \frac{1}{u}\right)^{-1}.$$
(4.119)

$$\therefore v = \frac{f}{u-f} u.$$
(4.120)

Now, since $y = -\frac{h}{u}x$ passes through $(v, -H)$, $-H = -\frac{h}{u}v$ and therefore

$$H = \frac{f}{u-f} h.$$
(4.121)

This implies a *vertical magnification factor*

$$M = \frac{H}{h} = \frac{f}{u-f}.$$
(4.122)

Note that we could apply these transformations to an *array* of object coordinates $\{x_i, y_i\}$ to result in image coordinates $\{X_i, Y_i\}$:

$$X_i = \frac{f}{x_i - f} x_i,$$
(4.123)

$$Y_i = -\frac{y_i}{x_i} X_i = -\frac{f}{x_i - f} y_i.$$
(4.124)

This is the basis of the MATLAB simulation in Fig. 4.13. The ray intersection example is illustrated using the path from the red spot on the object to the green spot in the image. However, a bitmap image of $\{x_i, y_i\}$ pixel locations are also transformed. This gives a clearer picture of both inversion and magnification, as well as distortion resulting from magnification depending on the distance x_i from the lens.

An alternative simulation is required when $x_i < f$. In this case, the object is behind the lens (positive x coordinate in Fig. 4.14) and an observer (negative x coordinate in Fig. 4.14) experiences diverging rays emanating from the biconvex lens. The *apparent source* of these rays in the location of the *virtual image*. As illustrated in Fig. 4.14, the virtual image is upright and enlarged, which explains how this system of object and lens can be used as a magnifying glass.

4.4.5　*A concave spherical mirror*

This is a different type of scenario involving *reflection* in a *mirror* rather than refraction via a lens system. As indicated by the curious 'flying cow' demonstration[23] at Winchester College, which comprises a toy cow/dinosaur suspended upside down in front of a large spherical mirror, a *real* and inverted image is manifested between the object and the mirror, giving the appearance of the bovine or lizard floating in space like a strange spectral form. Both the cow and dinosaur variants are illustrated in Fig. 4.15.

---

[23]The *Flying Cow* at Winchester College mysteriously disappeared to pastures unknown after many years of service, being dangled upside down (or udder-side up!) in front of a *concave spherical mirror*. We now have a brightly coloured flying dinosaur instead.

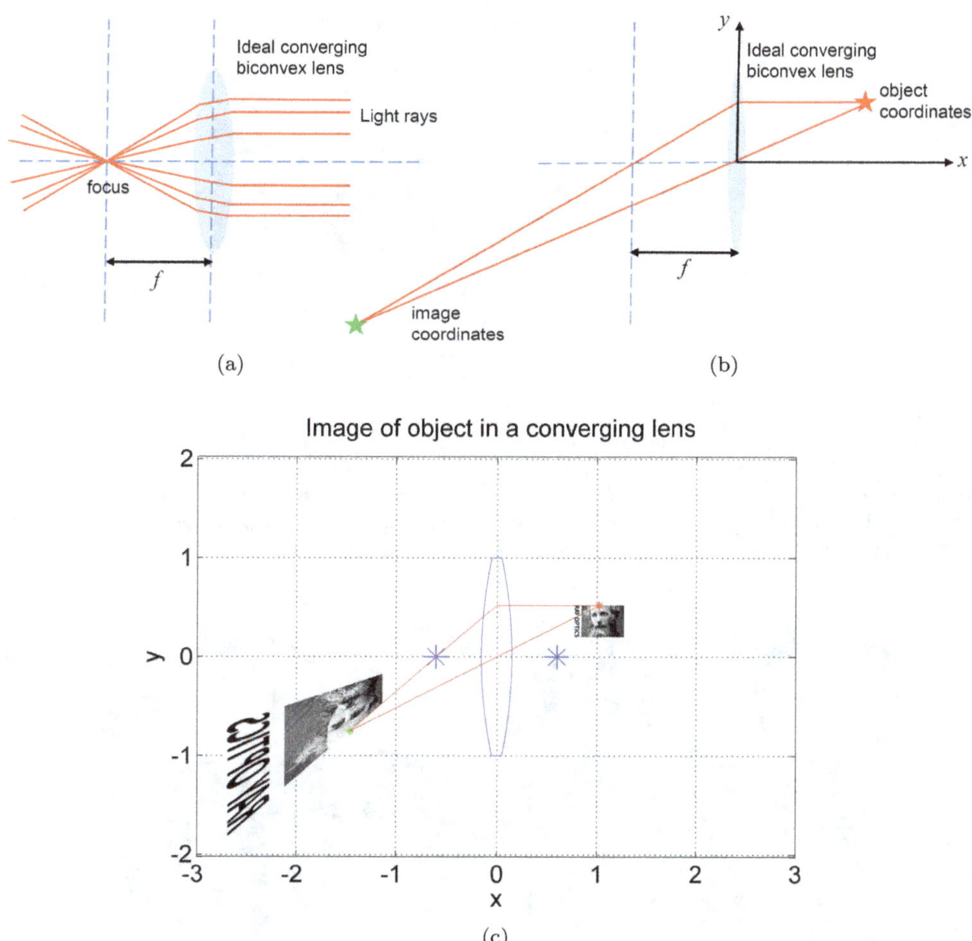

Fig. 4.13. An ideal, 'thin' biconvex lens has two rules for rays passing through it. 1: Rays through the centre of the lens *pass straight through*, and there is no angle of refraction. 2: Rays that are parallel to the lens normal are refracted such that they *all pass though the focus of the lens* at a perpendicular distance f from the lens. These two rules can be combined to determine the real image position (X_i, Y_i) of rays produced at coordinate (x_i, y_i) relative to the centre of the lens. In this simulation, $x_i > f$ and $X_i = \frac{f}{x_i - f} x_i$ and $Y_i = \frac{f}{x_i - f} y_i$, i.e. the lens is used as part of a projection system and creates a *real*, inverted image, rather than as a magnifying glass (which results in a virtual image). This *geometric transformation* concept of the physics of lenses is described in (a) and (b) and modelled in (c). Note the *inversion* and *magnification* of the image compared to the object, i.e. the ray source.

As in the previous examples, we can determine a transformation of object into image coordinates by finding the intersection of two rays that we can trace with mathematical certainty via the expression of Cartesian line segments $y(x)$. In this case (see Fig. 4.16), the first is the ray which passes through the centre of the circle and reflects straight back (since all circle radials are normal to the circle) and the second being a horizontal ray which is reflected off the circular surface at angle θ to the normal.

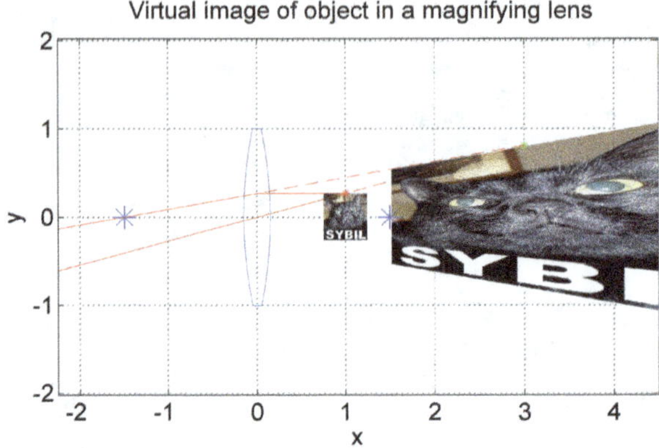

Fig. 4.14. When an object (the source of light) is placed within the focal range f of an idealised thin biconvex lens, an observer at the opposite side of the lens will experience diverging rays. The apparent source of these rays is the *virtual* image, i.e. it isn't actually a source of light, but merely what the observer interprets based on an assumption of straight-line ray propagation. A lens used in this configuration can be used as a magnifying glass since the virtual image is enlarged and also not inverted.

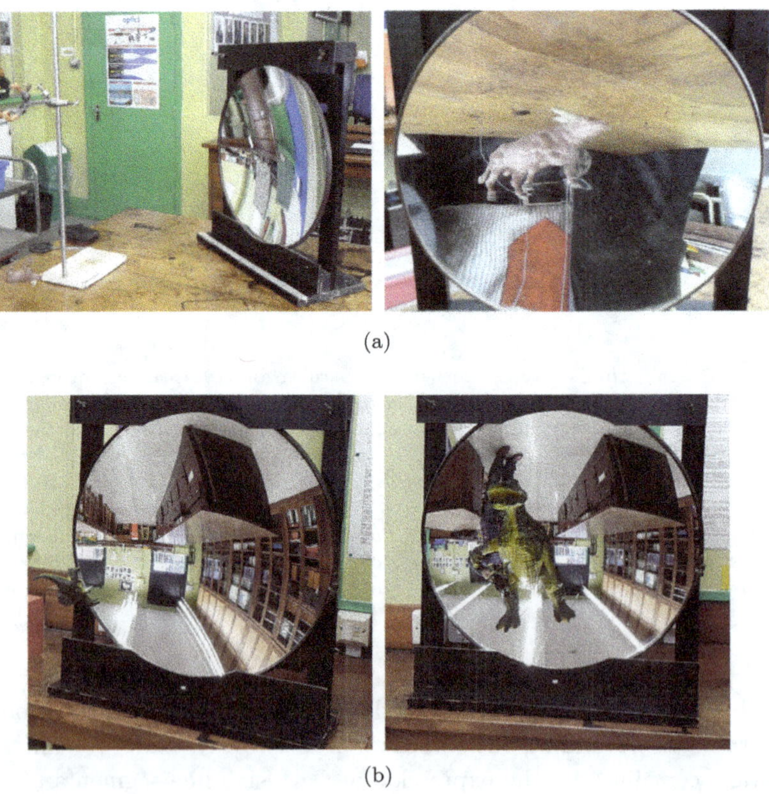

(a)

(b)

Fig. 4.15. A toy cow (a) or dinosaur (b) suspended upside down in front of a large spherical mirror can result in a *real* and inverted image between the object and the mirror, giving the appearance of the bovine or lizard floating in space like a strange spectral form. The effect is quite startling, especially if the toy is made to gently oscillate towards and away from the mirror, while the observer remains still.

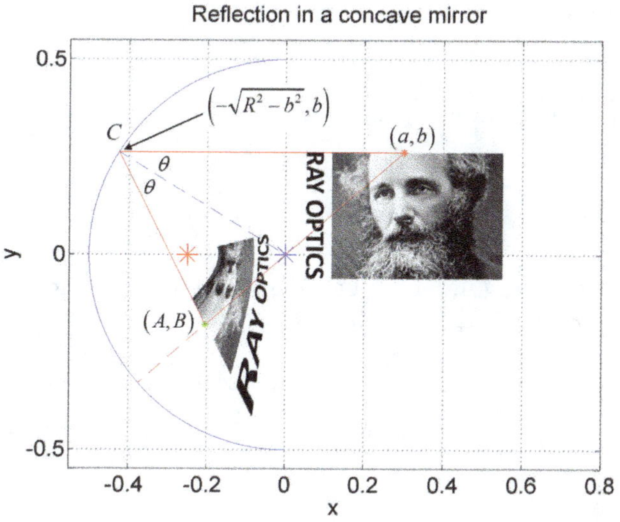

Fig. 4.16. Ray diagram illustrating the curious optical phenomenon of an inverted *real* image in a concave spherical mirror. The image will appear to float in mid-air! The geometric transformation of object coordinates (a,b) (the small red star) into image coordinates (A,B) (the small green star) is determined by finding the intersection of undeviated reflections along a normal to the mirror (i.e. the ray path which passes through the spherical mirror centre denoted by the blue star) and a reflection of horizontal rays. The latter obeys the condition angle of incidence $\theta =$ angle of reflection θ about the mirror normal at the point of reflection. Note that the 'half-radius' point (the large red star) is an *approximate* focus, in the sense that reflected rays (almost) pass through it.

Assume that the object (i.e. the light source) coordinates are (a,b) and the image coordinates are at (A,B).

Ray I: There-and-back reflection through the centre of a curved mirror, i.e. along a radial path. The angle of reflection from the mirror surface normal is $\theta = 0$:

$$y = \frac{b}{a}x. \tag{4.125}$$

Ray II: Horizontal ray from object. The angle of reflection from the mirror normal is θ:

$$y = -mx + c. \tag{4.126}$$

Let's use geometry and the law of reflection to find m and c. If the spherical mirror has radius R, the mirror surface coordinates (x,y) satisfy

$$R^2 = x^2 + y^2. \tag{4.127}$$

Therefore, the (x,y) coordinates of reflection point C are

$$\left(-\sqrt{R^2 - b^2}, b\right), \tag{4.128}$$

which means the angle of reflection θ from the normal line, which is a radial that connects C with $(0,0)$ is

$$\theta = \tan^{-1}\left(\frac{b}{\sqrt{R^2 - b^2}}\right). \tag{4.129}$$

The reflected ray has gradient

$$m = \tan 2\theta. \tag{4.130}$$

Using $\left(-\sqrt{R^2 - b^2}, b\right)$ in $y = -mx + c$,

$$b = -m\left(-\sqrt{R^2 - b^2}\right) + c. \tag{4.131}$$

$$\therefore c = b - m\sqrt{R^2 - b^2}.$$

The image coordinates (A, B) can be determined from the intersection of $y = \frac{b}{a}x$ and $y = -mx + c$, where $x = A$ and $y = B$:

$$\frac{b}{a}A = -mA + c \tag{4.132}$$

$$A\left(\frac{b}{a} + m\right) = c. \tag{4.133}$$

$$\therefore A = \frac{c}{\frac{b}{a} + m}.$$

Hence, since $B = \frac{b}{a}A$,

$$B = \frac{b}{a}\frac{c}{\frac{b}{a} + m}. \tag{4.134}$$

In summary, the transformation of object coordinates $\{x_i, y_i\}$ into image coordinates $\{X_i, Y_i\}$ by a concave spherical mirror is given by (as illustrated in Fig. 4.16)

$$X_i = -\frac{m_i\sqrt{R^2 - y_i^2} - y_i}{\frac{y_i}{x_i} + m_i}, \tag{4.135}$$

$$Y_i = -\frac{y_i}{x_i}\frac{m_i\sqrt{R^2 - y_i^2} - y_i}{\frac{y_i}{x_i} + m_i}, \tag{4.136}$$

$$m_i = \tan 2\theta_i, \tag{4.137}$$

$$\theta_i = \tan^{-1}\left(\frac{y_i}{\sqrt{R^2 - x_i^2}}\right). \tag{4.138}$$

4.4.6   *A convex cylindrical mirror*

For our final geometric optics curiosity, let us consider a *convex* cylindrical mirror. As illustrated in Fig. 4.17, light rays from an object at (x, y) will appear to originate from a virtual image location (X, Y). The virtual image is a smaller, slightly distorted and upright version of the object.

The line between (X, Y) and where rays from (x, y) pass the x-axis at $(k, 0)$ must pass through the reflection point $(R\cos\alpha, R\sin\alpha)$ on the surface of the cylinder, which has radius R. This is illustrated in Fig. 4.18. The angle α is defined by

$$\tan(2\alpha) = \frac{y}{x} = \frac{Y}{X}. \tag{4.139}$$

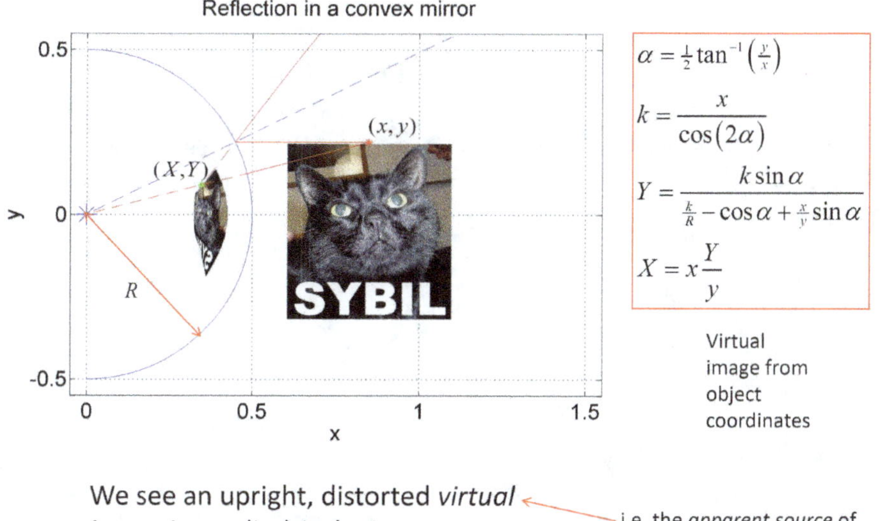

Fig. 4.17. Geometric optics construction to determine the virtual image (X, Y) (green star) of a object (x, y) (red star) resulting from reflection in a cylindrical convex mirror of radius R. The virtual image location is found from the intersection of a ray from object to circle centre and is the apparent source of a horizontal ray from object to mirror. The $(x, y) \to (X, Y)$ transformation is applied to every pixel location of the cat (object), resulting in the distorted, but upright, virtual image inside the circle.

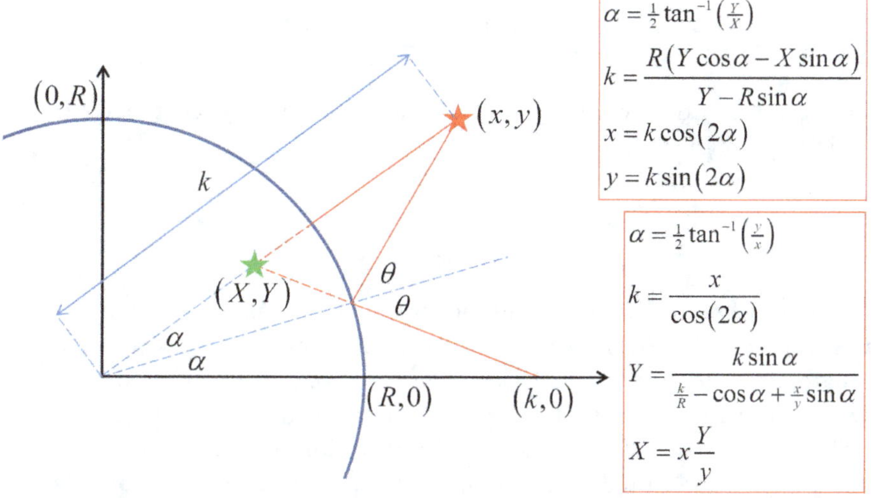

Fig. 4.18. Alternative geometric optics construction for reflection in a convex cylindrical mirror that is used to determine both the object-to-virtual-image transformation $(x, y) \to (X, Y)$ *and the* virtual-image-to-object transformation $(X, Y) \to (x, y)$. In this case, the observer is at $(k, 0)$. The *apparent path* of the rays are from (X, Y) to $(k, 0)$, whereas the actual ray path is from (x, y) to $(k, 0)$.

Hence,

$$\alpha = \tfrac{1}{2}\tan^{-1}\left(\frac{y}{x}\right) = \tfrac{1}{2}\tan^{-1}\left(\frac{Y}{X}\right). \tag{4.140}$$

If the line between (X, Y), the reflection point $(R\cos\alpha, R\sin\alpha)$ and $(k, 0)$ has equation $y = -mx + c$, the gradient $-m$ can be evaluated using the known points (X, Y) and $(R\cos\alpha, R\sin\alpha)$:

$$m = \frac{Y - R\sin\alpha}{R\cos\alpha - X}. \tag{4.141}$$

Hence, using (X, Y),

$$c = Y + mX \tag{4.142}$$

$$c = Y + \frac{Y - R\sin\alpha}{R\cos\alpha - X}X \tag{4.143}$$

$$c = \frac{RY\cos\alpha - XY + XY - RX\sin\alpha}{R\cos\alpha - X} \tag{4.144}$$

$$c = \frac{R\left(Y\cos\alpha - X\sin\alpha\right)}{R\cos\alpha - X}. \tag{4.145}$$

Substituting for m and c into $y = -mx + c$,

$$y = \frac{R\left(Y\cos\alpha - X\sin\alpha\right) - \left(Y - R\sin\alpha\right)x}{R\cos\alpha - X}. \tag{4.146}$$

Now, when $y = 0$, $x = k$.

$$\therefore k = \frac{R\left(Y\cos\alpha - X\sin\alpha\right)}{Y - R\sin\alpha}. \tag{4.147}$$

The object coordinates (x, y) can therefore be found from the virtual image coordinates (X, Y) using

$$y = k\sin(2\alpha), \tag{4.148}$$

$$x = k\cos(2\alpha). \tag{4.149}$$

This may at first be a strange scenario, but in fact it has significant importance in an artistic sense! The transformation of a *correctly proportioned* virtual image with coordinates (X_i, Y_i) to its *anamorphic image* (x_i, y_i) results in an object position that would, in turn, result in a correctly proportioned virtual image when viewed in a cylindrical mirror. If the object and virtual image heights are deemed to be the same, a 3D printer (and then perhaps a metal casting process) could be used to create a distorted sculpture that, when viewed in a cylindrical mirror, would look correctly proportioned. This idea has been used to spectacular effect by artists such as Jonty Hurwitz.[24]

---

[24] Anamorphic artist Jonty Hurwitz (https://jontyhurwitz.com/).

The object-to-image transformation can also be derived. In this case, the inputs are x, y, R, and X, Y are desired. From Fig. 4.18,

$$\frac{X}{Y} = \frac{x}{y} \tag{4.150}$$

since the line $y = x \tan(2\alpha)$ passes through both the object and image coordinates. Using $k = \frac{R(Y \cos \alpha - X \sin \alpha)}{Y - R \sin \alpha}$, $X = \frac{Yx}{y}$, $\alpha = \frac{1}{2} \tan^{-1}\left(\frac{y}{x}\right)$ and $k = \frac{x}{\cos(2\alpha)}$,

$$k = \frac{R\left(Y \cos \alpha - \frac{Yx}{y} \sin \alpha\right)}{Y - R \sin \alpha}. \tag{4.151}$$

$$\therefore Y \cos \alpha - \frac{Yx}{y} \sin \alpha = \frac{k}{R}(Y - R \sin \alpha). \tag{4.152}$$

$$\therefore k \sin \alpha = Y\left(\frac{x}{y} \sin \alpha - \cos \alpha + \frac{k}{R}\right). \tag{4.153}$$

Hence,

$$Y = \frac{k \sin \alpha}{\frac{x}{y} \sin \alpha - \cos \alpha + \frac{k}{R}}, \tag{4.154}$$

$$X = \frac{Yx}{y}. \tag{4.155}$$

4.4.7 *Anamorphic image in a cylindrical mirror*

A simplified 2D transformation can be used to map the pixel locations of a rectangular bitmap image onto an *anamorphic projection*. This is not truly three-dimensional as in Hurwitz's sculptures; however, the reflections of an anamorphic projection in a carefully placed cylindrical mirror do rather look like they are sourced from a virtual image which looks convincingly three-dimensional from a reasonable range of viewing angles. It is rather simpler in computation too than the exact geometric optics convex spherical mirror transformation in the previous section. The concept is as follows:

(1) Define a unit circle $x^2 + y^2 = 1$ centred on the origin of a coordinate system. Import a bitmap image into a computer program and scale this, such that (i) the image is centered at $(0,0)$ and (ii) the image pixel locations are also scaled so that each corner intersects with the unit circle. If pixel number is the original (x, y) coordinate of the imported image, then scale by $\sqrt{W^2 + H^2}$, where W is the pixel width of the image and H is the pixel height.

(2) Define a circular sector of angular width $\Delta\theta$. In Fig. 4.19, $\Delta\theta = 160°$. Let the radii of this sector run from $r = 1$ to $r = \rho$, where ρ is the *radial stretch factor*. In Fig. 4.19, $\rho = 3.00$. Finally, shift the origin of this circular sector such that the origin is at the bottom of the bitmap image, which has now been fitted into the unit circle. In Fig. 4.19, this point is represented by a red star at coordinate

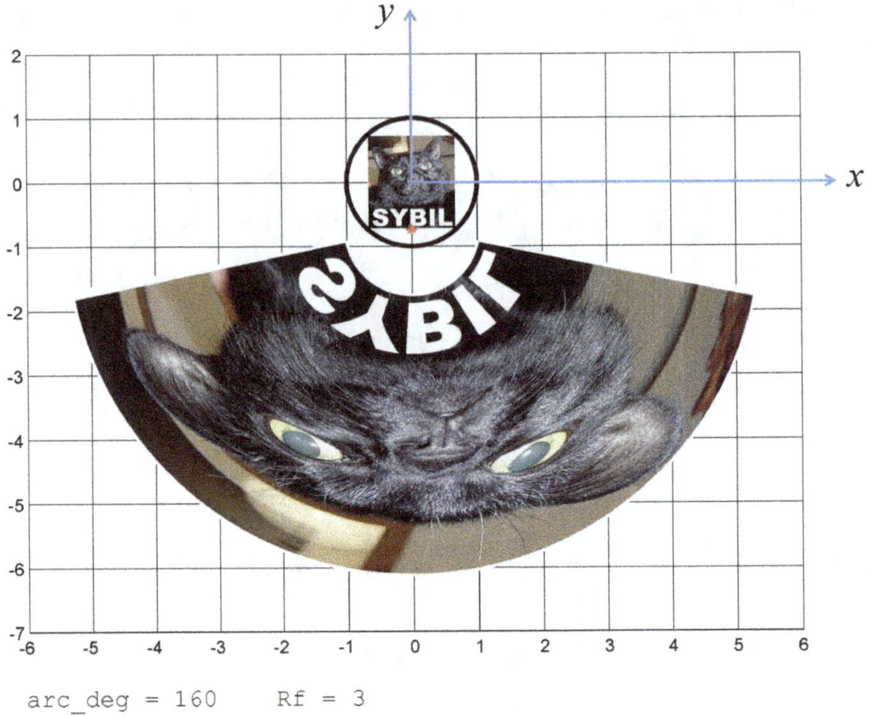

Fig. 4.19. Anamorphic projection of the cat image, which is imported into a MATLAB computer program and fitted into a unit circle. A circular sector is defined between radii $r = 1$ and $r = \rho = 3.0$, between polar angles $-\frac{1}{2}\Delta\theta \leq \theta \leq \frac{1}{2}\Delta\theta$ (with $\Delta\theta = 60°$) and centred at the red spot, which is the centre of the bottom edge of the encircled image. The sector is gridded to the same granularity as the pixels which comprise the image, and height y and width x pixel coordinates are mapped to range r and angle θ, resulting in the distorted feline in the sector.

$(0, -\beta)$, where

$$\beta = \tfrac{1}{2}\frac{H}{\sqrt{W^2 + H^2}}. \tag{4.156}$$

(3) Grid the circular sector to the same extent at the original bitmap. If the bitmap was 2,000 pixels wide by 3,000 pixels high, let there be 3,000 equally spaced radial steps between $r = 1$ to $r = \rho$ and 2,000 angular steps between $\theta = -\frac{1}{2}\Delta\theta$ and $\theta = \frac{1}{2}\Delta\theta$.

(4) Plot a coloured surface where the colours at location r_i, θ_i match those of circle-enclosed bitmap image (x_i, y_i). The coordinate transformations are

$$x_i \rightarrow r_i \sin\theta_i, \tag{4.157}$$

$$y_i \rightarrow -\beta - r_i \cos\theta_i. \tag{4.158}$$

Figure 4.19 illustrates the process, and Fig. 4.20 demonstrates how effective this approximate transformation can be. In these examples, a 2.5 cm diameter section of chrome curtain pole was cut to a height of about 10 cm. The anamorphic image (plus the unit circle) was created in MATLAB and displayed on a tablet PC, placed

(a)

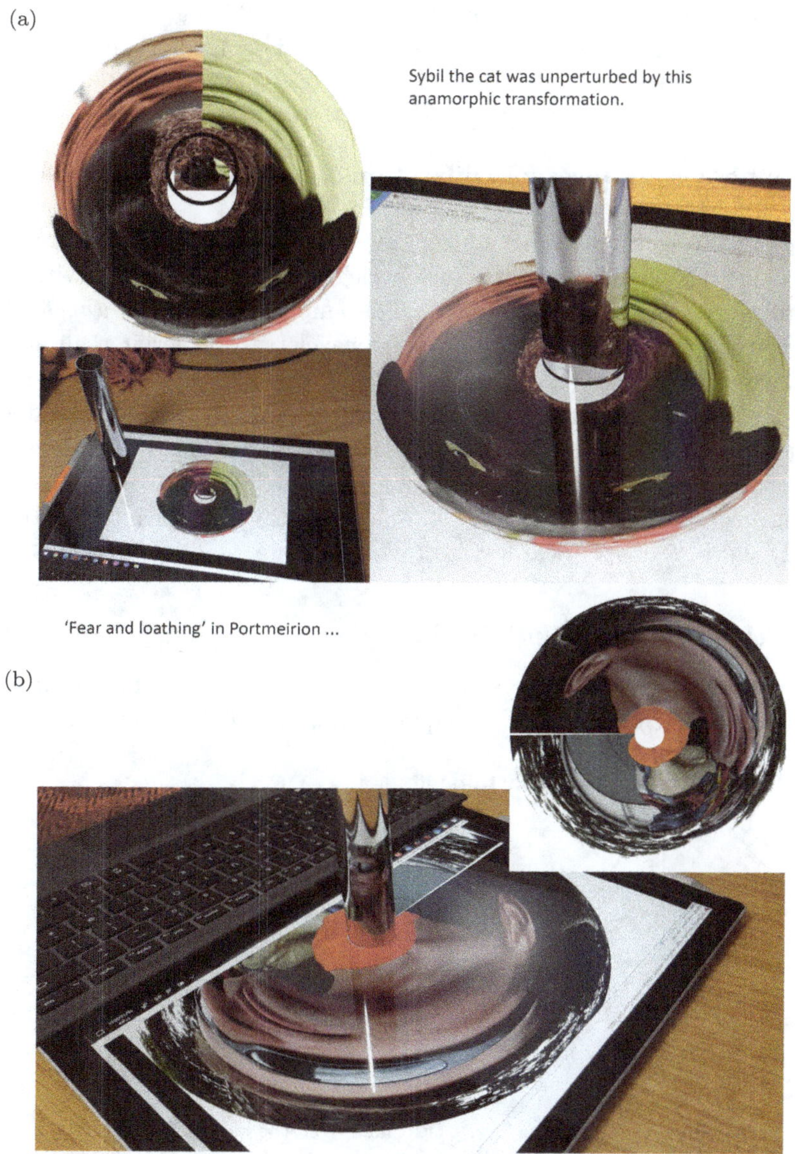

(b)

Fig. 4.20. A practical demonstration of an anamorphic projection. A 2D bitmap image is fitted inside a black circle, which, in Cartesian coordinate terms, is the unit circle $x^2 + y^2 = 1$. The image pixel centres are then mapped onto a sector, with x values mapping to within a defined angular range and y values to a corresponding radial extent r. This produces the anamorphic projection. In both examples (a) and (b), the angular width $\Delta\theta = 360°$. When a polished cylindrical mirror is placed within the black circle, the reflections of the anamorphic projection give the illusion of a 3D (virtual) image floating inside the mirror.

flat on the table. The image was zoomed until the curtain pole diameter fitted the circle. When viewed from about 30 cm away (and at an angle of about 30° to 40°), a convincing correctly proportioned virtual image can be seen. There is some distortion as you vary the viewing angle, so the transformation isn't perfect.

However, this approximate method enables an anamorphic image to be formed from a photograph within minutes, whereas a 3D-printed sculpture is likely to be a far more time-consuming, and expensive process!

4.5   Fraunhofer and Fresnel Diffraction

The *Huygens–Fresnel*[25] *principle* states: 'Every unobstructed point of a wavefront, at a given instant, serves as a source of spherical secondary wavelets (with the same frequency as that of the primary wave). The amplitude ψ of the *wavefield* at any point beyond is the *superposition* of all these wavelets (considering their amplitudes and relative phases).'

In other words, to determine the wavefield resulting from an illuminated slit or other such aperture, we need to add up the effect of spherical wave sources in the vicinity of the aperture. Let us build up such a model by first considering two slits of spacing d that are so thin (width $w \ll d$, and $w \ll \lambda$) that each can be considered as a point source[26] of spherical waves of wavelength λ.

4.5.1   *Two infinitesimally thin slits*

Using a Cartesian coordinate system (see Fig. 4.21), let the slits be at locations $(0, \frac{1}{2}d)$ and $(0, -\frac{1}{2}d)$ and both be equally illuminated such that they produce

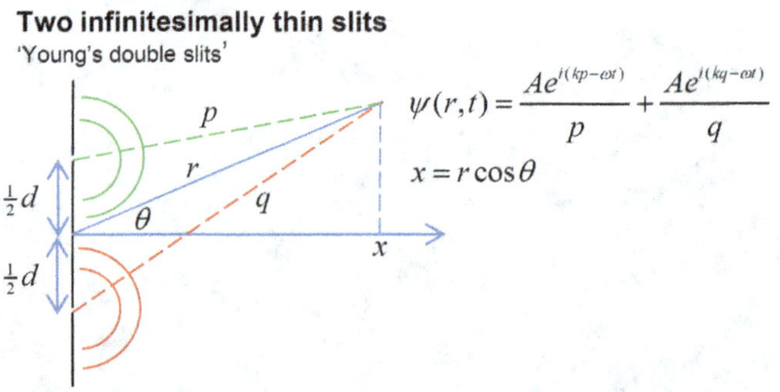

$$\psi(r,t) = \frac{Ae^{i(kp-\omega t)}}{p} + \frac{Ae^{i(kq-\omega t)}}{q}$$

$$x = r\cos\theta$$

Fig. 4.21. The interference pattern $\psi(r, \theta, t)$ formed from the *superposition* of a pair of waves of the same frequency and wavelength, with a fixed phase relationship (i.e. a *coherent* wave source) and vertical separation d, can be computed by considering the complex summation $\psi(r, \theta, t) = \frac{A}{p}e^{i(kp-\omega t)} + \frac{A}{q}e^{i(kq-\omega t)}$, where p, q are the respective distances from the source(s) to polar coordinate location (r, θ). In the *far field*, i.e. $p, q \gg d, \lambda$, we only need to think about angular, i.e. θ, variations in the phase term e^{ikp} rather than the $\frac{1}{p}$ amplitude term. k is the wavenumber $k = \frac{2\pi}{\lambda}$.

---

[25]Christiaan Huygens (1629–1695), Augustin-Jean Fresnel (1788–1827).

[26]Our analysis will be a two-dimensional one, i.e. constrained to a plane that includes the slit width and a direction normal to it. If the slit is long (i.e. $\gg \lambda$), then we can ignore any diffraction effects in the dimension perpendicular to our 2D plane. A laser-illuminated pair of slits or diffraction grating will produce a series of horizontal spots on a screen. Alternatively, a *two-dimensional grating* of small holes of radii $\ll \lambda$ and spacing $\approx \lambda$ will produce a grid of spots on a screen.

spherical waves of wavelength λ. The wave disturbance ψ at range r and polar angle θ from $(0,0)$ is

$$\psi(r,\theta,t) = \frac{A}{p}e^{i(kp-\omega t)} + \frac{A}{q}e^{i(kq-\omega t)}, \tag{4.159}$$

where A is a constant, which relates to the illumination of the slits. Note that the amplitude is divided by the distance (p, q respectively) of the wave source from (r,θ). This means the wave power $\propto |\psi|^2$ varies as the inverse square of distance, which is consistent with the idea of an isotropic radiation source of a given power L (or 'luminosity'), resulting in a power per unit area of $\frac{L}{4\pi r^2}$ at distance r, i.e. the total power divided by the surface area of a sphere of radius r.

Using the cosine rule,

$$p^2 = r^2 + \tfrac{1}{4}d^2 - rd\cos(90° - \theta) = r^2 + \tfrac{1}{4}d^2 - rd\sin\theta, \tag{4.160}$$

$$q^2 = r^2 + \tfrac{1}{4}d^2 - rd\cos(90° + \theta) = r^2 + \tfrac{1}{4}d^2 + rd\sin\theta. \tag{4.161}$$

Hence,

$$p = r\sqrt{1 + \frac{1}{4}\frac{d^2}{r^2} - \frac{d\sin\theta}{r}}, \tag{4.162}$$

$$q = r\sqrt{1 + \frac{1}{4}\frac{d^2}{r^2} + \frac{d\sin\theta}{r}}. \tag{4.163}$$

Now, if $r \gg d$, only the $e^{i(kp-\omega t)}$ and $e^{i(kq-\omega t)}$ terms will vary significantly. Also, in this limit, p, q will $\to r$. Hence, if $r \gg d$,

$$\psi(r,\theta,t) \approx \frac{Ae^{-i\omega t}}{r}\left(e^{ikp} + e^{ikq}\right). \tag{4.164}$$

In the $r \gg d$ limit, we can also perform a binomial expansion of the expressions for p and q:

$$p \approx r + \tfrac{1}{8}\frac{d^2}{r} - \tfrac{1}{2}d\sin\theta, \tag{4.165}$$

$$q \approx r + \tfrac{1}{8}\frac{d^2}{r} + \tfrac{1}{2}d\sin\theta. \tag{4.166}$$

Therefore,

$$\psi(r,\theta,t) \approx \frac{Ae^{i(kr-\omega t)}}{r}e^{\frac{ikd^2}{8r}}\left(e^{-k\frac{1}{2}d\sin\theta} + e^{k\frac{1}{2}d\sin\theta}\right) \tag{4.167}$$

$$\psi(r,\theta,t) \approx \frac{2Ae^{i(kr-\omega t)}}{r}e^{\frac{ikd^2}{8r}}\cos\left(\tfrac{1}{2}kd\sin\theta\right), \tag{4.168}$$

with the final expression making use of the identity $\cos x = \tfrac{1}{2}(e^{ix} + e^{-ix})$.

Can we also neglect the $e^{\frac{ikd^2}{8r}}$ term? If so, since wavenumber $k = \frac{2\pi}{\lambda}$,

$$\frac{2\pi}{\lambda}\frac{d^2}{8r} \ll 1. \tag{4.169}$$

$$\therefore r \gg \frac{\pi d^2}{4\lambda} \approx \frac{d^2}{\lambda}. \tag{4.170}$$

In this *Fraunhofer*[27] 'far-field' limit of $r \gg \frac{d^2}{\lambda}$ (and also $r \gg d$),

$$\psi(r,\theta,t) \approx \frac{2Ae^{i(kr-\omega t)}}{r} \cos\left(\tfrac{1}{2}kd\sin\theta\right). \tag{4.171}$$

At range r, the wave power $|\psi|^2$ varies as $\cos^2\left(\tfrac{1}{2}kd\sin\theta\right)$. This is the classic *Young's double slit*[28] interference pattern. Maxima occur when $\tfrac{1}{2}kd\sin\theta_n = n\pi$, where n is an integer, i.e.

$$d\sin\theta_n = n\lambda. \tag{4.172}$$

$$\therefore \theta_n = \sin^{-1}\left(\frac{n\lambda}{d}\right).$$

This is the same result if one considers two parallel rays at angle θ, emanating from a coherent[29] point–source pair of separation d. The *path difference* between the rays is $d\sin\theta_n$, so when this is a whole number of wavelengths, we have constructive interference. The geometric interpretation of the far-field pattern is, therefore, one of parallel rays, rather than spherically spreading wavefronts, as illustrated in Fig. 4.22.

If a screen is placed at distance D from a pair of slits and the slits are illuminated by a laser such that $r \gg \frac{d^2}{\lambda}$ and $r \gg d$, then the diffraction pattern will be one of bright maxima on a screen at distance y_n from the boresight direction (i.e. $\theta = 0°$). From trigonometry,

$$y_n = D\tan\theta_n. \tag{4.173}$$

Now, for small (radian) angles $\theta_n \ll 1$, $\tan\theta_n \approx \sin\theta_n \approx \theta_n$.

$$\therefore y_n \approx \frac{nD\lambda}{d}. \tag{4.174}$$

This is the result often quoted in A-Level formula sheets and is usually sufficiently precise for most situations. For larger angles, i.e. when $\theta_n \ll 1$ is not true, use the more exact result

$$y_n = D\tan\left(\sin^{-1}\left(\frac{n\lambda}{d}\right)\right). \tag{4.175}$$

---

[27] Joseph von Fraunhofer (1787–1826).

[28] Thomas Young (1773–1829).

[29] i.e. each point source in the pair produces the same frequency of radiation, and both are in phase.

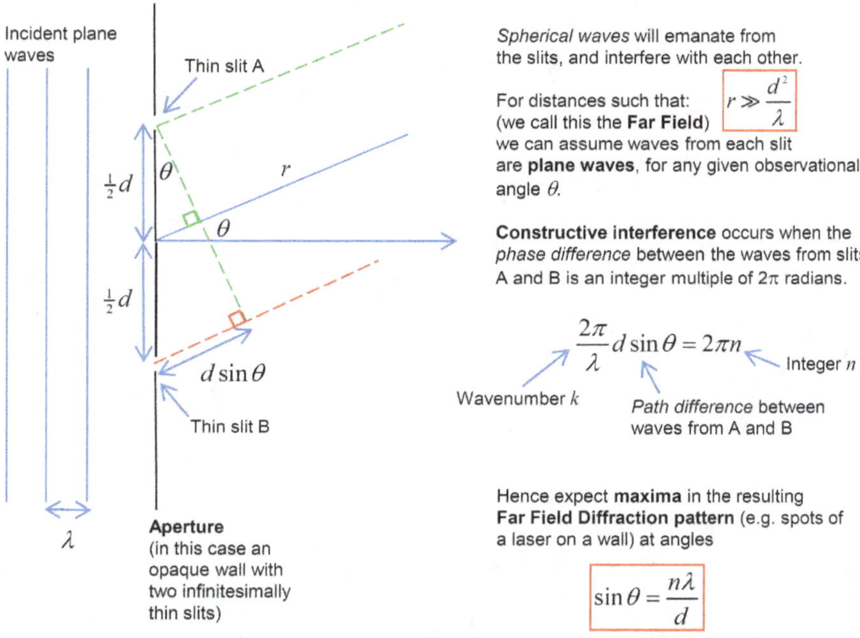

Fig. 4.22. Consider interference from two parallel rays at angle θ, emanating from a wave source of separation d. For the parallel wave (i.e. plane wave) approximation to hold, assume the distance from the source is $\gg$ wavelength λ. The *path difference* between the rays is $d\sin\theta_n$; therefore, when this is a whole number of wavelengths $n\lambda$, we have constructive interference. Alternatively, the *phase difference* $= 2\pi \times \frac{\text{path difference}}{\lambda}$ is an integer multiple of 2π radians. A detector that moves around polar angle θ would yield maxima at $\theta_n = \sin^{-1}\left(\frac{n\lambda}{d}\right)$. The maximum number of peaks in this pattern is $1 + 2 \times \text{floor}\left(\frac{d}{\lambda}\right)$. The floor function means 'round down to the nearest integer'.

4.5.2 *The far-field diffraction pattern of a finite width slit*

The analysis of the double slit in the previous section can be extended to include pairs of infinitesimal slits, which, when added up, cover a whole aperture of width a. In this case, let the aperture section pairing be at locations $(0, z)$ and $(0, -z)$, where $0 \leq z \leq \frac{1}{2}a$, i.e. change d to $2z$ in the expression for ψ. Define A/a in this case to be the illumination amplitude per unit length of the aperture so that the constant A becomes $\frac{A}{a}dz$ for our pairs of aperture–elements. Adapting the double-slit wavefield $\psi(r, \theta, t) \approx \frac{2Ae^{i(kr-\omega t)}}{r} e^{\frac{ikd^2}{8r}} \cos\left(\frac{1}{2}kd\sin\theta\right)$,

$$d\psi \approx \frac{2Adze^{i(kr-\omega t)}}{ar} e^{\frac{ikz^2}{2r}} \cos\left(kz\sin\theta\right). \tag{4.176}$$

The wavefield $\psi(r, \theta, t)$ is the sum of contributions $d\psi$ from pairs of sources that, together, constitute the aperture:

$$\psi(r, \theta, t) = \frac{2Ae^{i(kr-\omega t)}}{ar} \int_0^{\frac{1}{2}a} e^{\frac{ikz^2}{2r}} \cos\left(kz\sin\theta\right) dz. \tag{4.177}$$

We can simplify this integral by again considering a Fraunhofer far-field linear phase regime, $r \gg \frac{d^2}{\lambda}$, where we can assume that $e^{\frac{ikz^2}{2r}} \approx 1$:

$$\psi(r,\theta,t) \approx \frac{2Ae^{i(kr-\omega t)}}{ar} \int_0^{\frac{1}{2}a} \cos\left(kz\sin\theta\right) dz \tag{4.178}$$

$$= \frac{2Ae^{i(kr-\omega t)}}{ar} \left[\frac{\sin\left(kz\sin\theta\right)}{k\sin\theta}\right]_0^{\frac{1}{2}a} \tag{4.179}$$

$$= \frac{Ae^{i(kr-\omega t)}}{r} \frac{\sin\left(\frac{1}{2}ak\sin\theta\right)}{\frac{1}{2}ak\sin\theta}. \tag{4.180}$$

Therefore,

$$|\psi|^2 = \frac{A^2}{r^2}\frac{\sin^2\alpha}{\alpha^2}, \tag{4.181}$$

where

$$\alpha = \tfrac{1}{2}ak\sin\theta = \frac{\pi a\sin\theta}{\lambda}. \tag{4.182}$$

$|\psi|^2$ versus θ has a strongly tapering *envelope* (i.e. unlike the diffraction pattern for the infinitesimally thin slits in Fig. 4.23, which is has a constant maximum amplitude of the oscillations), with *zeros* at $\alpha = m\pi$, apart from when $m = 0$, where $|\psi|^2$ has

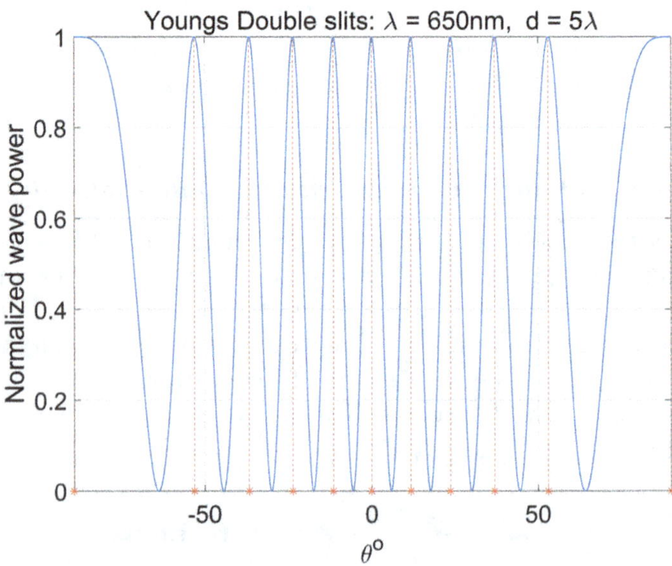

Fig. 4.23. Normalised wave power $|\psi|^2 \propto \cos^2\left(\frac{1}{2}kd\sin\theta\right)$ in the far field of two infinitesimally thin slits separated by distance d. Wavenumber $k = \frac{2\pi}{\lambda}$ of the incident plane waves. Maxima occur when $\theta_n = \sin^{-1}\left(\frac{n\lambda}{d}\right)$. In this example, $\frac{\lambda}{d} = \frac{1}{5}$, so $\theta_n = \sin^{-1}\left(\frac{1}{5}n\right) = 0°, \pm 11.5°, \pm 23.6°, \pm 36.9°, \pm 53.1°, \pm 90°$.

an overall maximum of $\frac{A^2}{r^2}$, i.e. for $m \neq 0$, $|\psi|^2 = 0$ at $a\sin\theta_m = m\lambda$.

$$\therefore \theta_m = \sin^{-1}\left(\frac{m\lambda}{a}\right). \tag{4.183}$$

An example of this tapering envelope is the red curve in Fig. 4.24(b). The physical fact that $|\psi(r,0,t)|$ is a maximum as a function of angle θ is perhaps

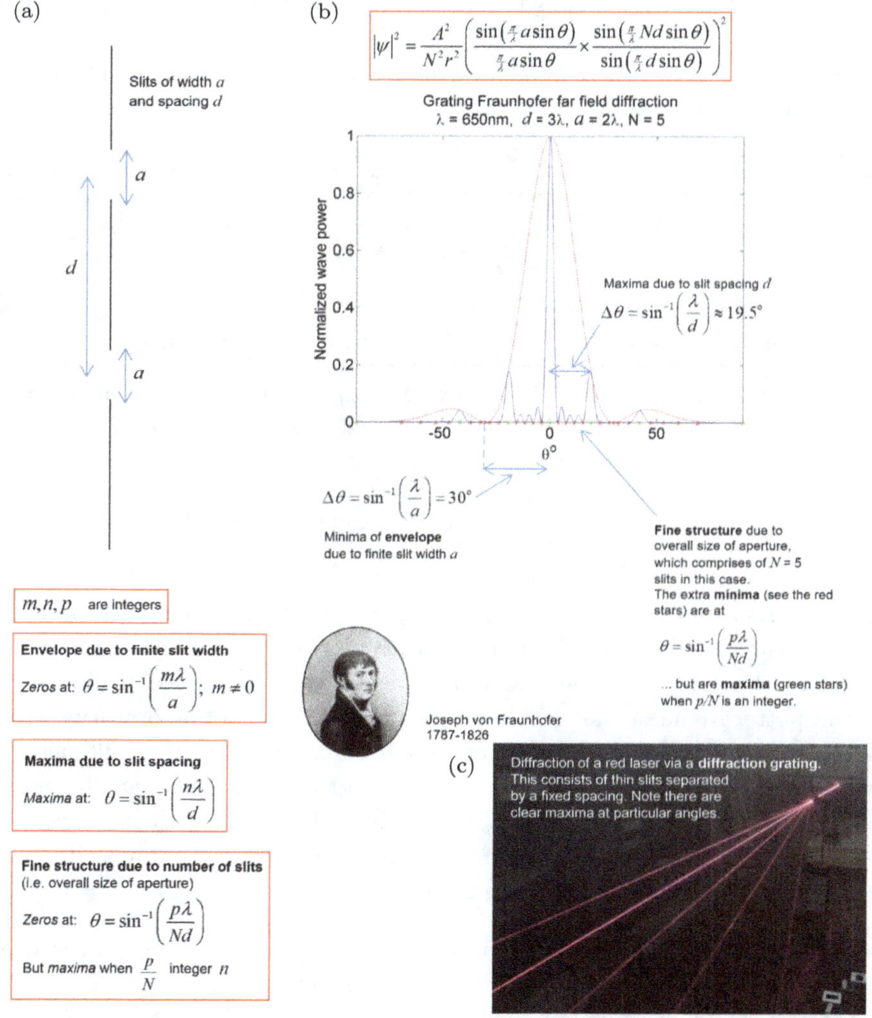

Fig. 4.24. Summary of the *Fraunhofer far field diffraction pattern* from N slits of width a, each separated by distance d. The slit width a (the *smallest* feature of a diffraction grating) gives rise to the *largest* feature of the diffraction pattern. The resulting *envelope* has zeros when $\theta = \sin^{-1}\left(\frac{m\lambda}{a}\right)$, apart from a maxima at boresight $\theta = 0$ radians. *Grating lobes* have maxima at $\theta = \sin^{-1}\left(\frac{n\lambda}{d}\right)$ (unless an envelope zero) and there are additional zeros at $\theta = \sin^{-1}\left(\frac{p\lambda}{Nd}\right)$, apart from when these correspond to a grating lobe. m, n, p are integers such that $\left|\frac{m\lambda}{a}\right| \leq 1$, $\left|\frac{n\lambda}{d}\right| \leq 1$, $\left|\frac{p\lambda}{Nd}\right| \leq 1$. (a) Slit spacing d and width a. (b) Mathematical model of a Fraunhofer far field diffraction pattern. (c) Laboratory demonstration of diffraction of a red laser using a grating. A smoke machine is used to see the fan of laser beams.

unremarkable, but the mathematical proof of why $y = \frac{\sin x}{x}$ (which is the essential functional form of ψ versus θ, using a small-angle approximation for θ) is a maximum of 1 at $x = 0$, is most definitely of interest since, at face value, it implies that $1 = \frac{0}{0}$ for the $x \to 0$ limit of $y = \frac{\sin x}{x}$. To prove why $\lim_{x \to 0} \left(\frac{\sin x}{x} \right) = 1$, we can use L'Hôpital's[30] rule:

$$\lim_{x \to 0} \left(\frac{\sin x}{x} \right) = \lim_{x \to 0} \left(\frac{\frac{d}{dx}(\sin x)}{\frac{d}{dx}(x)} \right) = \lim_{x \to 0} \left(\frac{\cos x}{1} \right) = 1. \tag{4.184}$$

Alternatively, one can use a Maclaurin[31] expansion of $\sin x$, divide this by x and then take the limit $x \to 0$:

$$\frac{\sin x}{x} = \frac{x - \frac{1}{3!}x^3 + \frac{1}{5!}x^5 - \frac{1}{7!}x^7 + \cdots}{x} = 1 - \frac{1}{3!}x^2 + \frac{1}{5!}x^4 - \frac{1}{7!}x^6 + \cdots \tag{4.185}$$

$$\therefore \lim_{x \to 0} \left(\frac{\sin x}{x} \right) = 1. \tag{4.186}$$

4.5.3 *Far field Fraunhofer diffraction pattern of a grating*

If an aperture consists of N slits of width a, each spaced by d, the far-field patterns from each slit can be summed[32] to yield an overall expression for the wavefield power:

$$|\psi|^2 = \frac{A^2}{N^2 r^2} \left(\frac{\sin\left(\frac{\pi}{\lambda} a \sin\theta\right)}{\frac{\pi}{\lambda} a \sin\theta} \times \frac{\sin\left(\frac{\pi}{\lambda} N d \sin\theta\right)}{\sin\left(\frac{\pi}{\lambda} d \sin\theta\right)} \right)^2. \tag{4.187}$$

This imparts *three* levels of structure to the resulting diffraction pattern when N slits of a diffraction grating are illuminated by a laser beam. Note that the smallest feature of the aperture (the slit width a) gives rise to the largest features of the diffraction pattern (the overall envelope). For a typical school laboratory setup, as illustrated in Fig. 4.24(c), a grating between 100 and 600 lines per millimetre will be illuminated by a laser beam one or two millimetres wide. So, N may be between 200 and 1,000 and d between 1.67×10^{-6} m and 1.99×10^{-5} m. If the slits widths are $a \approx \lambda$, this means about 532 nm for a green laser in air and imply a slit spacing range of 3.1λ to 37.4λ. With reference to the example in Fig. 4.24(b):

- The *largest* structure of the diffraction pattern (e.g. on a screen) is the overall envelope of $|\psi|^2$ versus θ. This is due to the *smallest* structure of the grating, the slit width a. $|\psi|^2$ has a maxima at $\theta = 0$, and a taper with zeros at angles

$$\theta_m = \sin^{-1}\left(\frac{m\lambda}{a} \right), \tag{4.188}$$

where integer $m \neq 0$.

---

[30]Guillaume de l'Hôpital (1661–1704).

[31]Colin Maclaurin (1698–1746).

[32]http://www.eclecticon.info/index_htm_files/Diffraction.pdf.

- The *intermediate* structure of the diffraction pattern comprises the 'grating lobes', which are due to the slit spacing d. The maxima are at

$$\theta_n = \sin^{-1}\left(\frac{n\lambda}{d}\right),\tag{4.189}$$

where n is an integer (unless θ_n corresponds to a zero of the envelope at θ_m).

- The *smallest* structure of the diffraction pattern are subsidiary maxima due to the number of slits N that are illuminated. A greater N will result in sharper grating lobes and more subsidiary maxima (albeit at much lower power than the grating lobes as N increases). The zeros are at

$$\theta_p = \sin^{-1}\left(\frac{p\lambda}{Nd}\right)\tag{4.190}$$

but maxima when p/N are integers m, i.e. corresponding to the 'grating lobes' of the intermediate structure.

4.5.4 *Diffraction limit and resolving power*

Almost all of the diffraction effects described so far result in a main lobe of angular width (in radians)

$$\Delta\theta \approx \frac{\lambda}{d},\tag{4.191}$$

where d is a characteristic length of the grating, slit etc.[33] For any optical instrument, the 'resolving power' is likely to be diffraction limited. So, this ratio effectively yields the minimum angular deviation that two objects could be resolved via an optical system. For the James Webb Space Telescope, launched on the 25 December 2021 and in operation since July 2022, the main reflector width $d = 6.5\,\text{m}$. This system uses infrared light. Taking a characteristic wavelength of $3.15\,\mu\text{m}$, this means $\Delta\theta$ is about $1.22 \times \frac{3.15\times 10^{-6}}{6.5} \times \frac{180}{\pi} \times 3600 \approx 0.12$ arc-seconds.[34]

4.5.5 *The Fresnel wavefield and a knife-edge diffraction pattern*

The *Fresnel*[35] *wavefield* is an alternative approach to the Fraunhofer far-field diffraction pattern calculation and investigates the region of *quadratic phase*, i.e. when $r \approx \frac{d^2}{\lambda}$. Consider a uniformly illuminated slit aperture extending from z_1 to z_2, as illustrated in Fig. 4.25. The wavefield $\psi(x,t)$ is to be computed from the sum of spherical waves that comprise the aperture:

$$\psi(r,t) = \int_{z_1}^{z_2} A\frac{e^{i(kp-\omega t)}}{p}\,dz.\tag{4.192}$$

---

[33] For circular apertures, the diffraction limit is approximately $1.22\,\lambda/d$.

[34] One arc-second is $1/3600$ of a degree. According to https://sci.esa.int/documents/34594/36271/1567258183173-JWST_scientific_cap.pdf (Accessed 24/8/23), the angular resolution of JWST is quoted as 0.068 arc-seconds at a wavelength of $2\,\mu$m.

[35] Augustin-Jean Fresnel (1788–1827).

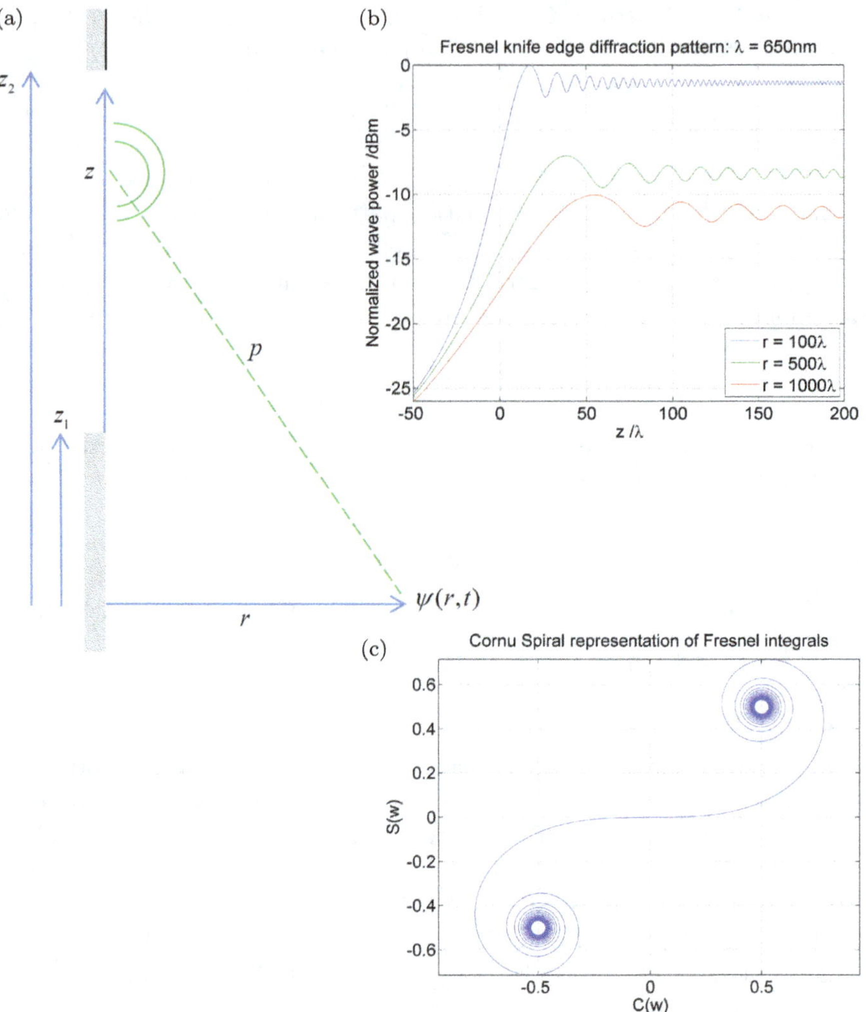

Fig. 4.25. (a) The *Fresnel* wavefield assumes a perpendicular displacement r from an aperture, which admits a coherent source of waves of wavelength λ. Unlike a *Fraunhofer* wavefield, we consider quadratic variations in the phase term, i.e. r is not so large than we can assume plane waves from the aperture. (b) Plots the wavefield power $10\log_{10}|\psi|^2$ in decibels vs. vertical distance z from the bottom edge of an aperture. In this 'knife-edge' case, there is no top edge, i.e. $z_2 \to \infty$. Note how the average power *and* the inter-peak distance vary as r is increased. (c) To calculate the Fresnel wave power requires the use of a *Cornu spiral*, which is a plot of the integrals $S(w) = \int_0^w \sin\left(\frac{1}{2}\pi t^2\right) dt$ vs. $C(w) = \int_0^w \cos\left(\frac{1}{2}\pi t^2\right) dt$.

p is the distance from a wave source to $(r, z = 0)$. Unlike the polar geometry of the Fraunhofer pattern, r is now defined to be the *perpendicular distance* from the aperture. Define z to be a distance in the plane of the aperture, so (r, z) forms a Cartesian coordinate system. Also, A is now a more general scaling constant which could be determined by measuring wave intensity at some r and comparing this with the predicted variation of $|\psi|^2$ calculated when $A = 1$.

If we assume $r \gg z$, we can use Pythagoras' theorem to show that

$$p = \sqrt{r^2 + z^2} = r\sqrt{1 + \frac{z^2}{r^2}} \approx r\left(1 + \tfrac{1}{2}\frac{z^2}{r^2}\right) = r + \tfrac{1}{2}\frac{z^2}{r}. \tag{4.193}$$

Since $r \gg z$, the quadratic term in z will only affect the phase $kp - \omega t$ in $e^{i(kp - \omega t)}$ significantly rather than p in the denominator. Hence, if $r \gg z$,

$$\psi(r, t) \approx \frac{A}{r} e^{i(kr - \omega t)} \int_{z_1}^{z_2} e^{\frac{ikz^2}{2r}} \, dz. \tag{4.194}$$

Using wavenumber $k = \frac{2\pi}{\lambda}$,

$$\psi(r, t) \approx \frac{A}{r} e^{i(kr - \omega t)} \int_{z_1}^{z_2} e^{\frac{i\pi z^2}{\lambda r}} \, dz. \tag{4.195}$$

To evaluate this integral expression, consider the substitution

$$\tfrac{1}{2}\pi t^2 = \frac{\pi z^2}{\lambda r}. \tag{4.196}$$

Therefore,

$$z = t\sqrt{\tfrac{1}{2}\lambda r}, \tag{4.197}$$

$$dz = dt\sqrt{\tfrac{1}{2}\lambda r}, \tag{4.198}$$

and define the (transformed) integral limits of t to be

$$w_1 = \frac{z_1}{\sqrt{\tfrac{1}{2}\lambda r}}, \tag{4.199}$$

$$w_2 = \frac{z_2}{\sqrt{\tfrac{1}{2}\lambda r}}. \tag{4.200}$$

Noting de Moivre's theorem $e^{i\theta} = \cos\theta + i\sin\theta$, the Fresnel wavefield expression $\psi(r, t) \approx \frac{A}{r} e^{i(kr - \omega t)} \int_{z_1}^{z_2} e^{\frac{i\pi z^2}{\lambda r}} \, dz$ becomes

$$\psi(r, t) \approx \frac{A}{r} e^{i(kr - \omega t)} \sqrt{\tfrac{1}{2}\lambda r} \int_{w_1}^{w_2} \left(\cos\left(\tfrac{1}{2}\pi t^2\right) + i\sin\left(\tfrac{1}{2}\pi t^2\right)\right) dt. \tag{4.201}$$

This representation means we can write the Fresnel wavefield power in terms of 'special' (integral) functions[36] $C(w)$ and $S(w)$:

$$C(w) = \int_0^w \cos\left(\tfrac{1}{2}\pi t^2\right) dt, \tag{4.202}$$

---

[36] *Special functions* such as C, S, erf and bessel, are often encoded efficiently as a numerical recipe in programming languages, such as MATLAB or Python. Indeed, these are similar to sine and cosine, which also have to be evaluated numerically, e.g. as a truncated polynomial series based on a Taylor expansion. For details of how to code up special functions from polynomials, see the excellent book *Numerical Recipes* [28].

$$S(w) = \int_0^w \sin\left(\tfrac{1}{2}\pi t^2\right) dt. \tag{4.203}$$

$$\therefore |\psi|^2 \approx \frac{A^2\lambda}{2r}\left|C(w_2) + iS(w_2) - C(w_1) - iS(w_1)\right|^2. \tag{4.204}$$

If S is plotted against C, the rather beautiful *Cornu spiral* is formed, as illustrated in Fig. 4.25(c). The variation of the Fresnel diffraction pattern is proportional to the magnitude of the complex number formed from the difference between coordinates in an Argand diagram of C, S space, with each $C + iS$ (complex number) coordinate corresponding to w_1 or w_2, respectively.

It is possible to show (and indeed graphically from the Cornu spiral) that

$$C(\infty) = \int_0^\infty \cos\left(\tfrac{1}{2}\pi t^2\right) dt = \tfrac{1}{2}, \tag{4.205}$$

$$S(\infty) = \int_0^\infty \sin\left(\tfrac{1}{2}\pi t^2\right) dt = \tfrac{1}{2}. \tag{4.206}$$

This means we can determine a special case of the Fresnel wavefield for a 'knife-edge' diffraction pattern. That is, we illuminate a sharp edge with a laser beam or a similarly coherent source of light and track the wave power at (x, y) coordinates from the tip of the edge. In this case, $z_1 = z$ and $z_2 = \infty$, i.e. there is no upper boundary to the aperture. This means

$$w_1 = w = \frac{z}{\sqrt{\tfrac{1}{2}\lambda r}}. \tag{4.207}$$

$$\therefore |\psi|^2 \approx \frac{A^2\lambda}{2r}\left|\tfrac{1}{2}(1 + i) - C(w) - iS(w)\right|^2. \tag{4.208}$$

As illustrated in Fig. 4.25(b), there will be some wave power within the geometric shadow of the knife edge, but this decreases rapidly with negative z. Above the knife edge, we observe decaying oscillations in wave power with z. Of course, $|\psi|^2$ will also reduce with r.

4.5.6   *A numeric calculation of a generalised wavefield*

Now that we have the analytical machinery to consider the Fraunhofer and Fresnel regimes, we are in a good position to make sense of[37] a purely numeric calculation which is not restricted to the Fresnel and Fraunhofer assumptions, as illustrated in Fig. 4.26. For a uniformly illuminated aperture of width a, define wavefield

$$\psi(r, t) = \frac{A}{a}e^{-i\omega t}\int_{\text{aperture}} \frac{e^{ikp}}{p}\,dz, \tag{4.209}$$

where coordinate z maps out the extent of the aperture of width a.

---

[37]The analytical calculations enable us to check the limiting cases of our numerics, i.e. in our case, the Fresnel and Fraunhofer regimes. It is *always wise to do a sanity check* of numerical models, as it is not always easy to spot erroneous code, inaccurate model inputs and, indeed, artefacts resulting from finite step sizes, grid sizes etc.

Modelling general diffraction effects from a finite width slit

We can use a computer to evaluate the wavefield in the vicinity of a finite width slit which is uniformly illuminated. We are therefore not restricted to the limitations of the Fraunhofer and Fresnel regimes

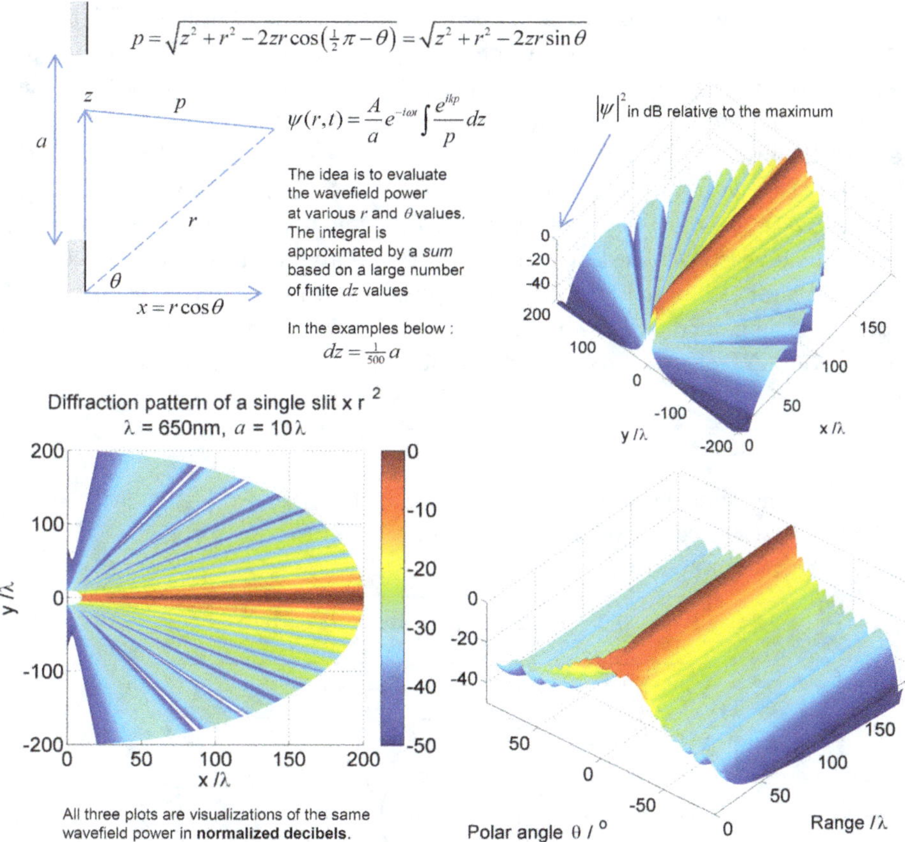

All three plots are visualizations of the same wavefield power in **normalized decibels**.

Fig. 4.26. Approximate numeric calculation of a generalised wavefield resulting from coherent illumination of an aperture of width a. The various views of the three-dimensional surface of wave power $|\psi|^2$ (in decibels) vs. (x, y) displacement from the aperture illustrate the transition between the *Fresnel* and *Fraunhofer* regimes. The latter is perhaps more clearly realised, indicating the fact that, in most situations, the Fraunhofer far-field pattern is a reasonably accurate model.

By applying the cosine rule,

$$p = \sqrt{z^2 + r^2 - 2zr\cos\left(\tfrac{1}{2}\pi - \theta\right)}. \tag{4.210}$$

$$\therefore p = \sqrt{z^2 + r^2 - 2zr\sin\theta}.$$

For a numeric method, let

$$\psi(r, t) \approx \sum_{\text{aperture}} \frac{e^{ikp}}{p}\delta z, \tag{4.211}$$

with δz being a small fraction of a such as $\delta z = \frac{1}{500}a$.

4.6   Doppler and Mach

4.6.1   *Derivation of the Doppler shift formula*

Consider a receding wave source of frequency f, as illustrated in Fig. 4.27. It crosses the x-axis of a Cartesian reference frame at angle θ with speed u. The receiver of the waves is stationary at the origin of the Cartesian frame. The speed of waves, relative to the observer, is w. Depending on the velocity $\mathbf{u}$ of the wave source relative to the observer, the observer will experience a *Doppler*[38] frequency shift Δf from f. If the source recedes, the frequency diminishes and the wavelength increases ('redshift'). If the source is approaching, the observed frequency will increase and the wavelength will decreases ('blueshift'). The ensuing analysis will derive the equations for Doppler shift Δf based on Fig. 4.27. However, perhaps a more intuitive understanding can be gained by a geometric approach, which is illustrated in Fig. 4.28.

The period ΔT of waves received by an observer (in the x direction) at the frame origin O is

$$\Delta T = \Delta t + \frac{r - x}{w}, \tag{4.212}$$

where Δt is the time between wave crests at source, $r - x$ is the *extra distance* travelled by the source between wave crests and w is the wave speed.

From the geometry in Fig. 4.27,

$$r = \sqrt{(x + u\Delta t \cos\theta)^2 + u^2\Delta t^2 \sin^2\theta} \tag{4.213}$$

$$= \sqrt{x^2 + u^2\Delta t^2 \cos^2\theta + 2ux\Delta t \cos\theta + u^2\Delta t^2 \sin^2\theta} \tag{4.214}$$

$$= \sqrt{x^2 + u^2\Delta t^2 + 2ux\Delta t \cos\theta} \tag{4.215}$$

$$= x\sqrt{1 + 2\cos\theta\frac{u\Delta t}{x} + \left(\frac{u\Delta t}{x}\right)^2}. \tag{4.216}$$

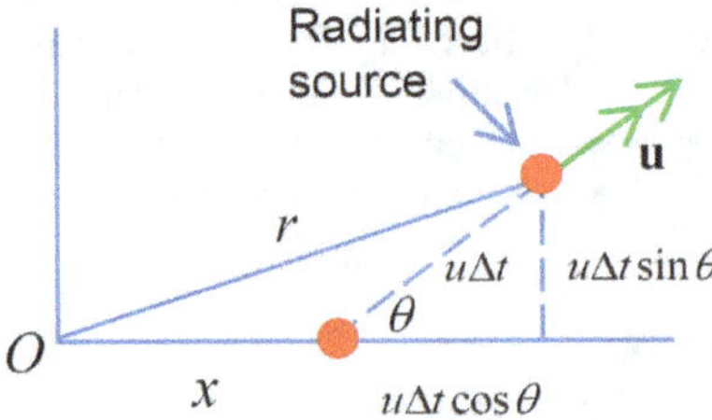

Fig. 4.27. A source of radiation moves at speed u from $(x, 0)$ to $(x + u\Delta t \cos\theta, u\Delta t \sin\theta)$ in time Δt. The period ΔT of waves received by an observer (in the x direction) at the frame origin O is $\Delta T = \Delta t + \frac{r-x}{w}$. The difference between ΔT and Δt results in a *Doppler frequency shift*: $\Delta f = -\frac{\cos\theta}{\frac{w}{u} + \cos\theta} f$, where w is the wave speed and f is the frequency of radiation emitted at source. In terms of wavelength, one can show that $u\cos\theta = \frac{\Delta\lambda}{\lambda_e} w$, where $\Delta\lambda = \lambda_o - \lambda_e$, i.e. the wavelength difference between observed and emitted radiation.

---

[38] Christian Doppler (1803–1853).

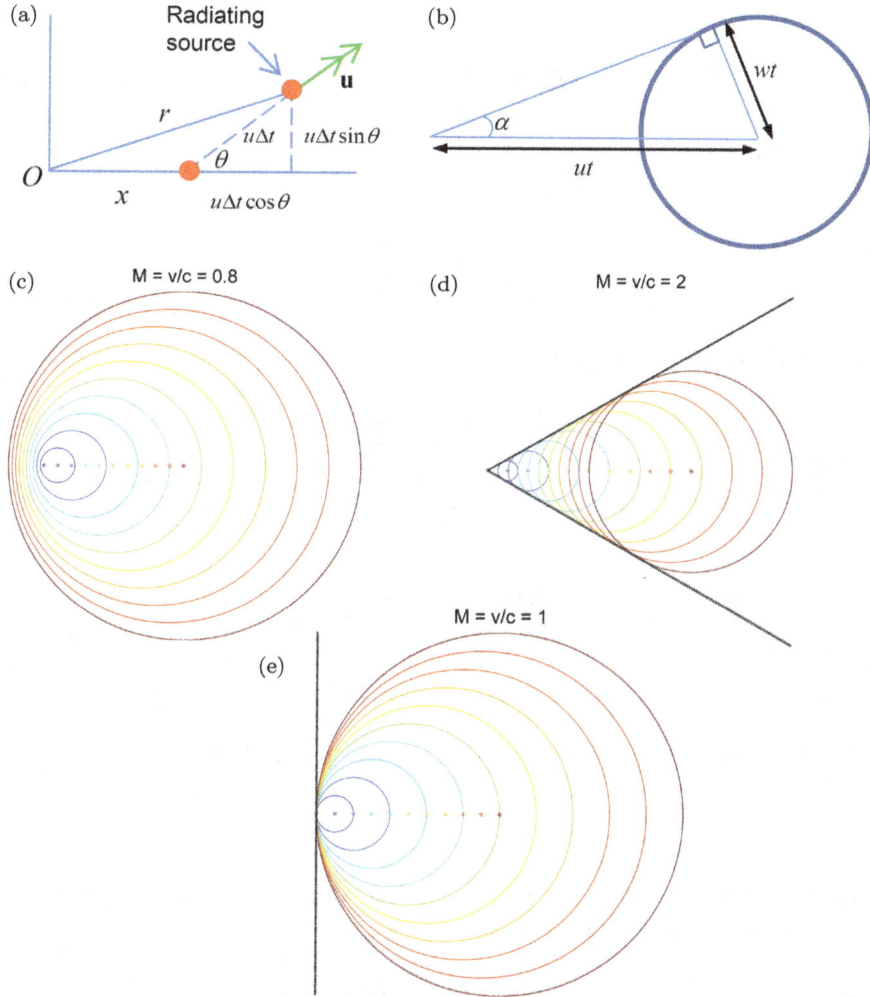

Fig. 4.28. (a) Geometric summary of the *Doppler effect*, when wave speed w (or c) is greater than source speed u, and a *Mach cone*, when wave speed is less than the source speed. In (c) to (e), a radiation source moves from right to left at speed u and wave speed is c. The coloured circles are the positions of the wavefront produced by the source when at the far right (dark red) star position and when the source is at the other star positions. The waves bunch up as the source approaches and spread out as the source recedes, which describes the Doppler effect. When wave speed is less than source speed a *shock front* appears, i.e. a *Mach cone* in (d). The angle of the cone α can be determined via the construction in (b), i.e. $\alpha = \sin^{-1}\left(\frac{w}{u}\right) = \sin^{-1}\left(\frac{1}{M}\right)$, where *Mach number* $M = \frac{\text{wave source speed}}{\text{wave speed}} = \frac{u}{w}$.

If $x \gg u\Delta t$,

$$r \approx x\left(1 + \cos\theta\frac{u\Delta t}{x}\right). \tag{4.217}$$

Therefore,

$$r - x \approx u\Delta t\cos\theta \tag{4.218}$$

and

$$\Delta T = \Delta t \left(1 + \frac{u \cos\theta}{w}\right). \tag{4.219}$$

Define the frequencies

$$f = \frac{1}{\Delta t}, \tag{4.220}$$

$$f + \Delta f = \frac{1}{\Delta T}, \tag{4.221}$$

i.e. a frequency shift Δf. Therefore,

$$f + \Delta f = \frac{f}{1 + \frac{u \cos\theta}{w}} \tag{4.222}$$

$$\Delta f = \frac{f}{1 + \frac{u \cos\theta}{w}} - f \tag{4.223}$$

$$\Delta f = \frac{1 - 1 - \frac{u \cos\theta}{w}}{1 + \frac{u \cos\theta}{w}} f. \tag{4.224}$$

$$\therefore \Delta f = -\frac{\cos\theta}{\frac{w}{u} + \cos\theta} f. \tag{4.225}$$

If $\theta = 0$, i.e. the source moves radially away from an observer,

$$\Delta f = -\frac{1}{\frac{w}{u} + 1} f. \tag{4.226}$$

Now, if the wave speed $w \gg u$ (e.g. in the case of an ambulance producing sound waves or a satellite broadcasting radio waves),

$$\Delta f \approx -\frac{u}{w} f, \tag{4.227}$$

which is the Doppler shift formula typically quoted in most textbooks.

We can also rewrite these formulae in terms of wavelength shift. Define

$$\lambda_e = \frac{w}{f}, \tag{4.228}$$

$$\lambda_o = \frac{w}{f + \Delta f}. \tag{4.229}$$

Therefore,

$$f + \Delta f = \frac{w}{\lambda_o}, \tag{4.230}$$

$$f = \frac{w}{\lambda_e}, \tag{4.231}$$

$$\implies \Delta f = \frac{w}{\lambda_o} - \frac{w}{\lambda_e} \tag{4.232}$$

$$\Delta f = w \left(\frac{\lambda_e - \lambda_o}{\lambda_o \lambda_e} \right) \tag{4.233}$$

$$\Delta f = \left(\frac{\lambda_e - \lambda_o}{\lambda_o} \right) f. \tag{4.234}$$

Substituting for Δf in $\Delta f = -\frac{\cos\theta}{\frac{w}{u} + \cos\theta} f$,

$$\frac{\lambda_e - \lambda_o}{\lambda_o} = -\frac{\cos\theta}{\frac{w}{u} + \cos\theta} \tag{4.235}$$

$$\frac{\lambda_e}{\lambda_o} = 1 - \frac{\cos\theta}{\frac{w}{u} + \cos\theta} \tag{4.236}$$

$$\frac{\lambda_e}{\lambda_o} = \frac{w}{w + u\cos\theta}. \tag{4.237}$$

$$\therefore \frac{\lambda_o}{\lambda_e} = 1 + \frac{u}{w}\cos\theta. \tag{4.238}$$

Define *redshift* as

$$z = \frac{\lambda_o - \lambda_e}{\lambda_e}. \tag{4.239}$$

Therefore,

$$1 + z = \frac{\lambda_o}{\lambda_e}. \tag{4.240}$$

Hence,

$$z = \frac{u}{w}\cos\theta. \tag{4.241}$$

If Doppler wavelength shift $\Delta\lambda = \lambda_o - \lambda_e$, then

$$u\cos\theta = \frac{\Delta\lambda}{\lambda_e}w. \tag{4.242}$$

$u\cos\theta = \frac{\Delta\lambda}{\lambda_e}w$ is probably the most useful formula of all those derived since it yields the 'radial' velocity component (i.e. in the x direction) in terms of fractional wavelength shift $\frac{\Delta\lambda}{\lambda_e}$ and wave speed w.

4.6.2 *Doppler shift in radar applications*

Radio detection and ranging, or radar, is a remote sensing technique where electromagnetic pulses are transmitted and reflected off targets, typically a metallic object such as an aircraft. At microwave frequencies (around 1–10 GHz), it is possible to detect aircraft at ranges of hundreds of kilometres using a modestly powerful transmitter of the sort that might be installed at an airport. To measure the range of an aircraft, the radar has a sensitive receiver which listens for the reflection(s) associated with the transmitted pulse. The time delay is the 'there-and-back' time Δt, and hence range R can be computed from

$$R = \tfrac{1}{2}c\Delta t, \tag{4.243}$$

where c is the speed of light. The received signal will be Doppler shifted if the target is moving. Indeed, this fact can be used to design a *Doppler filter* which will ignore reflections from unmoving 'clutter', such as mountains and other topographic obstacles, and hence enable a small aircraft to be readily identified in a situation where radar echoes might otherwise be swamped. In the radar scenario, we must note an important difference in our Doppler shift calculation: the 'observed' wavelength was transmitted from the (stationary) radar, reflected off the moving aircraft and then received back by the radar. From the perspective of the aircraft, the radar is a moving source; therefore, the radiation it receives will be Doppler shifted. The radiation it transmits towards the radar will then be Doppler shifted *again* by the same relative amount. The classical formula for radar Doppler shift can be derived as follows:

$$f_{\text{target}} = \frac{f}{1 + \frac{u\cos\theta}{c}}, \tag{4.244}$$

$$f_{\text{radar}} = \frac{f_{\text{target}}}{1 + \frac{u\cos\theta}{c}}. \tag{4.245}$$

$$\therefore f + \Delta f = \frac{f}{\left(1 + \frac{u\cos\theta}{c}\right)^2}. \tag{4.246}$$

$$\therefore \frac{\Delta f}{f} = \left(1 + \frac{u\cos\theta}{c}\right)^{-2} - 1. \tag{4.247}$$

If $\frac{u\cos\theta}{c} \ll 1$,

$$\Delta f \approx -\frac{2u\cos\theta}{c}f, \tag{4.248}$$

i.e. in a radar scenario, the Doppler shift is *twice* the value described earlier.

4.6.3 Mach's construction, and a geometric visualisation of Doppler shift

Mach number $M = \frac{u}{w}$ is the ratio between the wave source speed u and the speed of waves w. A Mach number greater than unity associated with sound waves implies a *supersonic* disturbance, which radiates out as a *shock front*, as illustrated in Figs. 4.28(d) and (e).

The Doppler shift described previously can be readily visualised for *Mach numbers of less than unity* if one plots a sequence of circular wavefronts being transmitted from the wave source. If the plot is set up for a wave source to be moving *towards* an observer, the wavelength shift formula $-u\cos\theta = \frac{\Delta\lambda}{\lambda_e}w \Rightarrow -\frac{u}{w} = \frac{\lambda_o - \lambda_e}{\lambda_e}$

and therefore, for $M < 1$,

$$\frac{\lambda_o}{\lambda_e} = 1 - M. \tag{4.249}$$

For Mach numbers *greater than unity*,[39] a shock front will result, beyond which the 'wave medium' has no knowledge of the waves produced by the source. The angle of the shock front (which, in 3D, will form a 'Mach cone') is the inverse sine of the reciprocal of the Mach number. To prove this, consider a wave source travelling at speed u towards an observer (see (b) of Fig. 4.28). After t seconds, it will have travelled a distance ut. In the same time, a spherical wavefront produced at time $t = 0$ will have radius wt. Repeating this process for various t values results in a shock front of angle α from the direction of travel between source and observer. Using basic trigonometry,

$$ut \sin \alpha = wt \tag{4.250}$$

$$\alpha = \sin^{-1}\left(\frac{w}{u}\right). \tag{4.251}$$

$$\therefore \alpha = \sin^{-1}\left(\frac{1}{M}\right). \tag{4.252}$$

The Mach cone becomes 'infinitely wide' (i.e. $\alpha = 90°$) when $M = 1$. This explains why a sonic boom is heard in all directions when an aircraft breaks the sound barrier. As it moves faster, the cone angle decreases. *Hypersonic* jets or missiles travelling at Mach 7 would have a cone angle of

$$\alpha = \sin^{-1}\left(\frac{1}{7}\right) \approx 8.2°. \tag{4.253}$$

4.7 Water Waves and the Kelvin Wedge

4.7.1 *Wave velocities and dispersion relationships for water waves*

A wave of frequency f and wavelength λ travels at speed $c = f\lambda$. Using $\omega = 2\pi f$ and $k = \frac{2\pi}{\lambda}$,

$$c = \frac{\omega}{k}. \tag{4.254}$$

This is called the *phase velocity* of a wave. Because water waves of different wavelengths can travel at different phase velocities, we also have the idea of a *group velocity* of a 'wave packet':

$$c_g = \frac{d\omega}{dk}. \tag{4.255}$$

Deep water waves have a *dispersion relationship* which relates ω and k:

$$\omega = \sqrt{gk}, \tag{4.256}$$

---

[39] $\frac{\lambda_o}{\lambda_e} = 1 - M$ is clearly unphysical for Mach number $M > 1$, as this would imply a *negative* observed wavelength λ_o.

where g is the strength of gravity. *Shallow water waves* have an alternative expression:

$$\omega = k\sqrt{gD},\qquad(4.257)$$

where D is the water depth. This means a phase (and group) velocity $\frac{\omega}{k}$ of

$$c = \sqrt{gD}.$$

This formula can be used to explain why ocean waves rise up and then break as they approach the shore. As the water depth D reduces, so does the wave speed. Since water is essentially incompressible, the waves must increase in height to conserve the same rate of flow of water. If wave height is h' at depth D' and h at depth D, since water volume ch will flow per second per metre width of wave, $c'h' = ch \Rightarrow h'\sqrt{D'} = h\sqrt{D}$ and therefore

$$h' = h\sqrt{\frac{D}{D'}}.$$

So, if water depth decreases from 10 to 2 m, then we anticipate the wave height to gain by a factor of $\sqrt{5} \approx 2.2$.

Ripples (where surface tension σ cannot be ignored) have a dispersion relationship:

$$\omega^2 = \left(\frac{\sigma k^3}{\rho} + gk\right)\tanh kD.\qquad(4.258)$$

Note that $\tanh kD \approx 1$ when $D \gg \lambda$, i.e. 'deep water ripples'.

All these fluid waves assume that the density ρ of water is much greater than the density of the upper fluid (air). If this is not the case, the general dispersion

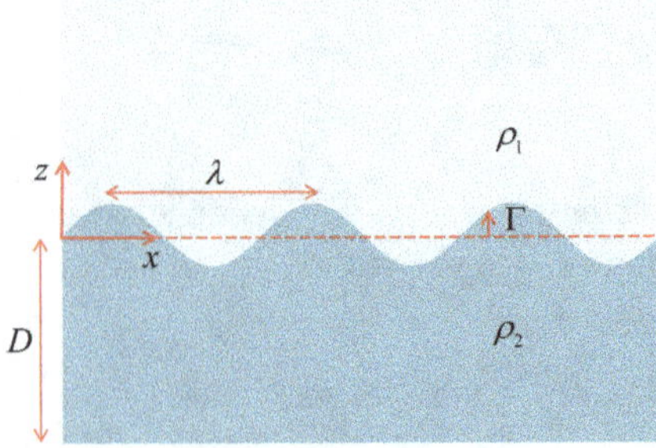

Fig. 4.29. Waves of wavelength λ on an interface between fluid of density ρ_2 and ρ_1, where the lower fluid is of depth D obeys the *dispersion relationship* $\omega^2 = \frac{\sigma k^3 + (\rho_2 - \rho_1)gk}{\rho_1 + \rho_2 \coth(kD)}$. σ is the surface tension of the interface. If $\rho_1 = \rho$, special cases for $\rho_2 \ll \rho_1$ are *ripples*, where $\omega^2 = \left(\frac{\sigma k^3}{\rho} + gk\right)\tanh kD$, *shallow water* waves (where we can ignore surface tension) $\omega^2 = gk^2 D$ and *deep water* gravity waves $\omega^2 = gk$. In all cases, the group velocity of wave packets is given by $c_g = \frac{d\omega}{dk}$, and the phase velocity of individual frequency components is $c_p = \frac{\omega}{k}$.

relationship[40] is

$$\omega^2 = \frac{\sigma k^3 + (\rho_2 - \rho_1)\, gk}{\rho_1 + \rho_2 \mathrm{cotanh}\,(kD)}, \tag{4.259}$$

where ρ_1 is the density of a fluid layer above another fluid layer of density ρ_2 of depth D. The parameters described in this general dispersion relationship are illustrated in Fig. 4.29. The amplitude of a small disturbance with wavelength λ is Γ. If $\rho_2 \gg \rho_1$, then we have $\omega^2 = \left(\frac{\sigma k^3}{\rho_2} + gk\right)\tanh kD$. A full derivation of this dispersion relationship (and indeed the special cases which result in deep water, shallow water and ripples) is provided in Appendix B.

4.7.2 *The Kelvin wedge*

A *wake* is an interference pattern of waves formed by the motion of a body through a fluid. Intriguingly, the angular width of the wake produced by ships (as well as kayaks, barges and ducks) in deep water *is the same* (about 38.9°). A mathematical explanation for this phenomenon was first proposed by Lord Kelvin (1824–1907), and the triangular envelope of the wake pattern has since been known as the *Kelvin wedge*. The derivation presented here is inspired by the reasoning in Faber's splendid *Fluid Dynamics for Physicists* [5].

Consider a ship moving at velocity v through deep water. It will create wave-like disturbances with a variety of wavelengths. i.e. the resulting waves have a spectrum of *Fourier components*. The waves of interest are those which are most pronounced, i.e. those which superpose to form the Kelvin wedge. Let these be travelling at angle α to the ship's velocity, as illustrated in Fig. 4.30(a).

Unlike the other waves, which will naturally disperse (and also attenuate due to the viscosity of water), the waves which form the Kelvin wedge waves continue to be sustained by the motion of the ship. To achieve this, the phase velocity c of these waves must equal the component of the ship's velocity parallel to their wavevector $\mathbf{k}_\alpha$. This is known as a *surf-riding condition*:

$$c = v \sin\alpha. \tag{4.260}$$

Now, if we assume deep water waves, the dispersion relationship is

$$\omega = \sqrt{gk}, \tag{4.261}$$

which means a phase velocity of

$$c = \frac{\omega}{k} = \sqrt{g} k^{-\frac{1}{2}}. \tag{4.262}$$

Hence,

$$\sqrt{g} k^{-\frac{1}{2}} = v \sin\alpha, \tag{4.263}$$

---

[40]Note: $\mathrm{cotanh}(x) = \frac{1}{\tanh(x)} = \frac{e^x + e^{-x}}{e^x - e^{-x}}$.

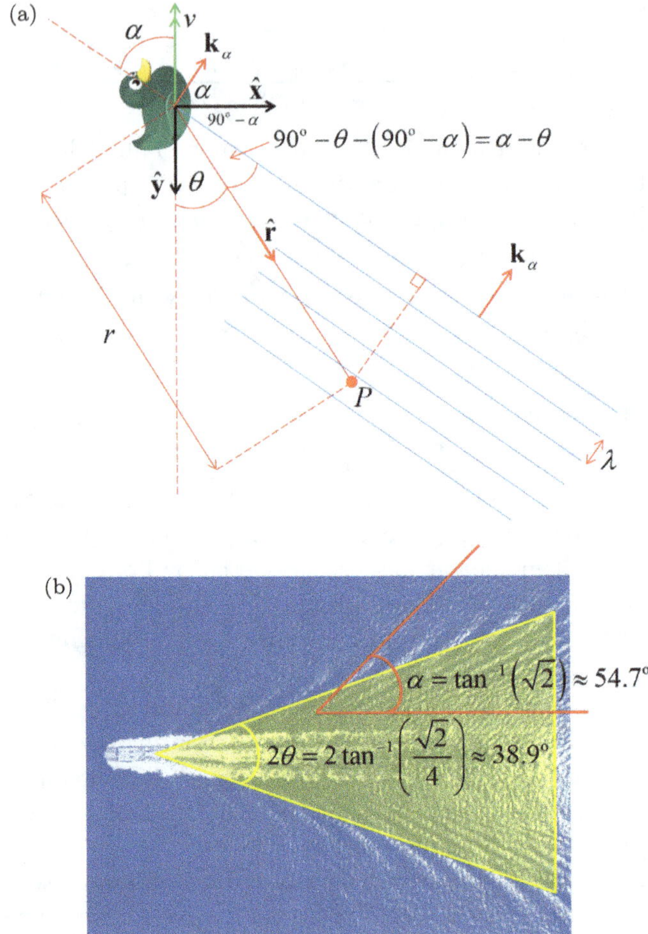

Fig. 4.30. (a) A wake is an interference pattern of waves formed by the motion of a body through a fluid. In this case a duck moving at velocity v. (b) Intriguingly, the angular width 2θ of the wake produced by ships (and ducks!) in deep water (depth $\gg$ wavelength) is the same (about 38.9°), whereas the angle of attack α of the waves is about 54.7°. A mathematical explanation for this phenomenon was first proposed by Lord Kelvin (1824–1907). The triangular envelope of the wake pattern has since been known as the *Kelvin wedge*. The Kelvin wedge is independent of the strength of gravity g as well as the speed v of the body, although the wavelength λ is affected by both parameters since $v = \sqrt{\frac{3g\lambda}{4\pi}}$.

$$gk^{-1} = v^2 \sin^2 \alpha. \tag{4.264}$$

$$\therefore k = \frac{g}{v^2 \sin^2 \alpha}. \tag{4.265}$$

Now, consider a coordinate system centred on the body moving through the fluid. Let $\hat{\mathbf{y}}$ point directly behind the body and $\hat{\mathbf{x}}$ to the right, as depicted in Fig. 4.30(a). The amplitude ψ of waves produced in the wake can be represented by the complex exponential

$$\psi \propto e^{i(\mathbf{k}\cdot\mathbf{r} - \omega t)}. \tag{4.266}$$

In the *surf-riding condition*, the wavevector $\mathbf{k} = \mathbf{k}_\alpha$ is

$$\mathbf{k} = k\left(\hat{\mathbf{x}}\cos\alpha - \hat{\mathbf{y}}\sin\alpha\right). \tag{4.267}$$

Now, define a position vector

$$\mathbf{r} = r\left(\hat{\mathbf{x}}\sin\theta + \hat{\mathbf{y}}\cos\theta\right) \tag{4.268}$$

from the wave source at $(0,0)$ to point P in the diagram, where θ is the angle from the $\hat{\mathbf{y}}$ direction. Hence, the spatial part of the wave phase $\phi = \mathbf{k}\cdot\mathbf{r}$ is, using $k = \frac{g}{v^2\sin^2\alpha}$,

$$\phi = kr\left(\cos\alpha\sin\theta - \sin\alpha\cos\theta\right) \tag{4.269}$$

$$= -\frac{gr}{v^2}\frac{\left(\sin\alpha\cos\theta - \cos\alpha\sin\theta\right)}{\sin^2\alpha} \tag{4.270}$$

$$= -\frac{gr}{v^2}\frac{\sin\left(\alpha - \theta\right)}{\sin^2\alpha}. \tag{4.271}$$

In the observed Kevin wedge, we see constructive interference perpetuated at a particular angle α. Another way of thinking about this is for the wave phase to remain the same at this angle, which means

$$\frac{\partial\phi}{\partial\alpha} = 0. \tag{4.272}$$

Using the expression above for ϕ,

$$\frac{\partial\phi}{\partial\alpha} = -\frac{gr}{v^2}\left\{\frac{\sin^2\alpha\cos\left(\alpha - \theta\right) - 2\sin\left(\alpha - \theta\right)\sin\alpha\cos\alpha}{\sin^4\alpha}\right\} \tag{4.273}$$

$$= -\frac{gr}{v^2}\left\{\frac{\sin\alpha\cos\left(\alpha - \theta\right) - 2\sin\left(\alpha - \theta\right)\cos\alpha}{\sin^3\alpha}\right\}, \tag{4.274}$$

which implies $\frac{\partial\phi}{\partial\alpha} = 0$ when

$$\sin\alpha\cos\left(\alpha - \theta\right) - 2\sin\left(\alpha - \theta\right)\cos\alpha = 0 \tag{4.275}$$

$$\tan\alpha = 2\tan\left(\alpha - \theta\right) \tag{4.276}$$

$$\tfrac{1}{2}\tan\alpha = \frac{\tan\alpha - \tan\theta}{1 + \tan\alpha\tan\theta} \tag{4.277}$$

$$\tan\alpha + \tan^2\alpha\tan\theta = 2\tan\alpha - 2\tan\theta \tag{4.278}$$

$$\tan\theta\left(2 + \tan^2\alpha\right) = \tan\alpha. \tag{4.279}$$

$$\therefore \tan\theta = \frac{\tan\alpha}{2 + \tan^2\alpha}. \tag{4.280}$$

Now, if the Kelvin Wedge is at a fixed value of θ and α we might also expect

$$\frac{d\theta}{d\alpha} = 0. \tag{4.281}$$

Using the expression $\tan\theta = \frac{\tan\alpha}{2+\tan^2\alpha}$ and differentiating both sides with respect to α,

$$\sec^2\theta\,\frac{d\theta}{d\alpha} = \frac{\left(2+\tan^2\alpha\right)\sec^2\alpha - 2\tan^2\alpha\,\sec^2\alpha}{\left(2+\tan^2\alpha\right)^2}. \tag{4.282}$$

$$\therefore \frac{d\theta}{d\alpha} = \frac{2-\tan^2\alpha}{\left(2+\tan^2\alpha\right)^2}. \tag{4.283}$$

So, $\frac{d\theta}{d\alpha} = 0$ when $2 - \tan^2\alpha = 0$, i.e.

$$\tan\alpha = \pm\sqrt{2}. \tag{4.284}$$

Therefore,

$$\alpha = \pm\tan^{-1}\sqrt{2} \approx \pm 54.7^\circ \tag{4.285}$$

and

$$\tan\theta = \frac{\tan\alpha}{2+\tan^2\alpha} = \pm\tfrac{1}{4}\sqrt{2}, \tag{4.286}$$

which means

$$\theta = \pm\tan^{-1}\left(\tfrac{1}{4}\sqrt{2}\right) \approx \pm 19.5^\circ. \tag{4.287}$$

So, in summary, $2\theta = 38.9^\circ$ is the angular width of the Kelvin wedge, and $\alpha = 54.7^\circ$ is the angle of the principal waves in the wake from the $\hat{\mathbf{y}}$ direction directly behind the wave source.

Returning to $k = \frac{2\pi}{\lambda} = \frac{g}{v^2\sin^2\alpha}$ from our surf-riding condition, this means a relationship between wavelength λ of (clearly visible) wave crests in the Kelvin wedge and the source speed v. This means one can, in principle, calculate the speed of a vessel from an aerial photo of its wake. This is rather a fun exercise with *Google Earth*, which in many areas will have sufficiently high-resolution imagery to resolve the wave crests in a Kelvin Wedge as well as the wedge itself.

Now, consider the basic Pythagorean identity $1 = \cos^2\alpha + \sin^2\alpha$. Dividing both sides by $\sin^2\alpha$ yields

$$\frac{1}{\sin^2\alpha} = \frac{1}{\tan^2\alpha} + 1. \tag{4.288}$$

So, if

$$\frac{2\pi}{\lambda} = \frac{g}{v^2\sin^2\alpha} \tag{4.289}$$

and $\tan^2\alpha = 2$ for the Kelvin wedge, this means

$$\frac{2\pi}{\lambda} = \frac{g}{v^2} \times \left(\tfrac{1}{2} + 1\right) = \frac{g}{v^2} \times \tfrac{3}{2}. \tag{4.290}$$

The vessel speed v is therefore proportional to the square root of the wavelength λ of the waves in the Kelvin wedge, and the constant of proportionality is $\sqrt{\frac{3g}{4\pi}}$.

$$\therefore v = \sqrt{\frac{3g\lambda}{4\pi}}. \tag{4.291}$$

The *Red Jet* catamaran, which sails between West Cowes on the Isle of Wight and Southampton in the UK mainland, has a maximum speed of 38 knots, or $v = 38/1.944 = 19.5$ ms^{-1}. This means it should produce a Kelvin wedge of maximum wavelength $\lambda = \frac{4\pi}{3g} v^2 = 163$ m, where $\frac{4\pi}{3g} = 0.427$ m^{-1}s^2 using $g = 9.81$ ms^{-2}. The maximum speed[41] at the Inner Harbour at Cowes is 6 knots, which means one should observe (from the shore) an approaching Red Jet producing Kelvin wedges with maximum wavelengths of

$$\lambda = 0.427 \times (6/1.944)^2 \text{ m} \tag{4.292}$$

$$= 4.1 \text{ m}. \tag{4.293}$$

4.8 Simple Harmonic Motion and Resonance in a Driven Mass–Spring System

4.8.1 *Simple harmonic motion in a mass–spring system*

As illustrated in Fig. 4.31(a), consider an object of mass m suspended vertically from a fixed point by a light elastic string (or a light coiled spring) of unextended length l that obeys Hooke's law,[42] i.e. the *restoring force is proportional to the extension of the spring*. The constant of proportionality between force and extension is the *spring constant* k. The equilibrium extension z of the string can be calculated by equating the Hookean restoring force kz with the weight of the object mg, i.e.

$$z = \frac{mg}{k}. \tag{4.294}$$

Now, assume that the mass is pulled down a further x_0 beyond the equilibrium extension z and then released from rest. The subsequent motion $x(t)$, with positive x being downwards beyond the equilibrium extension, is described by Newton's Second Law, *mass $\times$ acceleration = vector sum of force*:

$$m\ddot{x} = mg - k(x + z). \tag{4.295}$$

Since $kz = mg$,

$$\ddot{x} = -\frac{k}{m}x. \tag{4.296}$$

---

[41] https://www.cowes.co.uk/safety-navigation/navigation/ (Accessed 29/5/24).
[42] Robert Hooke (1635–1703).

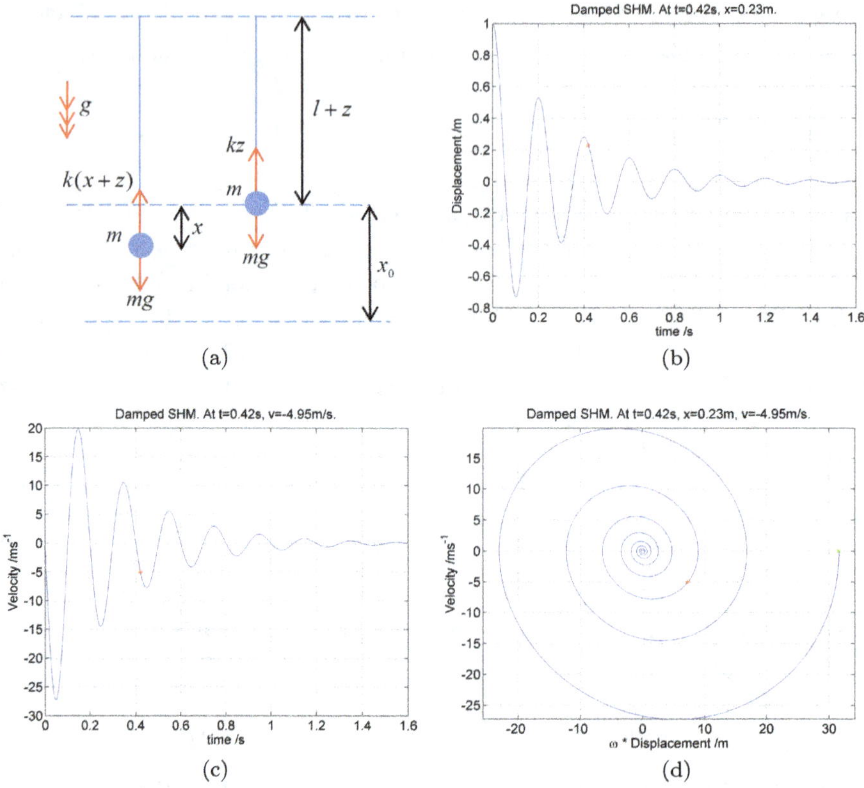

Fig. 4.31. (a) A mass m is suspended by a light *Hookean* elastic string of unextended length l, which provides a restoring force $k(x + z)$ that is proportional to extension $z + x$. In equilibrium, where $x = 0$, the weight mg balances the extension z, so $z = \frac{mg}{k}$. Displacement x is the *extra extension* beyond equilibrium. When released from rest at $x = x_0$, the spring will oscillate with frequency $f = \frac{1}{2\pi}\sqrt{\frac{k}{m}}$. In most practical situations, there is also a damping force which opposes motion. We shall assert a force of magnitude $2m\gamma\,|\dot{x}|$, where $\dot{x}$ is the velocity of mass m. This results in an exponential decay of oscillation amplitude, as described in (b), (c) and (d). The red star is the position (or velocity) at time $t = 0.42$ s. (d) describes the motion in a graph of velocity $\dot{x}$ vs. ωx, i.e. a parametric plot, with both axes' quantities being time varying, resulting in an inspiralling trajectory. Angular speed $\omega = \sqrt{\frac{k}{m}}$.

This is an example of the differential equation of *simple harmonic motion* (SHM):

$$\ddot{x} = -\omega^2 x. \tag{4.297}$$

So, we can compare the two equations to relate spring constant and mass to the angular frequency ω of the oscillations of the mass about the equilibrium position:

$$\omega^2 = \frac{k}{m}. \tag{4.298}$$

Solutions to $\ddot{x} = -\omega^2 x$ are of the general form

$$x(t) = A\cos(\omega t - \phi), \tag{4.299}$$

i.e. cosinusoidal oscillations of period $T = \frac{2\pi}{\omega}$ with constants A and ϕ fixed by initial conditions $x(0) = x_0$ and $\dot{x}(0) = u$. (If the system is released from rest, the initial velocity $u = 0$.) From $x(t) = A\cos(\omega t - \phi)$ and hence $x(0) = x_0 = A\cos(-\phi) = A\cos\phi$,

$$A = \frac{x_0}{\cos\phi}. \tag{4.300}$$

Velocity $\dot{x}(t) = \frac{dx}{dt}$ is

$$\dot{x}(t) = -\omega A \sin(\omega t - \phi). \tag{4.301}$$

Therefore, noting $\sin(-\phi) = -\sin\phi$ and using $A = \frac{x_0}{\cos\phi}$,

$$u = \dot{x}(0) = \omega A \sin\phi = \omega x_0 \tan\phi. \tag{4.302}$$

$$\therefore \phi = \tan^{-1}\left(\frac{u}{\omega x_0}\right). \tag{4.303}$$

Now, since $u = \omega x_0 \tan\phi$,

$$u^2 + \omega^2 x_0^2 = \omega^2 x_0^2 \left(\tan^2\phi + 1\right) \tag{4.304}$$

$$= \frac{\omega^2 x_0^2}{\cos^2\phi} \tag{4.305}$$

$$= \omega^2 A^2. \tag{4.306}$$

So, a more elegant formula for A involving inputs u, ω, x_0 is

$$A = \frac{\sqrt{u^2 + \omega^2 x_0^2}}{\omega}. \tag{4.307}$$

Note that we have asserted $A > 0$ and, therefore, only adopt the positive root of $u^2 + \omega^2 x_0^2 = \omega^2 A^2$.

In summary, if $x(0) = x_0$ and $\dot{x}(0) = u$,

$$x(t) = \frac{\sqrt{u^2 + \omega^2 x_0^2}}{\omega} \cos\left(\omega t - \tan^{-1}\left(\frac{u}{\omega x_0}\right)\right). \tag{4.308}$$

If the system is released from rest, $u = 0$, which means

$$x(t) = x_0 \cos\omega t. \tag{4.309}$$

In the case of our mass–spring system

$$\omega^2 = \frac{k}{m}, \tag{4.310}$$

so the period of oscillations is expected to be

$$T = \frac{2\pi}{\omega} = 2\pi\sqrt{\frac{m}{k}} \tag{4.311}$$

and their frequency to be

$$f = \frac{1}{T} = \frac{1}{2\pi}\omega = \frac{1}{2\pi}\sqrt{\frac{k}{m}}. \tag{4.312}$$

As exemplified here, in many problems which result in SHM models (which could be electrical, fluidic, mechanical etc.), the general idea is to derive a $\ddot{x} = -\omega^2 x$ expression (e.g. from the application of Newton's second law or by differentiating an expression of total energy) and then determine the period from $T = \frac{2\pi}{\omega}$.

4.8.2 *Resonance in a driven mass–spring system*

Now consider adding a damping device to our mass–spring system which provides a force $-2\gamma m\dot{x}$, i.e. which always opposes motion. A force that is proportional to speed is similar to that of viscous drag in fluid dynamics and, therefore, deemed appropriate at low oscillation speeds. For higher speeds, a force proportional to speed^2 is more appropriate, although this will not allow us to make much analytical progress, so we shall not pursue this dynamics regime. However, if a numerical solution method[43] is used (e.g. using an approximation of constant acceleration between small but finite time steps Δt), then a $\dot{x}^2$ force is recommended.

In addition, let the whole system be driven by a sinusoidal driving force

$$F = mA_0\omega_0^2 \sin \omega t, \tag{4.313}$$

where

$$\omega_0^2 = \frac{k}{m}. \tag{4.314}$$

$\omega_0 = \sqrt{\frac{k}{m}}$ is the 'free oscillation' or 'natural (angular) frequency' of the system. For a laboratory demonstration, a driven mechanical oscillator could be implemented via some form of electromagnetic vibrator which moves a rod (connected to the mass and spring perhaps via a string and a pulley or suspended from above if the equipment can withstand it), although this system can be applied in a very general way to almost all driven oscillations of modest amplitude[44] and speed.

By Newton's second law,

$$m\ddot{x} = mg - k(x + z) - 2\gamma m\dot{x} + F. \tag{4.315}$$

$$\therefore \ddot{x} + \omega_0^2 x + 2\gamma\dot{x} = A_0\omega_0^2 \sin \omega t. \tag{4.316}$$

There are three variants of general solution of the form

$$x(t) = X(t) + A \sin (\omega t - \phi), \tag{4.317}$$

where A is a positive constant with dimensions of $x(t)$.

---

[43] A discussion of numerical methods for the solution of SHM equations, as a incentive for the learning of calculus techniques, is in the *Physics Education* paper 'Numerical methods as an introduction to calculus' [10].

[44] Oscillations of an 'immodest' amplitude, e.g. a large-angle pendulum swing, are often *nonlinear* and only obey SHM for small angles, i.e. $\ll 1$ radian.

The *transient* part $X(t)$, which is the general solution to $\ddot{x} + \omega_0^2 x + 2\gamma\dot{x} = 0$, is

$$X(t) = \begin{cases} A_1 e^{-\gamma t} \cos\left(t\sqrt{\omega^2 - \gamma^2} - \Phi\right) & \omega > \gamma \quad \text{`under damped'} \\ e^{-\gamma t}\left(A_1 e^{t\sqrt{\gamma^2 - \omega^2}} + A_2 e^{-t\sqrt{\gamma^2 - \omega^2}}\right) & \omega < \gamma \quad \text{`over damped'} \\ e^{-\gamma t}\left(A_1 + A_2 t\right) & \omega = \gamma \quad \text{`critically damped'} \end{cases} \tag{4.318}$$

where the constant pairs A_1, A_2 or A_1, Φ can be found[45] from initial conditions $x(0) = x_0$ and $\dot{x}(0) = u$. The term *transient* refers to the exponentially decaying factor $e^{-\gamma t}$. This means for $t \gg \frac{1}{\gamma}$, the transient term $X \to 0$. The *steady-state* dynamics of a driven mass–spring system are, therefore, *oscillations of the same frequency as the driving force but out of phase by* ϕ. We can solve for A and ϕ by substituting $x = A\sin(\omega t - \phi)$ into $\ddot{x} + \omega_0^2 x + 2\gamma\dot{x} = A_0\omega_0^2 \sin\omega t$. This can be done, albeit rather tediously, with the key mathematical step of comparing the coefficients of $\sin\omega t$ and $\cos\omega t$ in the resulting expression. This yields 'two equations from one' and allows us to solve for A and ϕ. A more elegant method is to start with a complex-number representation of the driven, damped SHM equation of the system

$$\ddot{z} + \omega_0^2 z + 2\gamma z = A_0\omega_0^2 e^{i\omega t} \tag{4.319}$$

and seek steady-state solutions of the form

$$z(t) = Ae^{i(\omega t - \phi)}. \tag{4.320}$$

$\ddot{x} + \omega_0^2 x + 2\gamma\dot{x} = A_0\omega_0^2 \sin\omega t$ is the *imaginary* part of $\ddot{z} + \omega_0^2 z + 2\gamma z = A_0\omega_0^2 e^{i\omega t}$, and $x = A\sin(\omega t - \phi)$ is the imaginary part of $z = A\cos(\omega t - \phi) + iA\sin(\omega t - \phi)$. Substituting $z = Ae^{i(\omega t - \phi)}$ into $\ddot{z} + \omega_0^2 z + 2\gamma z = A_0\omega_0^2 e^{i\omega t}$ yields

$$\left(-\omega^2 + 2i\gamma\omega + \omega_0^2\right) Ae^{i(\omega t - \phi)} = A_0\omega_0^2 e^{i\omega t} \tag{4.321}$$

$$\left(-\omega^2 + 2i\gamma\omega + \omega_0^2\right) Ae^{-i\phi} = A_0\omega_0^2 \tag{4.322}$$

$$\left|-\omega^2 + 2i\gamma\omega + \omega_0^2\right| e^{i\tan^{-1}\left(\frac{2\gamma\omega}{\omega_0^2 - \omega^2}\right)} Ae^{-i\phi} = A_0\omega_0^2. \tag{4.323}$$

$$\therefore \left|-\omega^2 + 2i\gamma\omega + \omega_0^2\right| A = A_0\omega_0^2. \tag{4.324}$$

$$\therefore A = \frac{A_0\omega_0^2}{\sqrt{\left(\omega_0^2 - \omega^2\right)^2 + 4\gamma^2\omega^2}}. \tag{4.325}$$

The variation of steady-state amplitude with frequency is therefore

$$\frac{A}{A_0} = \frac{1}{\sqrt{\left(1 - \frac{\omega^2}{\omega_0^2}\right)^2 + 4\left(\frac{\gamma}{\omega_0}\right)^2\left(\frac{\omega}{\omega_0}\right)^2}}. \tag{4.326}$$

---

[45] http://www.eclecticon.info/index_htm_files/Mechanics%20-%20Forced%20SHM.pdf.

To determine the frequency variation of steady-state phase ϕ, note that $\left|-\omega^2 + 2i\gamma\omega + \omega_0^2\right| e^{i\tan^{-1}\left(\frac{2\gamma\omega}{\omega_0^2-\omega^2}\right)} Ae^{-i\phi} = A_0\omega_0^2$ implies the exponential term $e^{i\tan^{-1}\left(\frac{2\gamma\omega}{\omega_0^2-\omega^2}\right)} e^{-i\phi}$ must be unity. Hence,

$$e^{i\tan^{-1}\left(\frac{2\gamma\omega}{\omega_0^2-\omega^2}\right)} e^{-i\phi} = e^0. \tag{4.327}$$

$$\therefore \phi = \tan^{-1}\left(\frac{2\gamma\omega}{\omega_0^2 - \omega^2}\right). \tag{4.328}$$

$$\therefore \phi = \tan^{-1}\left(\frac{2\left(\frac{\gamma}{\omega_0}\right)\left(\frac{\omega}{\omega_0}\right)}{1 - \left(\frac{\omega}{\omega_0}\right)^2}\right). \tag{4.329}$$

A plot of A and ϕ versus $\frac{\omega}{\omega_0}$ is illustrated in Fig. 4.32, with different coloured curves corresponding to the values of $\frac{\gamma}{\omega_0}$ within the range of 0.1 to 0.7. A screenshot of a MATLAB program is also included, which compares the steady-state amplitude versus ω with a time variation $x(t)$, which includes the transient $X(t)$ as well as steady-state $A\sin(\omega t - \phi)$ components. The program calculates the curves automatically as the various inputs are changed. The $+/-$ buttons allow this to occur in a dynamic way.

An alternative (and perhaps more geometrically intuitive) derivation for $\phi(\omega)$ and $A(\omega)$ starts with $\left(-\omega^2 + 2i\gamma\omega + \omega_0^2\right) Ae^{-i\phi} = A_0\omega_0^2$, which implies

$$Ae^{-i\phi} = \frac{A_0\omega_0^2}{-\omega^2 + 2i\gamma\omega + \omega_0^2}. \tag{4.330}$$

Using the complex number identity $\frac{1}{a+ib} = \frac{a-ib}{(a+ib)(a-ib)} = \frac{a-ib}{a^2+b^2}$,

$$Ae^{-i\phi} = \frac{A_0\omega_0^2}{\left(\omega_0^2 - \omega^2\right)^2 + 4\gamma^2\omega^2} \left(\omega_0^2 - \omega^2 - 2i\gamma\omega\right). \tag{4.331}$$

$z = Ae^{-i\phi}$ can be drawn on an *Argand diagram* (or 'phasor' diagram) as a downward sloping vector of length A and angle ϕ below the real (horizontal) axis. From basic trigonometry and Pythagoras' theorem, we can immediately write down

$$\phi = \tan^{-1}\left(\frac{2\gamma\omega}{\omega_0^2 - \omega^2}\right) \tag{4.332}$$

and

$$A = \frac{A_0\omega_0^2}{\left(\omega_0^2 - \omega^2\right)^2 + 4\gamma^2\omega^2} \sqrt{\left(\omega_0^2 - \omega^2\right)^2 + 4\gamma^2\omega^2}, \tag{4.333}$$

which means

$$A = \frac{A_0\omega_0^2}{\sqrt{\left(\omega_0^2 - \omega^2\right)^2 + 4\gamma^2\omega^2}}. \tag{4.334}$$

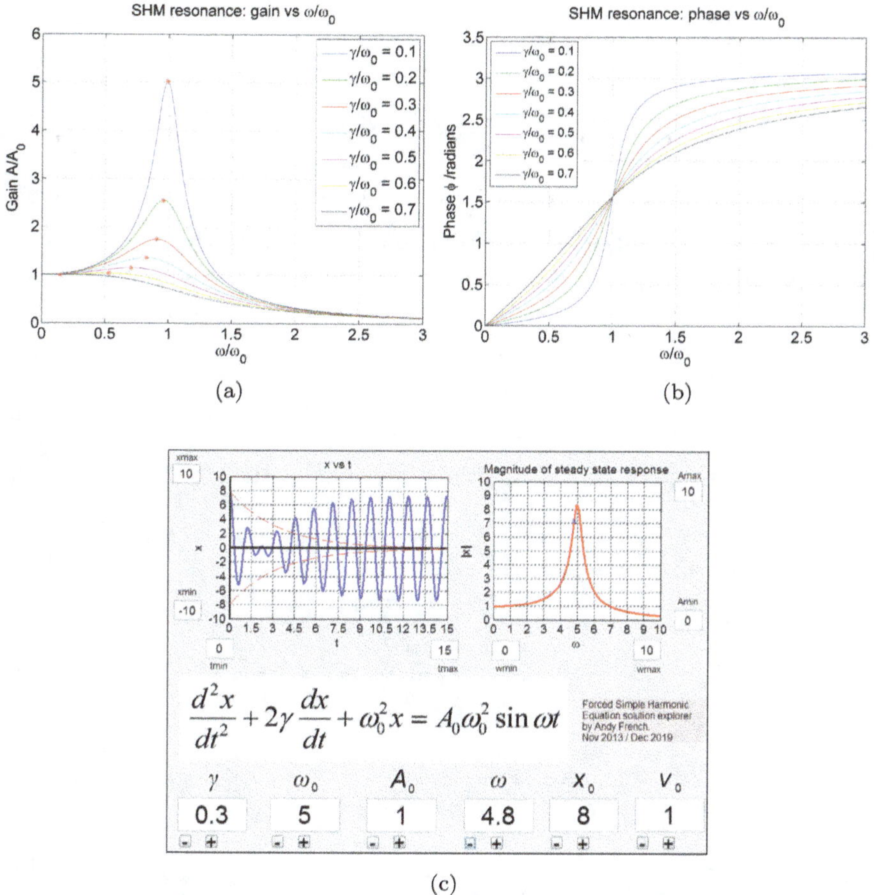

Fig. 4.32. (a) *Gain* and (b) *phase* curves for driven *simple harmonic motion* (SHM). For example, a mass suspended from a Hookean spring that experiences drag proportional to velocity and is driven by a sinusoidally time-varying force of frequency $\frac{\omega}{2\pi}$. (c) A MATLAB app is also presented, which illustrates exponentially decaying *transient* behaviour as well as the *steady-state* response of the system at the driving frequency. In general, an SHM system is modelled by the equation $\ddot{z} + \omega_0^2 z + 2\gamma z = A_0 \omega_0^2 e^{i\omega t}$, which has steady-state solutions of the form

$$z(t) = A e^{i(\omega t - \phi)}, \text{ where } \frac{A}{A_0} = \frac{1}{\sqrt{\left(1 - \frac{\omega^2}{\omega_0^2}\right)^2 + 4\left(\frac{\gamma}{\omega_0}\right)^2 \left(\frac{\omega}{\omega_0}\right)^2}} \text{ and } \phi = \tan^{-1}\left(\frac{2\left(\frac{\gamma}{\omega_0}\right)\left(\frac{\omega}{\omega_0}\right)}{1 - \left(\frac{\omega}{\omega_0}\right)^2}\right). \text{ The}$$

peak gain is $\frac{A_{\max}}{A_0} = \frac{\omega_0^2}{2\gamma\sqrt{\omega_0^2 - \gamma^2}}$ at $\frac{\omega}{\omega_0} = \sqrt{1 - 2\left(\frac{\gamma}{\omega_0}\right)^2}$.

The curve $A(\omega) = \dfrac{A_0 \omega_0^2}{\sqrt{\left(\omega_0^2 - \omega^2\right)^2 + 4\gamma^2 \omega^2}}$ defines the *resonance* curve and has a

maxima of $A_{\max} = \dfrac{A_0 \omega_0^2}{2\gamma\sqrt{\omega_0^2 - \gamma^2}}$ when $\omega = \omega_* = \sqrt{\omega_0^2 - 2\gamma^2}$. It is intriguing that the damping factor γ *reduces* the resonance peak angular frequency from the *natural frequency* ω_0. You can see this effect quite clearly in Fig. 4.32(a). Let's differentiate $A(\omega)$ to determine the coordinates of its peak at $(\omega_*, A_{\max})$:

$$\frac{dA}{d\omega} = -\frac{1}{2} A_0 \omega_0^2 \left(\left(\omega_0^2 - \omega^2\right)^2 + 4\gamma^2 \omega^2\right)^{-\frac{3}{2}} \left(2\left(\omega_0^2 - \omega^2\right)(-2\omega) + 8\gamma^2 \omega\right). \quad (4.335)$$

If $\frac{dA}{d\omega} = 0$ and $\omega > 0$, when $\omega = \omega_*$.

$$\therefore 2\left(\omega_0^2 - \omega_*^2\right)(-2\omega_*) + 8\gamma^2\omega_* = 0, \tag{4.336}$$

$$\Rightarrow \omega_0^2 - \omega_*^2 = 2\gamma^2. \tag{4.337}$$

$$\therefore \omega_* = \sqrt{\omega_0^2 - 2\gamma^2}. \tag{4.338}$$

So, as the damping factor γ gets closer to ω_0, we expect the angular frequency ω_* at the resonance peak to reduce from ω_0. In fact, there will only be a peak at all when

$$\omega_0^2 - 2\gamma^2 > 0. \tag{4.339}$$

$$\therefore \gamma < \frac{\omega_0}{\sqrt{2}}. \tag{4.340}$$

At the resonance peak $(\omega_*, A_{\max})$ the amplitude of the steady-state oscillation is

$$A_{\max} = \frac{A_0\omega_0^2}{\sqrt{\left(\omega_0^2 - \omega_*^2\right)^2 + 4\gamma^2\omega_*^2}} \tag{4.341}$$

$$= \frac{A_0\omega_0^2}{\sqrt{\left(\omega_0^2 - \omega_0^2 + 2\gamma^2\right)^2 + 4\gamma^2\left(\omega_0^2 - 2\gamma^2\right)}} \tag{4.342}$$

$$= \frac{A_0\omega_0^2}{\sqrt{4\gamma^4 + 4\gamma^2\omega_0^2 - 8\gamma^4}} \tag{4.343}$$

$$= \frac{A_0\omega_0^2}{2\gamma\sqrt{\omega_0^2 - \gamma^2}}. \tag{4.344}$$

4.9 A Selection of Books on Waves

My personal standard text on waves and optics is the beautifully illustrated *Optics* by Hecht [17]. The *Feynman Lectures on Physics* [6] is also highly recommended. For a discussion of water waves, there are many excellent fluid dynamics texts, and I would recommend starting with Faber's *Fluid Dynamics for Physicists* [5]. For seismic waves, I would suggest the wonderful and nominatively deterministic Shearer's *Introduction to Seismology* [33], and Fowler's *The Solid Earth* [12]. For a comprehensive discussion of Fourier transforms applied to signal processing, I would highly recommend Richard's *Fundamentals of Radar Signal Processing* [31], although the engineering context is of course electromagnetic waves rather than acoustic waves. In terms of online resources, there is a huge amount that can be Googled, but I would personally start with R. Nave's amazing *Hyperphysics* (http://hyperphysics.phy-astr.gsu.edu/hbase/hframe.html). Much of the material presented here derives from the notes on my website http://www.eclecticon.info/physics_notes_waves.htm, which includes downloads of the ray optic simulations, anamorphic images and much more. There are also plenty of experiment videos too.

I would highly recommend the online simulations at https://phet.colorado.edu/. In my classroom teaching, I use *Geometric Optics* and *Bending Light* on a very regular basis for all year groups between ages 13 and 18. PhET *Wave on a String* and *Waves Intro* are also among my favourites personal favourites. For problems at the pre-university level, see http://www.eclecticon.info/physics_questions.htm, *Mastering Pre-University Physics* [23], *Isaac Physics* at https://isaacphysics.org/ and the website of the British Physics Olympiad https://www.bpho.org.uk/.

Chapter 5

Electromagnetism

One could sensibly argue that the harnessing of electricity represents a technological phase transition between the ancient and modern human worlds. Whereas the application of thermodynamics in the form of heat engines underpinned the 'first industrial revolution' in Europe between about 1760 and 1840, the use of *electromagnetism* was the key paradigm shift associated with the 'second industrial revolution' between about 1870 and the late 1940s. The mass development and use of digital electronics (the 'third industrial revolution') and indeed our current epoch of artificial intelligence, fifth-generation wireless communications and quantum computing (perhaps a 'fourth industrial revolution'), while certainly absolutely transformative, might also perhaps be viewed as a logical consequence of just two key ideas that we associate with a comprehension of the fundamentals of electricity: first, that all matter is comprised of *atoms* and, second, that these atoms have components that carry the property of *charge*. The forces between these charges (mediated by *electric* and *magnetic fields*) are *far* stronger than the force of gravity, which binds together mass in stars and planets.[1] It is electromagnetism that binds *us* and indeed sculpts the molecular and atomic structures that comprise the intricate details of our Universe. The practical essence of electromagnetism is to harness the energies associated with electric and magnetic fields using materials which allow for the movement and temporary separation of charges from their positive and negative pairings within neutral atoms. We will commence this chapter by building up simple models of resistor networks and power cables, i.e. metal conductors which enable the flow of charge, and then apply the ideas of charge conservation and energy transformation in an electrical circuit context.

James Clark Maxwell's[2] four equations (which are presented at the end of this chapter) describe how (vector) *electric fields* $\mathbf{E}$ and *magnetic flux densities*

---

[1] Both the gravitational and electric forces between objects have the same inverse-square law dependence on separation r. However, their respective strengths vary enormously. The ratio between the electrostatic repulsion between two electrons and their gravitational attraction is $\frac{F_E}{F_G} = \frac{e^2}{4\pi\varepsilon_0 r^2} \Big/ \frac{Gm_e^2}{r^2} \approx 4.2 \times 10^{42}$.

[2] James Clerk Maxwell (1831–1879).

195

B can be calculated from, respectively, the presence and motion of electric charges. Using Lorentz's[3] equation

$$\mathbf{F} = q\,(\mathbf{E} + \mathbf{v} \times \mathbf{B}),$$

one can then calculate the force $\mathbf{F}$ on an object which carries charge q coulombs, moving at velocity $\mathbf{v}$ in the vicinity of vector fields $\mathbf{E}$ and $\mathbf{B}$. This equation can then be combined with Newton's second law (mass $\times$ acceleration = vector sum of forces) to incorporate electromagnetic (EM) effects into mechanics. We will continue our chapter with models of the fields of *electric and magnetic dipoles* and then extend the fields-from-charges concept to *capacitors*, i.e. practical systems for separating charges and storing energy. We will then consider the microscopic origins of magnetism and develop *Curie's model of paramagnetism* and a numerical recipe for the *Ising/Metropolis model* of the more powerful *ferromagnetism*, which gives rise to permanent magnets and the highly useful temporary magnetic behaviour of iron. We will then conduct a classic tour of models systems that employ electric and magnetic effects: the *toroidal inductor* and *Rogowski coil*, the *Hall probe*, the *mass spectrometer* and *fine beam tube*, the *velocity selector* and, finally, the *cyclotron*. Our focus then moves to *EM induction* using a model of the *frequency response of a transformer* and then *resonance in an LCR (inductor, capacitor, resistor) circuit*, which lies at the heart of electronic devices for radio transmission and reception. We then derive the *Fresnel equations* of transmission and reflection of polarised EM waves upon a boundary between two media and, finally, conclude the chapter with an exposition of *Maxwell's equations* and some of the key derivations, such as the emergence of EM waves (that are necessarily transverse and travel at $c = 2.998 \times 10^8$ ms^{-1} in a vacuum), continuity equations for electric and magnetic fields on boundaries and *Poynting's formula* for the flow of radiative energy.

5.1   Power Cables and Continued Fraction Resistor Networks

5.1.1   *Resistors in series*

Consider a *series* circuit with applied voltage (i.e. *electromotive force*, or EMF) V_0 where current I flows through a pair of fixed resistors R_1 and R_2. That is, the voltage source and the resistors form a single closed loop, connected by electrically conductive wires, as illustrated in Fig. 5.1(a). From the definition of resistance, the potential differences across each resistor are

$$V_1 = IR_1, \tag{5.1}$$

$$V_2 = IR_2. \tag{5.2}$$

---

[3] Hendrik Lorentz (1853–1928).

Since potential difference is the electrical energy expended per unit charge, conservation of energy (i.e. *Kirchhoff's*[4] *second law*[5] when applied to loops in electrical circuits) means

$$V_0 = V_1 + V_2. \tag{5.3}$$

Now, if total resistance is R,

$$V_0 = IR. \tag{5.4}$$

Hence,

$$IR = IR_1 + IR_2. \tag{5.5}$$

$$\therefore R = R_1 + R_2, \tag{5.6}$$

i.e. *resistances add up arithmetically when wired in series.*

5.1.2 *Resistors in parallel*

If the pair of resistors were now wired in a *parallel* arrangement, each resistor is now separately connected to the applied voltage source V_0, as illustrated in Fig. 5.1(b). Current I_1 passes through the upper loop, and current I_2 passes through the lower loop. From the definition of resistance,

$$V_0 = I_1 R_1, \tag{5.7}$$

$$V_0 = I_2 R_2. \tag{5.8}$$

The total current draw from the voltage source is

$$I = I_1 + I_2, \tag{5.9}$$

i.e. a *statement of continuity of charge*, which is otherwise known as *Kirchhoff's first law*, i.e. the *sum of currents entering a point in an electrical circuit = sum of currents exiting this point.* If the total resistance of the circuit is R and therefore $V_0 = IR$,

$$\underbrace{\frac{V_0}{R}}_{I} = \underbrace{\frac{V_0}{R_1}}_{I_1} + \underbrace{\frac{V_0}{R_2}}_{I_2}. \tag{5.10}$$

$$\therefore \frac{1}{R} = \frac{1}{R_1} + \frac{1}{R_2}, \tag{5.11}$$

i.e. *reciprocals of resistances add up when wired in parallel.*[6]

---

[4] Gustav Kirchhoff (1824–1887).

[5] *Kirchhoff's second law* states that the sum of the EMFs $\sum_j \varepsilon_j$ around any closed loop of an electrical circuit equals the sum of the potential differences around the same loop. This is essentially a statement of conservation of energy. Using the definition of resistance $R_i = V_i/I_i$, a sum of potential differences equates to a sum of products of current and resistance, i.e. $\sum_j \varepsilon_j = \sum_i V_i = \sum_i I_i R_i$.

[6] The addition of quantities in reciprocals is also known as a *harmonic* addition. The *arithmetic mean* of 1,2,3 is $\frac{1}{3}(1+2+3) = 2$, whereas the *harmonic mean* is $\left(\frac{1^{-1}+2^{-2}+3^{-1}}{3}\right)^{-1} = 1\frac{7}{11} \approx 1.636$.

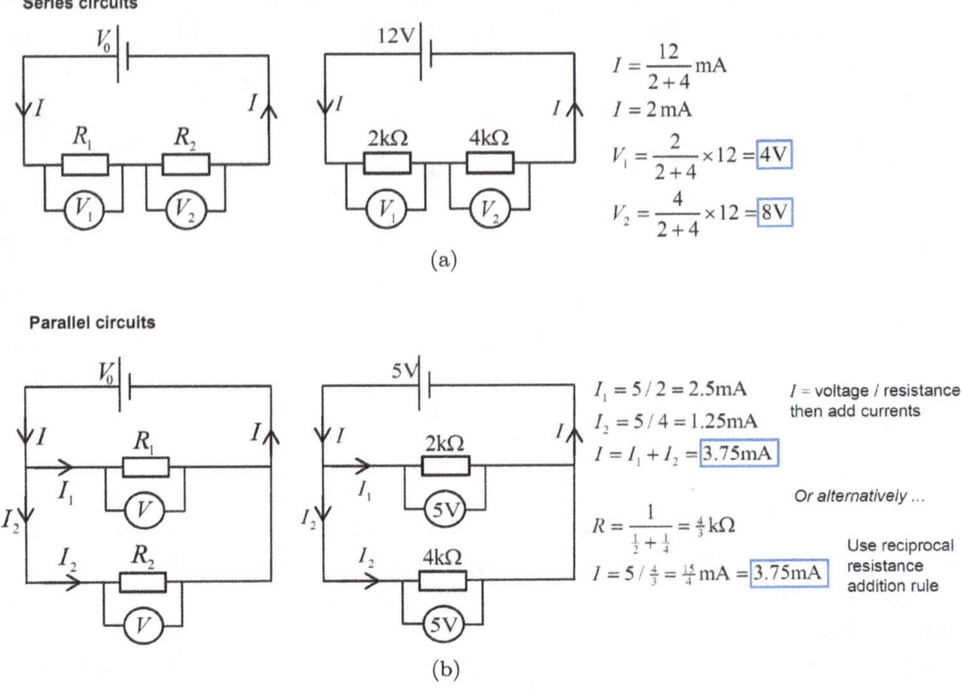

Fig. 5.1. Series (a) and parallel (b) arrangements of electrical resistors. For a series arrangement, continuity of current means the same current I flows through both resistors R_1 and R_2. Conservation of energy means the sum of the energy per unit charge (voltage) dissipated as heat by the resistors, i.e. $V_1 + V_2$, must equate to the applied voltage V_0. Putting these together, combined with the definition of resistance $R = V/I$, implies that $R_1 + R_2$ is the total resistance. For a parallel arrangement, each loop has the same applied voltage. Continuity of current means $I = I_1 + I_2$, which leads to parallel resistances adding in reciprocals, i.e. $\frac{1}{R_{\text{total}}} = \frac{1}{R_1} + \frac{1}{R_2}$.

### 5.1.3	*Resistance of a wire from electric sausages, and how to calculate the diameter of a power cable*

The rules for series and parallel resistance combinations help to explain the formula for the resistance of a wire of (uniform) cross-section A and length l, which is illustrated in Fig. 5.2. Since resistances sum up (in the conventional arithmetic sense) for resistors in series, the total resistance of a wire must be proportional to l. We could imagine a wire cut up into 'wire sausages', each with resistance δR_i. Adding them in series yields the total resistance $R = \sum_i \delta R_i$. Using the same idea, we could bunch up the electric sausages into a parallel arrangement (like a bunch of dynamite sticks from a Road Runner vs. Wile E. Coyote cartoon) to increase the overall cross-section. The cross-section A is proportional to the number of sausages we bunch up. Since parallel resistors add up in reciprocals, we therefore expect total resistance R to be proportional to $\frac{1}{A}$. Let's explore this with a very simple example. If a resistor is connected to a 1.0 V supply and draws 1.0 A of current, it will have a resistance of 1.0 Ω. If you add another identical resistor in parallel, the potential difference across each resistor is still 1.0 V; therefore, a total current of 2.0 A is now

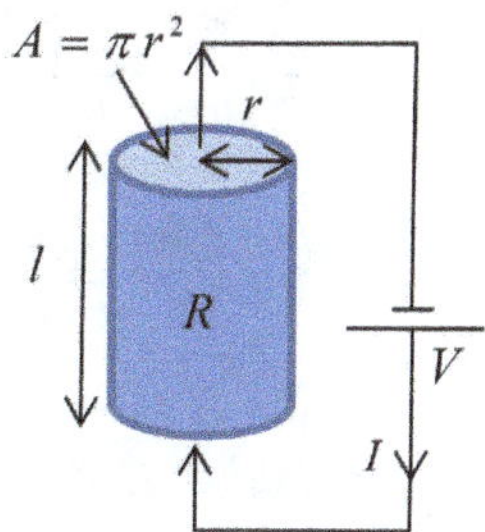

Fig. 5.2. The resistance R of a uniform conductor of resistivity ρ, length l and cross-section A is $R = \frac{\rho l}{A}$.

drawn. The total resistance is now $R = \frac{1.0\,\text{V}}{2.0\,\text{A}} = 0.5\,\Omega$. Extending this argument to N identical resistors in parallel, $R = \frac{1.0\,\Omega}{N}$. So, if the cross-sectional area A of our electric sausages is proportional to the number of sausages N wired in parallel, then $R \propto \frac{1}{A}$.

In summary, the resistance of a wire of radius r is

$$R = \frac{\rho l}{\pi r^2}, \tag{5.12}$$

and

$$R = \frac{\rho l}{A} \tag{5.13}$$

for a generic prism-shaped resistor of cross-section A. The *resistivity* ρ is a bulk material property. The resistivity of aluminium is about $2.8 \times 10^{-8}\,\Omega\text{m}$ and ρ for copper is about $1.7 \times 10^{-8}\,\Omega\text{m}$.

If the resistor doesn't have a uniform cross-section, then add up thin salami-slice sections in series to calculate the total resistance. This will result in an *integration* exercise if a formula is known, which relates cross-section to distance along the resistor's length. A relatively straightforward example is a *truncated* conical[7] resistor (of uniform resistivity ρ) of (untruncated) height l and base radius r. Let the first a of the cone be removed (from the apex) to create the resistor, i.e. the height of the truncated conical resistor is $h = l - a$. If the cone is placed on its side, with the apex being the origin of a Cartesian x, y coordinate system, the cone will form a volume or revolution about the x-axis of the line $y = \frac{r}{l}x$ from $0 \le x \le l$. The resistance of the (truncated) cone is

$$R = \int_a^l \frac{\rho dx}{\pi y^2}$$

$$= \frac{\rho l^2}{\pi r^2} \int_a^l \frac{1}{x^2} dx$$

$$= \frac{\rho l^2}{\pi r^2} \left[-\frac{1}{x} \right]_a^l$$

---

[7]You need to truncate the conical resistor since a resistor of zero cross-section (i.e. the 'first salami slice') would have infinite resistance!

$$= \frac{\rho l^2}{\pi r^2}\left(\frac{1}{a} - \frac{1}{l}\right).$$

$$\therefore R = \frac{\rho l}{\pi r^2}\left(\frac{l}{a} - 1\right).$$

We can use the cylindrical resistor formula $R = \frac{\rho l}{\pi r^2}$ to estimate the radius of a high-voltage power cable (more likely a bundle of cables) that conveys electrical power from a generating station to a city. Assuming that a $V = 765$ kV transmission line, carrying $P = 1,000$ MW of power, runs at a $\alpha = 1\%$ loss over $l = 160$ km, what is r? We solve this problem algebraically so that our model will be applicable for any inputs of V, P, l and α.

The current I transmitted along the line can be determined from the formula

$$P = VI. \tag{5.14}$$

$$\therefore I = \frac{P}{V}. \tag{5.15}$$

In our case,

$$I = \frac{1,000 \times 10^6}{765 \times 10^3}\ \text{A} = 1,307\ \text{A}. \tag{5.16}$$

The power loss in the cable of resistance R is

$$P_{\text{loss}} = \alpha P = I^2 R. \tag{5.17}$$

Therefore, using $I = \frac{P}{V}$, the cable resistance is

$$R = \frac{\alpha P}{\left(\frac{P}{V}\right)^2}. \tag{5.18}$$

$$\therefore R = \frac{\alpha V^2}{P}. \tag{5.19}$$

In our case, the power cable resistance is

$$R = \frac{0.01 \times \left(765 \times 10^3\right)^2}{1,000 \times 10^6}\ \Omega = 5.85\ \Omega. \tag{5.20}$$

Using the formula $R = \frac{\rho l}{\pi r^2}$ for the resistance of a wire of radius r and length l,

$$\frac{\rho l}{\pi r^2} = \frac{\alpha V^2}{P}. \tag{5.21}$$

Therefore, the power cable radius r is

$$r = \sqrt{\frac{P \rho l}{\pi \alpha V^2}}. \tag{5.22}$$

If we assume that the cable is made of aluminium with $\rho = 2.8 \times 10^{-8}$ Ωm,

$$r = \sqrt{\frac{1,000 \times 10^6 \times 2.8 \times 10^{-8} \times 160 \times 10^3}{\pi \times 0.01 \times \left(765 \times 10^3\right)^2}}\ \text{m} \tag{5.23}$$

$$= 15.6\ \text{mm}, \tag{5.24}$$

which seems physically reasonable, although impressively small if you think about the city-powering electrical energy the power cable transmits.

Note that 'full depth uniform conduction' is an approximation. For alternating current (AC), conduction (in magnitude) diminishes inwards from the surface, and the majority occurs within a *skin depth* δ. This is because *eddy currents* induced in a conductor impede current flow in the interior and reinforce it on the surface. A formula for skin depth is derived by Bleaney & Bleaney (see [2], p. 236):

$$\frac{\delta}{\text{mm}} \approx 1000 \times \sqrt{\frac{\rho}{\pi \mu_0 \mu f}}. \tag{5.25}$$

$\mu_0 = 1.257 \times 10^{-6} \text{NA}^{-2}$ is *vacuum magnetic permeability* and μ is the *relative permeability* of the conductor, which is a quantitative measure of how easily magnetisable it is. Aluminium, air and water all have a μ value of about unity. However, nickel has a relative permeability of about 100, while μ of pure iron is about 5,000. f is the frequency of the AC. Using $\rho = 2.8 \times 10^{-8}$ Ωm, $f = 50$ Hz and $\mu = 1$,

$$\delta \approx 11.9 \text{ mm}. \tag{5.26}$$

So, although it might be possible to reduce the radius of our power cable by a few millimetres if we use AC transmission, full-depth conduction (at a low frequency, such as 50 Hz) is not such a bad approximation. Since skin depth $\delta \propto \frac{1}{\sqrt{f}}$, higher-frequency signals would have a smaller skin depth.

5.1.4 *Continued fraction resistor networks: The golden circuit and the circuit of Deep Thought*

The series and parallel resistor combination rules allow us to compute the resistance $R_{\text{total}} = xR$ of a ladder network of the design described in Fig. 5.3. The total fractional resistance $\frac{R_{\text{total}}}{R}$ takes the form of a *continued fraction*

$$x = a_0 + \cfrac{1}{a_1 + \cfrac{1}{a_2 + \cfrac{1}{a_3 + \cfrac{1}{a_4 + \cfrac{1}{a_5 + \cdots}}}}}. \tag{5.27}$$

The resistance of the first resistor is $a_0 R$. To construct a *golden circuit*, let all the resistors take the same value:

$$a_i = 1, \tag{5.28}$$

i.e.

$$x = 1 + \cfrac{1}{1 + \cfrac{1}{1 + \cfrac{1}{1 + \cfrac{1}{1 + \cdots}}}}. \tag{5.29}$$

The *infinite* continued fraction z that divides 1 in $x = 1 + \frac{1}{z}$ is *also* x. We can, therefore, express x *in terms of itself*, i.e. $x = 1 + \frac{1}{x}$. This is a piece of mathematical magic that should bring a smile to all students, especially as it results in a quadratic

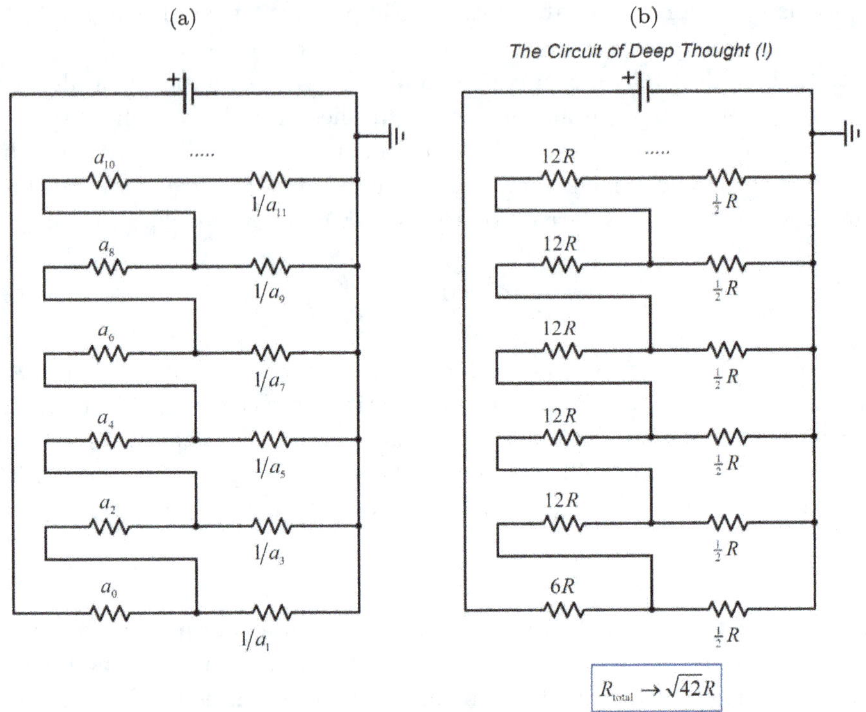

Fig. 5.3. (a) Infinite resistor 'ladder' with left-branch resistances $R_{2n} = a_{2n} R$ and right-branch resistances $R_{2n+1} = \dfrac{R}{a_{2n+1}}$, where $n = 0, 1, 2, 3, 4, 5, \ldots$. This yields a total fractional resistance of $\dfrac{R}{R_{\text{total}}} = a_0 + \cfrac{1}{a_1 + \cfrac{1}{a_2 + \cfrac{1}{a_3 + \cfrac{1}{a_4 + \cfrac{1}{a_5 + \cdots}}}}}$. If $a_n = 1$, then $\dfrac{R}{R_{\text{total}}} = \frac{1}{2}\left(1 + \sqrt{5}\right)$, i.e. the *golden ratio*. If the a values are as shown in (b), the total resistance ratio is $\sqrt{42}$. Since this is the root of 'the answer to the ultimate question', perhaps this circuit is at the heart of Douglas Adam's übercomputer *Deep Thought*.

equation for x, which can be easily solved analytically:

$$x = 1 + \frac{1}{x} \tag{5.30}$$

$$x^2 = x + 1 \tag{5.31}$$

$$x^2 - x - 1 = 0 \tag{5.32}$$

$$\left(x - \tfrac{1}{2}\right)^2 - \tfrac{1}{4} - 1 = 0 \tag{5.33}$$

$$x = \tfrac{1}{2} \pm \tfrac{1}{2}\sqrt{5}. \tag{5.34}$$

Since $x > 1$, we can take the positive root

$$x = \tfrac{1}{2}\left(1 + \sqrt{5}\right). \tag{5.35}$$

x is the *golden ratio*[8] $\phi \approx 1.618$, the asymptotic ratio of consecutive terms of the *Fibonacci sequence* and the proportions of ancient Greek columns and aesthetically pleasing rectangles.

Another similarly continued fraction circuit has resistances, for $n > 0$, set by the sequence terms

$$a_0 = 6, \tag{5.36}$$

$$a_{2n-1} = 2, \tag{5.37}$$

$$a_{2n} = 12, \tag{5.38}$$

i.e. a total resistance $R_{\text{total}} = xR$, where

$$x = 6 + \cfrac{1}{2 + \cfrac{1}{12 + \cfrac{1}{2 + \cfrac{1}{12 + \cfrac{1}{2 + \cdots}}}}}. \tag{5.39}$$

Like the *golden circuit*, we can express this continued fraction *recursively*, i.e. in terms of itself:

$$x + 6 = 12 + \cfrac{1}{2 + \cfrac{1}{12 + \cfrac{1}{2 + \cfrac{1}{12 + \cfrac{1}{2 + \cdots}}}}} \tag{5.40}$$

$$x + 6 = 12 + \cfrac{1}{2 + \cfrac{1}{x+6}} \tag{5.41}$$

$$x = 6 + \cfrac{x + 6}{2x + 13} \tag{5.42}$$

$$x = \cfrac{12x + 78 + x + 6}{2x + 13} \tag{5.43}$$

$$2x^2 + 13x = 13x + 84 \tag{5.44}$$

$$x^2 = 42. \tag{5.45}$$

$$\therefore x = \sqrt{42}.$$

After seven and a half million years of silent computation, the übercomputer *Deep Thought* in Douglas Adam's *The Hitchhiker's Guide to the Galaxy* returned the answer of 'forty two' to the 'Ultimate Question of Life, the Universe and Everything'. So, I would like to think the $a_0 = 6$, $a_{2n-1} = 2$, $a_{2n} = 12$ continued fraction circuit must be 'at the root' of its processor core! (You can groan now and observe the tumbleweed rolling past for the next seven and a half million years.)

---

[8] http://www.eclecticon.info/index_htm_files/Numbers%20-%20Fibonacci%20and%20Golden%20Ratio.pdf.

5.2   The Maximum Power Theorem

Consider a series electrical circuit comprising a cell of electromotive force (EMF) ε volts, with *internal resistance* r ohms and a *load* of resistance R ohms. The internal resistance of the cell can be modelled as a fixed resistor wired in series with the load. The total current I drawn from the cell is

$$I = \frac{\varepsilon}{r + R} \tag{5.46}$$

and the power dissipated by load R is

$$P = I^2 R. \tag{5.47}$$

$$\therefore P = \frac{\varepsilon^2 R}{(r + R)^2}.$$

The curve of P versus R is plotted in Fig. 5.4(d) and has a maximum at $\left(r, \frac{\varepsilon^2}{4r}\right)$ before decaying to zero as $\frac{1}{R}$ when $R \gg r$. The (R, P) coordinates of the maximum can be determined by differentiating P with respect to R. Using the *quotient rule,*

$$\frac{dP}{dR} = \varepsilon^2 \frac{(r + R)^2 - 2(r + R)R}{(r + R)^4} \tag{5.48}$$

$$= \varepsilon^2 \frac{r + R - 2R}{(r + R)^3} \tag{5.49}$$

$$= \varepsilon^2 \frac{r - R}{(r + R)^3}. \tag{5.50}$$

So, $\frac{dP}{dR} = 0$ when $R = r$ and $P = \frac{\varepsilon^2}{4r}$.

This *maximum power theorem* result can be determined in at least two further ways, both of which yield additional insight into the system. The first involves the curve of P versus I rather than R. The potential difference across the load is

$$V = \varepsilon - Ir \tag{5.51}$$

and therefore the power dissipated in the load is

$$P = IV = I\left(\varepsilon - Ir\right). \tag{5.52}$$

This is an *inverted quadratic variation* (see Fig. 5.4(d)) of P with I, with roots of $I = 0$ and $I = \frac{\varepsilon}{r}$. By symmetry, the peak of the parabola $P(I)$ is at $I = \frac{\varepsilon}{2r}$, which corresponds to a maximum power of

$$P = \frac{\varepsilon}{2r}\left(\varepsilon - \frac{\varepsilon}{2r}r\right) \tag{5.53}$$

$$= \frac{\varepsilon^2}{2r}\left(1 - \tfrac{1}{2}\right). \tag{5.54}$$

$$\therefore P = \frac{\varepsilon^2}{4r}.$$

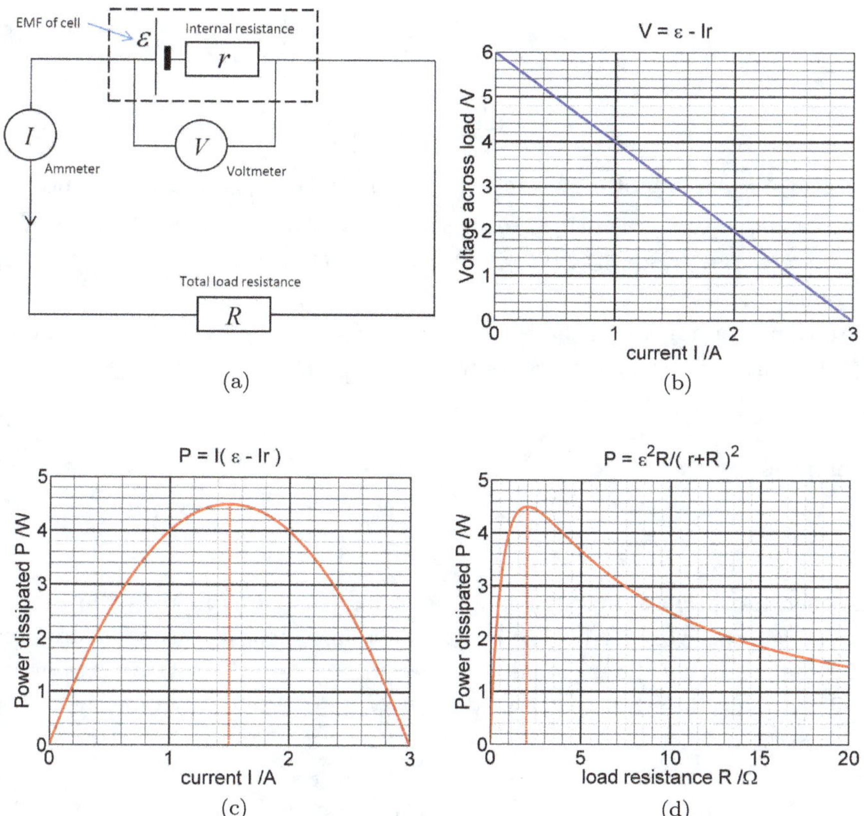

Fig. 5.4. (a) A load of resistance R is connected to a cell of EMF ε volts and internal resistance r ohms. Current I is drawn from the cell and potential difference V is across the load (or indeed across the terminals of the cell). (b) Load potential difference V is plotted against current I. By Kirchhoff's second law, this is a linear relationship: $V = \varepsilon - Ir$. Experimentally, r and ε can therefore be determined from a line of best fit of a graph of measurements of I vs. V, which can be changed by using a variable resistor R. (c) Power dissipated by the load R is $P = IV = I(\varepsilon - Ir)$. This is an *inverted parabola* in P vs. I, with peak $P_{\max} = \frac{\varepsilon^2}{4r}$ at $I = \frac{\varepsilon}{2r}$. (d) Alternatively, we can plot P vs. R, where the maximum of P is at $R = r$. In plots (b)–(d), $r = 2.00\ \Omega$ and $\varepsilon = 6.00$ V.

If we equate $P = I^2 R$ using $P = \frac{\varepsilon^2}{4r}$ and $I = \frac{\varepsilon}{2r}$, then

$$\frac{\varepsilon^2}{4r} = \left(\frac{\varepsilon}{2r}\right)^2 R. \tag{5.55}$$

$$\therefore r = R. \tag{5.56}$$

An alternative and perhaps a more direct way of showing this result is to express the already factorised $P = I\,(\varepsilon - Ir)$ in *completed square form* instead:

$$P = I\,(\varepsilon - Ir) \tag{5.57}$$

$$= I\varepsilon - I^2 r \tag{5.58}$$

$$= -r\left\{I^2 - \frac{\varepsilon}{r}I\right\} \tag{5.59}$$

$$= -r \left\{ \left(I - \frac{\varepsilon}{2r} \right)^2 - \frac{\varepsilon^2}{4r^2} \right\}. \tag{5.60}$$

$$\therefore P = \frac{\varepsilon^2}{4r} - r \left(I - \frac{\varepsilon}{2r} \right)^2. \tag{5.61}$$

So, P has a maximum of $\frac{\varepsilon^2}{4r}$ since $\left(I - \frac{\varepsilon}{2r} \right)^2$ must be a positive quantity. The smallest amount we can subtract from $\frac{\varepsilon^2}{4r}$ is zero, which occurs when $I - \frac{\varepsilon}{2r} = 0$ $\Rightarrow I = \frac{\varepsilon}{2r}$. The $r = R$ condition then follows using $P = I^2 R \Rightarrow \frac{\varepsilon^2}{4r} = \left(\frac{\varepsilon}{2r} \right)^2 R$, which can only be true if $r = R$.

Note that you don't even have to evaluate $P_{\max} = \frac{\varepsilon^2}{4r}$ to derive the criteria of $r = R$. If the maximum power occurs when $I = \frac{\varepsilon}{2r}$, then since $I = \frac{\varepsilon}{r+R}$,

$$\frac{\varepsilon}{2r} = \frac{\varepsilon}{r + R}, \tag{5.62}$$

which implies

$$2r = r + R. \tag{5.63}$$

$$\therefore r = R. \tag{5.64}$$

Another and perhaps even more illuminating way of explaining the maximum power theorem was explained to me by my friend and mentor Anson Cheung. In this case, let's start with a dimensional argument that power P must be proportional[9] to $\frac{\varepsilon^2}{r}$. What multiplies this (which will involve r and R) must be a purely numerical factor. Starting with

$$P = \frac{\varepsilon^2 R}{(r + R)^2}, \tag{5.65}$$

let us factor out $\frac{\varepsilon^2}{r}$.

$$\therefore P = \frac{\varepsilon^2}{r} \frac{rR}{(r + R)^2}. \tag{5.66}$$

Now,

$$(r + R)^2 - (r - R)^2 = r^2 + 2rR + R^2 - \left\{ r^2 - 2rR + R^2 \right\} \tag{5.67}$$

$$= 4rR. \tag{5.68}$$

$$\therefore rR = \tfrac{1}{4} \left((r + R)^2 - (r - R)^2 \right).$$

This means

$$P = \frac{\varepsilon^2}{4r} \frac{(r + R)^2 - (r - R)^2}{(r + R)^2} \tag{5.69}$$

$$= \frac{\varepsilon^2}{4r} \left\{ 1 - \left(\frac{r - R}{r + R} \right)^2 \right\}. \tag{5.70}$$

Now, in Chapter 4, we derived the *power reflection coefficient* to be $\left| \frac{Z_1 - Z_2}{Z_1 + Z_2} \right|^2$, where $Z_{1,2}$ is the *impedance* of media (labelled '1' and '2') joined by a boundary.

---

[9]Note that we could also factor out $\frac{\varepsilon^2}{R}$ instead, but $\frac{\varepsilon^2}{r}$ yields a more interesting result!

Waves will be reflected (and transmitted across) this boundary in power proportions based on the value of $\left|\frac{Z_1 - Z_2}{Z_1 + Z_2}\right|^2$ between 0 and 1. For electrical waves, the impedance of a fixed resistor is its resistance. Therefore, $1 - \left(\frac{r-R}{r+R}\right)^2$ represents the *transmission coefficient* of power from a medium of resistance r to a medium of resistance R if we can assume a DC limit and ignore any capacitative or inductive effects resulting from a time-varying electric current. For the transmission coefficient to be unity (i.e. maximised), $r = R$. This concept is of great practical importance to sound and telecommunications engineers. For maximum transmission, you must *impedance match* connected devices such as microphones, audio cables and amplification systems.

5.3   Modelling Electric Fields

Coulomb's[10] law, as illustrated in Fig. 5.5, defines the force $\mathbf{F}$ between two charges q and Q, separated by distance r. The force has magnitude $|\mathbf{F}| = \frac{qQ}{4\pi\varepsilon_0 r^2}$, where *permittivity of free space* $\varepsilon_0 = 8.85 \times 10^{-12}$ Fm^{-1}. The force per unit charge $\mathbf{E} = \frac{\mathbf{F}}{q}$ is the *electric field*. At distance r from charge Q,

$$\mathbf{E} = \frac{Q}{4\pi\varepsilon_0 r^2}\hat{\mathbf{r}},$$

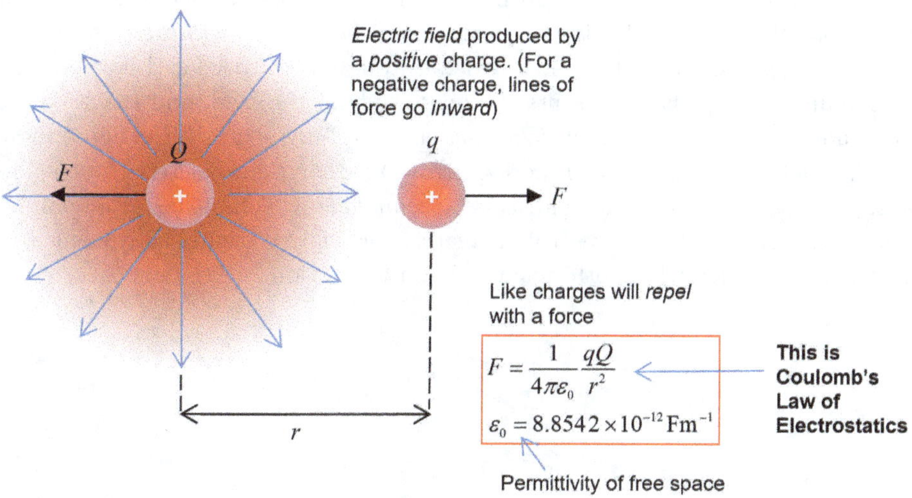

Fig. 5.5. A point charge Q will create an electric field of strength E, which acts radially. The force on another charge q at separation r is $F = qE$, where $E = \frac{Q}{4\pi\varepsilon_0 r^2}$. The permittivity of free space $\varepsilon_0 = 8.8542 \times 10^{-12}$ Fm^{-1}.

---

[10] Charles-Augustin de Coulomb (1736–1806).

where $\hat{\mathbf{r}}$ is a radial unit vector. Note that $\mathbf{E} = -\nabla V$, where ∇ is the *gradient operator*.[11] This means that the *electric potential* resulting from charge Q is

$$V = \frac{Q}{4\pi\varepsilon_0 r}.$$

Applications for different configurations of charges are illustrated in Figs. 5.7 and 5.9. In the ensuing sections, we derive analytic expressions for electric (vector) fields and their associated (scalar) potentials for several classic systems of charge arrangement. We start with *dipoles* (a pair of charges of opposing sign) and then move on to charged conductors separated by an insulator, i.e. *capacitors*. We will consider parallel plate, spherical and cylindrical geometries. Figures 5.7 and 5.9 contain the outputs of a MATLAB electric field plotter. The electric field $\mathbf{E}$ is calculated relative to a 2D Cartesian x, y coordinate system, although it could readily be generalised to 3D. Using the geometry described in Fig. 5.6 and a liberal application of Pythagoras' theorem, if charge Q_i is at location (X_i, Y_i), then the electric field at (x, y) is

$$\mathbf{E}(x,y) = \frac{1}{4\pi\varepsilon_0} \sum_i \frac{Q_i}{(x - X_i)^2 + (y - Y_i)^2} \frac{\hat{\mathbf{x}}(x - X_i) + \hat{\mathbf{y}}(y - Y_i)}{\sqrt{(x - X_i)^2 + (y - Y_i)^2}},$$

where $\hat{\mathbf{x}}, \hat{\mathbf{y}}$ are unit vectors of a Cartesian coordinate system. To generate the plots, an array (matrix) of x, y values are created, and then the horizontal $\hat{\mathbf{x}}$ and vertical $\hat{\mathbf{y}}$ components of $\mathbf{E}(x, y)$ are evaluated for each array element. The plots in Figs. 5.7 and 5.9 are a hybrid of an interpolated coloured surface of $\log_{10}|\mathbf{E}|$ and a *streamline* plot which indicates the direction of $\mathbf{E}$. I think *both* are important, as one needs to appreciate that the field exists throughout a space and not just along the streamlines. $\log_{10}|\mathbf{E}|$ represents the magnitude of the field, and the streamline arrows indicate direction. These *field lines* are helpfully spaced directional indicators (like contour lines, but with added arrows) rather than discrete, independent quantities which represent the field. $\mathbf{E}$ is a *vector field*, which means at all points in space (regardless of granuality—this is merely for computational convenience), we can assign a quantity with both magnitude *and* direction.

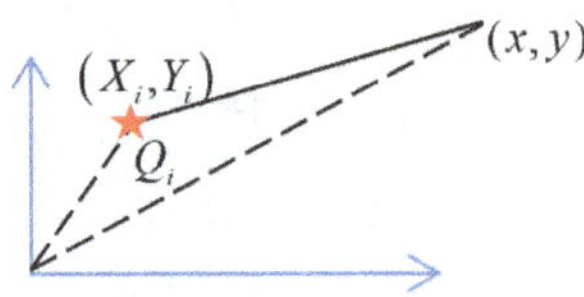

Fig. 5.6. Geometry for a 2D electric field calculator based on Coulomb's law. Charge Q_i is at location (X_i, Y_i), and the field $\mathbf{E}$ is computed at (x, y).

---

[11]If an electric field $\mathbf{E}$ acts in one direction, i.e. $\mathbf{E} = E(x)\hat{\mathbf{x}}$, where $\hat{\mathbf{x}}$ is a Cartesian unit vector in the direction of the x-axis, then $\mathbf{E} = -\nabla V = -\hat{\mathbf{x}}\frac{\partial V}{\partial x} - \hat{\mathbf{y}}\frac{\partial V}{\partial y} - \hat{\mathbf{z}}\frac{\partial V}{\partial z}$ becomes $E(x)\hat{\mathbf{x}} = -\frac{dV}{dx}\hat{\mathbf{x}}$, which means $E = -\frac{dV}{dx}$, i.e. 'field equals negative potential gradient'.

5.4 Electric and Magnetic Dipoles

A *dipole* is a pair of opposing charges $q, -q$ separated by a distance d. In many scenarios in electromagnetism, it is useful to calculate the electric field $\mathbf{E}$ due to one or more sets of dipoles (e.g. radio antennae, polar molecules) rather than individual charges. We derive the *far-field* results for the electric potential V and electric field. 'Far field' means the radial distance from the centre of the dipole is much larger than the charge separation d.

Let charge q be at Cartesian coordinate $(0, \frac{1}{2}d)$ and charge $-q$ be at $(0, -\frac{1}{2}d)$. Consider a location (x, y), and define polar coordinates r, θ such that

$$x = r\cos\theta, \tag{5.71}$$

$$y = r\sin\theta. \tag{5.72}$$

These are illustrated in Fig. 5.7(a). Define polar coordinate *unit vectors*[12] in terms of their Cartesian counterparts:

$$\hat{\mathbf{r}} = \hat{\mathbf{x}}\cos\theta + \hat{\mathbf{y}}\sin\theta, \tag{5.73}$$

$$\hat{\boldsymbol{\theta}} = -\hat{\mathbf{x}}\sin\theta + \hat{\mathbf{y}}\cos\theta. \tag{5.74}$$

Let distances p and s, respectively, be the distance from the charges q and $-q$ to (x, y). Using the cosine rule,

$$p^2 = r^2 + \tfrac{1}{4}d^2 - rd\cos(\tfrac{\pi}{2} - \theta). \tag{5.75}$$

$$\therefore p^2 = r^2 + \tfrac{1}{4}d^2 - rd\sin\theta. \tag{5.76}$$

$$s^2 = r^2 + \tfrac{1}{4}d^2 - rd\cos(\tfrac{\pi}{2} + \theta). \tag{5.77}$$

$$\therefore s^2 = r^2 + \tfrac{1}{4}d^2 + rd\sin\theta. \tag{5.78}$$

Hence,

$$p = r\sqrt{1 + \tfrac{1}{4}\frac{d^2}{r^2} - d\sin\theta}, \tag{5.79}$$

$$s = r\sqrt{1 + \tfrac{1}{4}\frac{d^2}{r^2} + d\sin\theta}. \tag{5.80}$$

Using Coulomb's law, the electric potential at (r, θ) is the sum of the electric potentials due to each charge:

$$V = \frac{q}{4\pi\varepsilon_0}\frac{1}{p} + \frac{-q}{4\pi\varepsilon_0}\frac{1}{s}. \tag{5.81}$$

$$\therefore V = \frac{q}{4\pi\varepsilon_0}\frac{(s - p)}{ps}.$$

---

[12]i.e. the magnitude of the unit vectors are, well, unity! They are also defined to be *orthonormal,* which means they are *mutually perpendicular* as well as having unit length.

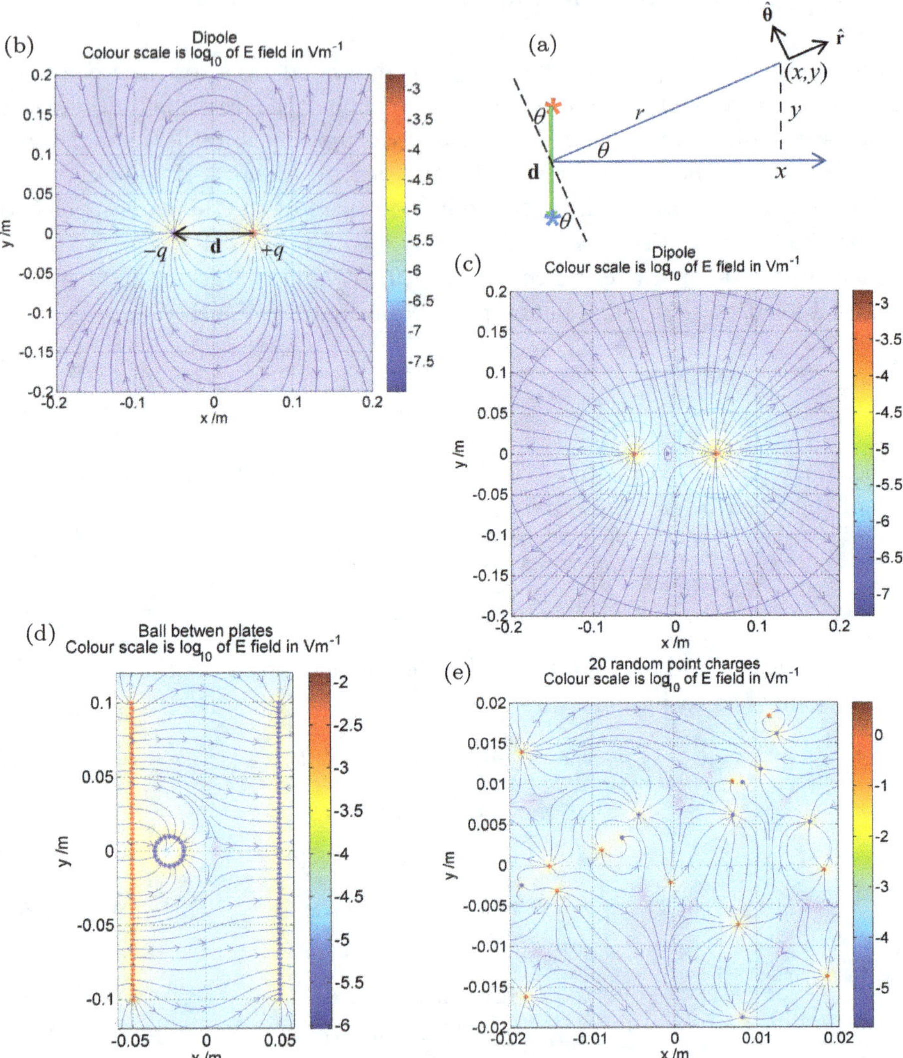

Fig. 5.7. (a) and (b): The electric field $\mathbf{E}$ resulting from two static charges of opposite sign and equal magnitude q, separated by displacement vector $\mathbf{d}$ is given by $\mathbf{E} = \frac{q}{4\pi\varepsilon_0 r^3}\left(2\hat{\mathbf{r}}d\sin\theta - \hat{\boldsymbol{\theta}}d\cos\theta\right) = \frac{3q(\mathbf{d}\cdot\hat{\mathbf{r}})\hat{\mathbf{r}} - q\mathbf{d}}{4\pi\varepsilon_0 r^3}$, where $\hat{\mathbf{r}}$ and $\hat{\boldsymbol{\theta}}$ are the polar coordinate unit vectors. Note that the field strength decays as an *inverse-cube law* with range r. (c) is a repelling dipole, with charges q and $2q$. (d) illustrates the field between a capacitor plate model of positive (red) and negative (blue) charges and a uniformly charged circle of negative charges placed between them. This is a crude model of a charged sphere in a uniform electric field, as, on a real conductive sphere, one would expect the charges to move, i.e. be *polarised* by the field until there are no $\mathbf{E}$ fields tangential to the sphere surface. The charge distribution would therefore *not* be uniform. (e) illustrates the electric fields between a random selection of positive and negative (static) charges.

Now,

$$ps = r^2 \sqrt{\left(1 + \tfrac{1}{4}\frac{d^2}{r^2} - d\sin\theta\right)\left(1 + \tfrac{1}{4}\frac{d^2}{r^2} - d\sin\theta\right)} \tag{5.82}$$

$$= r^2 \sqrt{\left(1 + \tfrac{1}{4}\frac{d^2}{r^2}\right)^2 - \frac{d^2}{r^2}\sin^2\theta}. \tag{5.83}$$

If we assume $r \gg d$, then

$$ps \approx r^2. \tag{5.84}$$

Using the same condition, we can binomially expand the expressions for p and s:

$$p \approx r + \tfrac{1}{8}\frac{d^2}{r^2} - \tfrac{1}{2}d\sin\theta, \tag{5.85}$$

$$s \approx r + \tfrac{1}{8}\frac{d^2}{r^2} + \tfrac{1}{2}d\sin\theta, \tag{5.86}$$

which means

$$s - p \approx d\sin\theta. \tag{5.87}$$

Hence, in the limit $r \gg d$,

$$V \approx \frac{q}{4\pi\varepsilon_0}\frac{d\sin\theta}{r^2}. \tag{5.88}$$

This is immediately interesting since the dipole potential decays as $\frac{1}{r^2}$, whereas the potential of an isolated charge decays as $\frac{1}{r}$.

If we define the dipole separation vector as

$$\mathbf{d} = \hat{\mathbf{y}}d = d\left(\hat{\mathbf{r}}\sin\theta + \hat{\boldsymbol{\theta}}\cos\theta\right), \tag{5.89}$$

then, since $d\sin\theta = \mathbf{d} \cdot \hat{\mathbf{r}}$ (see Fig. 5.7), we can therefore express the electric potential V as

$$V \approx \frac{q}{4\pi\varepsilon_0}\frac{\mathbf{d} \cdot \hat{\mathbf{r}}}{r^2}. \tag{5.90}$$

This formulation may be helpful when considering systems of many dipoles of different orientations expressed by $\mathbf{d}$. One would add up a sum of $\frac{q_i}{4\pi\varepsilon_0}\frac{\mathbf{d}_i \cdot \hat{\mathbf{r}}_i}{r_i^2}$ potentials to find the overall spatially varying potential $V(x, y, z)$. Note that the radial unit vector $\hat{\mathbf{r}}$ and radial distance r would have to be computed separately for each dipole, but this is a calculation a computer programme could easily perform. The quantity $q\mathbf{d}$ is called the *electric dipole moment*.

We can compute the electric field of a single isolated dipole by using the vector calculus definition of $\mathbf{E}$ in terms of potential V. We will need to use the plane polar-coordinate version[13] of the gradient operator ∇ to evaluate this in terms of r and θ:

$$\mathbf{E} = -\nabla V \tag{5.91}$$

$$= -\frac{\partial V}{\partial r}\hat{\mathbf{r}} - \frac{1}{r}\frac{\partial V}{\partial \theta}\hat{\boldsymbol{\theta}}. \tag{5.92}$$

Using $V = \frac{q}{4\pi\varepsilon_0}\frac{d\sin\theta}{r^2}$,

$$\frac{\partial V}{\partial r} = -\frac{2q}{4\pi\varepsilon_0}\frac{d\sin\theta}{r^3}, \tag{5.93}$$

$$\frac{1}{r}\frac{\partial V}{\partial \theta} = \frac{q}{4\pi\varepsilon_0}\frac{d\cos\theta}{r^3}. \tag{5.94}$$

Therefore,

$$\mathbf{E} = \frac{qd}{4\pi\varepsilon_0 r^3}\left(2\hat{\mathbf{r}}\sin\theta - \hat{\boldsymbol{\theta}}\cos\theta\right). \tag{5.95}$$

Now, since $\mathbf{d} = \hat{\mathbf{r}}d\sin\theta + \hat{\boldsymbol{\theta}}d\cos\theta$ and $\mathbf{d}\cdot\hat{\mathbf{r}} = d\sin\theta$,

$$\hat{\boldsymbol{\theta}}d\cos\theta = \mathbf{d} - (\mathbf{d}\cdot\hat{\mathbf{r}})\,\hat{\mathbf{r}}.$$

Hence,

$$\mathbf{E} = \frac{q}{4\pi\varepsilon_0 r^3}\left(2\hat{\mathbf{r}}d\sin\theta - \hat{\boldsymbol{\theta}}d\cos\theta\right) \tag{5.96}$$

$$= \frac{q}{4\pi\varepsilon_0 r^3}\left(2\left(\mathbf{d}\cdot\hat{\mathbf{r}}\right)\hat{\mathbf{r}} - \mathbf{d} + \left(\mathbf{d}\cdot\hat{\mathbf{r}}\right)\hat{\mathbf{r}}\right). \tag{5.97}$$

$$\therefore \mathbf{E} = \frac{3q\left(\mathbf{d}\cdot\hat{\mathbf{r}}\right)\hat{\mathbf{r}} - q\mathbf{d}}{4\pi\varepsilon_0 r^3}. \tag{5.98}$$

The same model can be used to determine the magnetic flux density $\mathbf{B}$ in the far field of a *magnetic dipole* (e.g. a bar magnet or, indeed, many planets such as Earth[14]):

$$\mathbf{B} = \frac{\mu_0 m}{4\pi r^3}\left(2\hat{\mathbf{r}}\sin\theta - \hat{\boldsymbol{\theta}}\cos\theta\right). \tag{5.99}$$

The mathematical difference between the electric and magnetic dipole fields is the quantity $\mu_0 m$. In the electric dipole field, this is $\frac{qd}{\varepsilon_0}$. For a magnetic dipole formed from a small current loop of radius a, carrying current I, the *magnetic dipole moment m* is

$$m = I\pi a^2. \tag{5.100}$$

The quantity $\mu_0 = 1.257 \times 10^{-6}$ NA^{-2} (to 4.s.f) is the *vacuum permeability*, which we encountered earlier in our discussion of skin depth in an electrical

---

[13] The Cartesian version of the gradient operator upon scalar field V would be $\nabla V = \hat{\mathbf{x}}\frac{\partial V}{\partial x} + \hat{\mathbf{y}}\frac{\partial V}{\partial y} + \hat{\mathbf{z}}\frac{\partial V}{\partial z}$. However, since $V = \frac{q}{4\pi\varepsilon_0}\frac{d\sin\theta}{r^2}$ is defined in terms of plane polar coordinates r, θ it makes sense to use the plane polar incarnation of $\nabla V = -\frac{\partial V}{\partial r}\hat{\mathbf{r}} - \frac{1}{r}\frac{\partial V}{\partial \theta}\hat{\boldsymbol{\theta}}$.

[14] The magnetic field of the Earth can be approximated by a dipole at large distances, but on the scale of the Earth's core itself, the field is much more complicated in structure [12, p. 34, Fowler].

conductor. The meaning of μ_0 is perhaps best illustrated via *Ampère's*[15] *theorem*, which, if expressed for a non-magnetisable 'magnetic circuit',[16] relates magnetic flux density $\mathbf{B}$ to the current I enclosed by a loop with a loop direction path element vector $d\mathbf{l}$. Ampère's theorem states that

$$\oint_{\text{loop}} \mathbf{B} \cdot d\mathbf{l} = \mu_0 I. \qquad (5.101)$$

This means the magnetic flux density $B = |\mathbf{B}|$ at (radial) distance r from a long current carrying wire is

$$B = \frac{\mu_0 I}{2\pi r} \qquad (5.102)$$

since $\mathbf{B}$ forms (tangential) loops around the wire and therefore $\mathbf{B} \cdot d\mathbf{l} = B \times 2\pi r$.

A magnetic flux density of 2.00×10^{-7} Tesla is therefore produced at a distance of 1.00 m from a wire carrying a current of 1.00 A.

5.5 Capacitors

5.5.1 *A device for separating charge and storing energy*

A *capacitor* is a fundamental electrical component (like a resistor and an inductor) which, in essence, is a sandwich of two conductors and an insulator, as illustrated in Fig. 5.8(c). The structure allows for charge to be transferred from one conductor to the other, and this separation is to be maintained if the capacitor is disconnected from a voltage source. If the insulator can be taken as ideal, no current can flow between the conductors; therefore, a charged capacitor can be regarded as a sort of charge storage system. I say 'sort of' since in a capacitor consisting of two metal plates and an air gap, one plate will have *excess* charge Q and the other a *loss* of charge Q. The *total* charge excess in the system is *zero*. So, charge is moved, or separated, rather than added overall. However, if we connect the opposing conductors of a charged capacitor via conductive wires and some form of resistive load, a current will flow. So, it is nonetheless quite natural to think of a capacitor as a source of charge and therefore electric current. Since work must be done to separate the charges in a capacitor, perhaps a better description is as a *source of energy*. A capacitor, like a reservoir of water behind a dam in a hydroelectric power station, is a store of potential energy.

Capacitance C (which is measured in *farads* (F) in honour of Michael Faraday,[17]) is defined as the charge Q transferred to one of the capacitor conductors divided

---

[15] André-Marie Ampère (1775–1836).

[16] For a further discussion of the application of *Ampère's theorem,* and its more general expression $\oint_{\text{loop}} \mathbf{H} \cdot d\mathbf{l} = I_{\text{enclosed}}$, see Section 5.7. $\mathbf{H}$ is the *magnetic field strength*, which for isotropic media, is related to *magnetic flux density* $\mathbf{B}$ via the relationship $\mathbf{B} = \mu\mu_0 \mathbf{H}$. The dimensionless constant μ is the *relative permeability*. For a vacuum, or air to a very good approximation (with nothing to magnetise), $\mu = 1$.

[17] Michael Faraday (1791–1867).

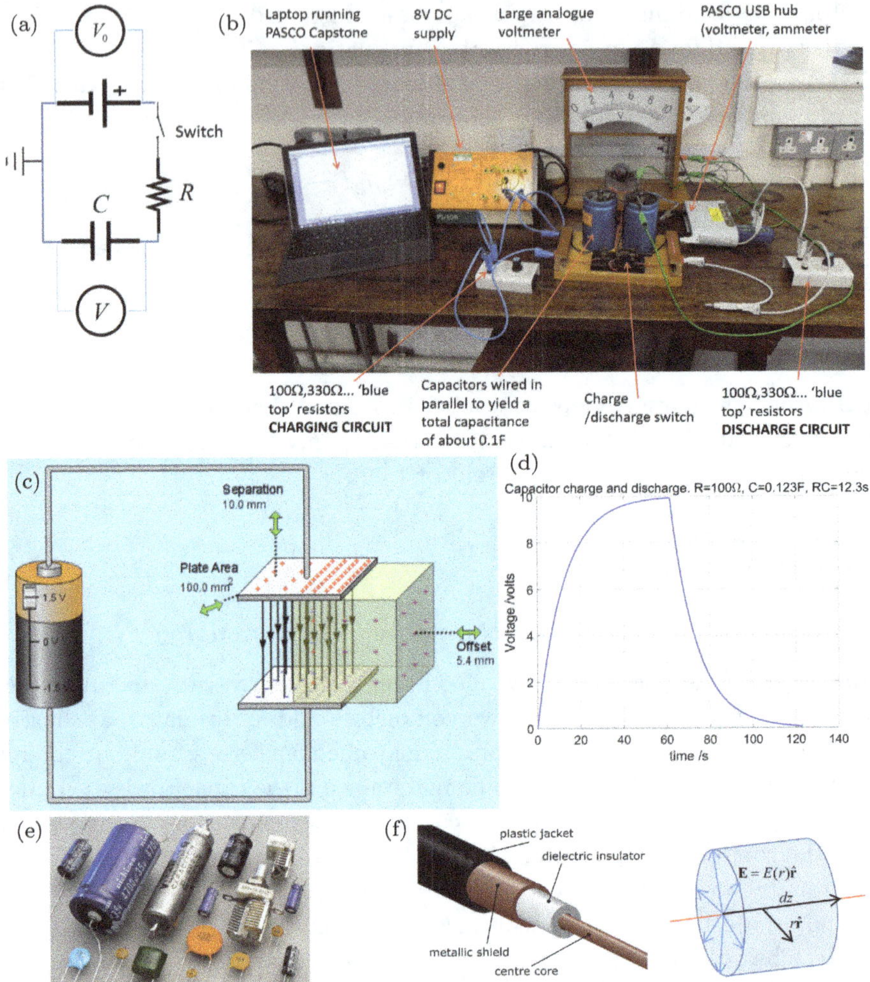

Fig. 5.8. (a) Circuit diagram of a capacitor of capacitance C being charged via connection to a constant voltage source V_0 via resistor R. The potential difference across the capacitor during charging is $V = V_0(1 - e^{-t/RC})$. (b) Experimental setup for a large (0.123 F) capacitor, which can be alternately charged and discharged by throwing a switch. (c) PhET screenshot showing electric field lines and charges in a parallel plate capacitor with a dielectric partially inserted. (d) Charging and discharging curves for a capacitor with time constant $RC = 12.3$ s. (e) Photo (from Wikipedia) of a variety of capacitors. (f) Schematic diagram of a coaxial cable and electric field associated with a cylindrical conductor.

by the potential difference (i.e. the energy per unit charge) across the conductor–insulator–conductor sandwich:

$$C = \frac{Q}{V}. \tag{5.103}$$

$$\therefore Q = CV. \tag{5.104}$$

The capacitance is typically a constant for a capacitor and depends on geometrical parameters, such as the conductor areas and their separation. It also depends

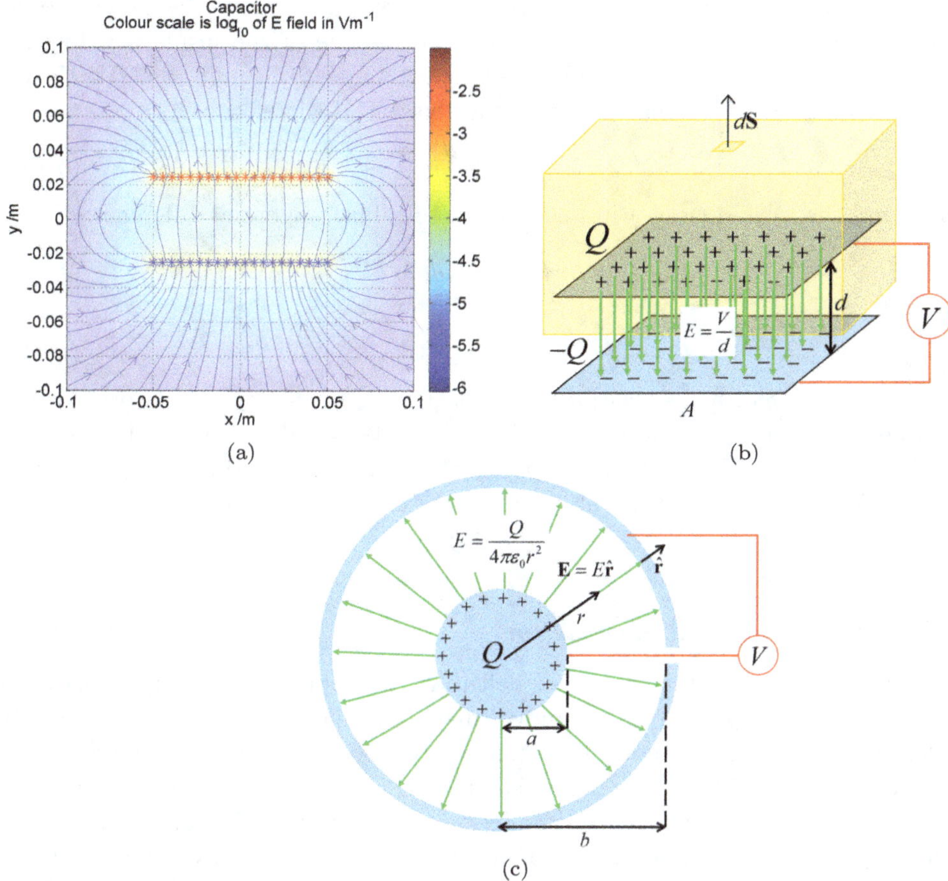

Fig. 5.9. (a) Coulomb's law for point charges can be added up, in a vector sense, to determine the field between two lines of positive (red) and negative (blue) charges. The field between them becomes increasingly uniform as one moves from the edges inwards. (b) Schematic diagram of a vacuum-separated parallel plate capacitor of plate area A and separation d. The orange 'Gaussian pillbox' (with surface vector area element $d\mathbf{S}$) encloses the top plate. Since $\int \mathbf{E} \cdot d\mathbf{S} = \frac{Q}{\varepsilon_0}$, $EA = \frac{Q}{\varepsilon_0}$, i.e. $Q = \varepsilon_0 E A$. The uniform field strength $E = \frac{V}{d}$ and the capacitance $C = \frac{Q}{V}$. This means $C = \frac{\varepsilon_0 E A}{V} = \frac{\varepsilon_0 \left(\frac{V}{d}\right) A}{V} = \frac{\varepsilon_0 A}{d}$. (c) Schematic diagram of a spherical capacitor comprising a charged sphere placed within a hollow conductor. The electric field strength between the conductors is $E = \frac{Q}{4\pi\varepsilon_0 r^2}$, and hence the potential difference is $V = \frac{Q}{4\pi\varepsilon_0}\left(\frac{1}{a} - \frac{1}{b}\right)$. This means the capacitance of the system is $C = 4\pi\varepsilon_0 \left(\frac{1}{a} - \frac{1}{b}\right)^{-1}$. In the limit of $b \gg a$, this yields the capacitance of an isolated charged sphere as $C = 4\pi\varepsilon_0 a$.

on the ability of the insulator (known as a *dielectric*) to be polarised by the electric field set up across it via the oppositely charged conductors. In the following section, we derive an expression for the capacitance of a parallel plate capacitor, and later in this chapter, we consider the capacitance of conductive spheres and cylinders.

A capacitor might be charged by connecting its conductors to a voltage source via conductive wires and a resistive load, as in Figs. 5.8(a), (b) and (c). The instantaneous power required to establish voltage V across the capacitor plates is $P = VI$, where I is the current flowing onto the capacitor (and indeed the same current removing charge from the conductor the opposite side of the insulator). In time interval dt, the amount of charge added to the capacitor is $dQ = CdV$. Hence, current $I = \frac{dQ}{dt} = C\frac{dV}{dt}$. The instantaneous power P is therefore

$$P = CV\frac{dV}{dt},\tag{5.105}$$

and hence the work done W (in t seconds) to establish potential difference V is

$$W = \int Pdt = \int CV\frac{dV}{dt}dt = C\int_0^V V'dV'.$$

$$\therefore W = \tfrac{1}{2}CV^2.\tag{5.106}$$

An even more direct derivation is to note that $dW = VdQ$ is the work done to separate charge dQ. Therefore, using $dQ = CdV$, $W = \int dW = \int VCdV = \tfrac{1}{2}CV^2$.

So, the *energy E stored in a capacitor* with a potential difference V is $E = \tfrac{1}{2}CV^2$. Using $Q = CV$, we can also express the energy stored as

$$E = \tfrac{1}{2}QV.\tag{5.107}$$

5.5.2 *The parallel plate capacitor*

A parallel plate capacitor can be modelled as two (oppositely) charged plates of area A, separated by distance d, sandwiching an insulating dielectric and typically rolled up like a Swiss cheese into a compact cylindrical packaging. A screenshot from the fantastic PhET[18] dynamic simulation illustrates the fields and charges associated with an (unrolled-up) parallel plate capacitor in Fig. 5.8(c) and cylindrically packaged capacitors in Fig. 5.8(e). Consider an imaginary rectangular box enclosing (only) the positively charged plate, which contains excess charge Q. In this first model, let's assume that the insulator between the plates is a vacuum, i.e. contains no molecules whose constituent charges can be polarised and therefore affect the electric field between the plates. *Gauss's law*[19] (which is one of the four fundamental laws of electromagnetism, known as *Maxwell's equations*; see the final section of this chapter for details) states that, in an unpolarised medium, the flux of an electric field $\mathbf{E}$ through a closed surface S is equal to the charge enclosed divided by the permittivity of free space $\varepsilon_0 = 8.85 \times 10^{-12}$ Fm^{-1}:

$$\int_S \mathbf{E} \cdot d\mathbf{S} = \frac{Q}{\varepsilon_0}.\tag{5.108}$$

Note that Gauss' law implies no electric field can exist within a hollow conductor (of any shape), which encloses zero charge. This is the basis of the *Faraday cage*

---

[18] https://phet.colorado.edu/en/simulations/capacitor-lab.

[19] Actually, the Maxwell equation for Gauss' law is $\int_S \mathbf{D} \cdot d\mathbf{S} = \int_V \rho dV$, where *electric displacement field* $\mathbf{D} = \varepsilon_0\mathbf{E} + \mathbf{P}$ incorporates both an applied electric field $\mathbf{E}$ and a *polarisation field* $\mathbf{P}$ resulting from *bound* charges in a medium which may be separated by $\mathbf{E}$ and, therefore, contribute to $\mathbf{D}$. The density of *free charges* is ρ, hence total charge enclosed within volume V is $Q = \int_V \rho dV$. If $\mathbf{P}$ is aligned with $\mathbf{E}$ (i.e. a medium has no preferred polarisation direction), then $\mathbf{D} = \varepsilon_r\varepsilon_0\mathbf{E}$ where ε_r is the relative permittivity. So, $\int_S \mathbf{D} \cdot d\mathbf{S} = \int_V \rho dV = Q$ implies $\int_S \mathbf{E} \cdot d\mathbf{S} = \frac{Q}{\varepsilon_r\varepsilon_0}$.

concept. A person-shaped conductor of chain mail can enable a technician to work safely on live power lines since there is no possibility of a potential difference across the body of the technician, and hence no dangerous currents will flow through them.

Returning to the parallel plate capacitor, let us assume its area A is much larger than d^2, which means we can assume the electric field between the capacitor plates is uniform, i.e. $\mathbf{E}$ is a constant and points between the plates, perpendicular to their surface, as illustrated in Fig. 5.8(c). We might justify this argument on the basis that any electric field parallel to the capacitor plate surface would cause charges to move. If the charge distribution on the plates is assumed to be static (and indeed uniform), then this is implies that the electric field must only act normal to the plate surface and have the same field strength over the plates (until you get rather close to the edges). Since we can extend our imaginary box as far as we like from the capacitor, $\int_S \mathbf{E} \cdot d\mathbf{S} = \frac{Q}{\varepsilon_0}$ reduces to $EA = \frac{Q}{\varepsilon_0}$.

The magnitude of the (constant) electric field between the capacitor plates is the gradient of the potential difference V between them.[20]

$$\therefore E = \frac{V}{d}. \tag{5.109}$$

So, if $E = \frac{Q}{\varepsilon_0 A}$ and $E = \frac{V}{d}$,

$$\frac{Q}{\varepsilon_0 A} = \frac{V}{d}.$$

$$\therefore Q = \frac{\varepsilon_0 A}{d} V. \tag{5.110}$$

Since capacitance $C = \frac{Q}{V}$, the capacitance of the parallel plate capacitor is therefore

$$C = \frac{\varepsilon_0 A}{d}. \tag{5.111}$$

It makes intuitive sense for C to be proportional to A, since a capacitor of greater surface area will be able to transfer more charge between its plates. The fact that C is inversely proportional to plate separation d is perhaps less intuitive, although the fact that field strength E is also inversely proportional to d and that Q is proportional to E, should hopefully remove any mystery.

Note that we have not yet considered the effect of a dielectric. If the insulator is a vacuum (and, to a good approximation, air), then $C = \frac{\varepsilon_0 A}{d}$ is a reasonable model. However, the charges which are bound to the molecular structures of the insulator are not unaffected by the electric field established between the plates. The molecules will have a tendency to polarise, which means positive charges will move towards the negative plate and vice versa, effecting the overall electric field. Polarisation of a dielectric within a parallel plate capacitor is illustrated in Fig. 5.8(c). The net result is that *more* charge can be stored on the capacitor plates by a factor of ε_r,

---

[20] Actually, $\mathbf{E} = -\nabla V$, but since our capacitor field acts from a high to low potential, we can simply express the field as *magnitude* $E = \frac{V}{d}$.

which is called the *relative permittivity*. For a dielectric which can polarise in any direction (i.e. the direction of $\mathbf{E}$),

$$C = \frac{\varepsilon_r \varepsilon_0 A}{d},$$
(5.112)

which means Gauss's law for an (isotropic) medium is

$$\int_S \mathbf{E} \cdot d\mathbf{S} = \frac{Q}{\varepsilon_r \varepsilon_0}.$$

The relative permittivity $\varepsilon_r = 1$ for a vacuum and $\varepsilon_r \approx 1$ for air, whereas water at 298 K has $\varepsilon_r \approx 78$. Barium strontium titanate has a relative permittivity of about 500, which makes it a useful material for constructing a dielectric between capacitor plates. Since capacitance is proportional to ε_r and the energy stored in a charged capacitor is $E = \frac{1}{2}CV^2$, this means a larger ε_r results in a proportionally higher energy density for a capacitor of a given geometry.

Note that *permittivity*, or 'dielectric constant' ε, is defined as

$$\varepsilon = \varepsilon_r \varepsilon_0.$$
(5.113)

ε has dimensions of Fm^{-1}, whereas ε_r is dimensionless.

The formula $C = \frac{\varepsilon_r \varepsilon_0 A}{d}$ gives some immediate insight into the anticipated sizes of capacitances. C is $\frac{\varepsilon_r A}{d}$ multiplied (in SI units) by $\varepsilon_0 = 8.85 \times 10^{-12}$ Fm^{-1}, so unless $\frac{\varepsilon_r A}{d}$ is a truly enormous number, we might expect capacitances for standard electrical components to be of the order of pico, nano or micro farads.[21]

5.5.3   *Charging and discharging a capacitor*

5.5.3.1   *Discharging a capacitor*

Consider a capacitor with charge Q_0 excess on one of its conductive plates, which is separated from another plate (with a charge *deficit* of Q_0) via an insulating dielectric material. Let both plates be connected via conducting wires through a resistive component. The wires, connectors and component have a total resistance of R. If C is the capacitance of the capacitor, the potential difference across the resistor is initially $V_0 = \frac{Q_0}{C}$ and will decay to zero as the charge flows from one capacitor plate to another until charge neutrality is achieved.

The rate of flow of capacitor charge Q is the current I through the resistor:

$$I = -\frac{dQ}{dt}.$$
(5.114)

There is a minus sign since I is a positive quantity and Q is reducing. From the definition of resistance, the potential difference across the resistor is $V = IR$. Since

---

[21]Although, at the time of writing, a new generation of *supercapacitors* are being developed, which apparently have capacitances of several *thousand* farads!

$Q = CV$, we can therefore write $I = -\frac{dQ}{dt}$ as a differential equation for $V(t)$:

$$\frac{V}{R} = -\frac{d(CV)}{dt}. \tag{5.115}$$

$$\therefore -\frac{dt}{RC} = \frac{1}{V}dV \tag{5.116}$$

$$\therefore -\int_0^t \frac{dt'}{RC} = \int_{V_0}^V \frac{1}{V'}dV' \tag{5.117}$$

$$\therefore -\frac{t}{RC} = \ln V - \ln V_0 = \ln\left(\frac{V}{V_0}\right) \tag{5.118}$$

$$\therefore V = V_0 e^{-\frac{t}{RC}}. \tag{5.119}$$

So, we expect an *exponential decay* of potential difference across the resistor, with *time constant RC*. When $t = RC$, $V = \frac{V_0}{e} \approx 0.37V_0$. Since $Q = CV$, the same exponential decay is also the time variation of the excess charge on the capacitor plate:

$$Q = Q_0 e^{-\frac{t}{RC}}. \tag{5.120}$$

Now, since $e^{\frac{t}{RC}} = \frac{V_0}{V}$

$$\implies t = RC\ln\left(\frac{V_0}{V}\right). \tag{5.121}$$

A time constant RC can therefore easily be determined from measurements of V and t. In the laboratory demonstration equipment illustrated in Fig. 5.8(b), a datalogger is used to achieve this. The data is copied from the PASCO Capstone software into a spreadsheet to perform the line-of-best-fit analysis. If t is plotted against $\ln\left(\frac{V_0}{V}\right)$, then this should be a straight-line graph through the origin, with gradient RC. A comically enormous analogue voltmeter is of course also essential for the demonstration.

It is also instructive, by analogy to radioactive decay (and indeed Newton's law of cooling from Chapter 3) to express $V(t)$ in terms of a half-life, $t_{1/2}$, which is the amount of time for the potential difference across the capacitor to decay from V_0 to $\frac{1}{2}V_0$. Since $t = RC\ln\left(\frac{V_0}{V}\right)$,

$$t_{1/2} = RC\ln 2 \approx 0.69RC. \tag{5.122}$$

The exponential decay $V = V_0 e^{-\frac{t}{RC}}$ can also be written in a perhaps more intuitive way using the half-life instead of RC:

$$V = \frac{V_0}{2^{\frac{t}{t_{1/2}}}}, \tag{5.123}$$

i.e. V is the initial voltage V_0 divided by two to the power of the number of half-lives.

5.5.3.2 *Charging a capacitor*

A capacitor of capacitance C, with no charge separation between its conductive plates, is connected to a constant voltage source V_0 via conductive wires and a resistive component. The wires, connectors and component have a total resistance of R. As charge flows round the circuit, depositing charge Q on the plate connected to the positive terminal of the voltage source and removing charge Q from the other plate, a potential difference of $V = \frac{Q}{C}$ is established across the capacitor. This means the potential difference across the resistor is $V_0 - V$ and the current flowing is $I = \frac{V_0 - V}{R}$. Since $I = \frac{dQ}{dt}$ and $Q = CV$,

$$\frac{V_0 - V}{R} = C\frac{dV}{dt}. \tag{5.124}$$

$$\therefore \frac{1}{RC}\int_0^t dt' = \int_0^V \frac{dV'}{V_0 - V'}. \tag{5.125}$$

Using the standard integral $\int \frac{1}{y}\frac{dy}{dx}dx = \int \frac{1}{y}dy = \ln|y| + c$,

$$\int_0^V \frac{dV'}{V_0 - V'} = -\int_0^V \frac{(-1)dV'}{V_0 - V'} = -\left[\ln(V_0 - V')\right]_0^V$$

$$= -\ln\left(\frac{V_0 - V}{V_0}\right) = \ln\left(\frac{V_0}{V_0 - V}\right).$$

Hence, since $\frac{1}{RC}\int_0^t dt' = \frac{t}{RC}$,

$$t = RC\ln\left(\frac{V_0}{V_0 - V}\right). \tag{5.126}$$

Rearranging to make capacitor voltage V the subject,

$$-\frac{t}{RC} = \ln\left(\frac{V_0 - V}{V_0}\right) \tag{5.127}$$

$$e^{-\frac{t}{RC}} = \frac{V_0 - V}{V_0} \tag{5.128}$$

$$\therefore V = V_0\left(1 - e^{-\frac{t}{RC}}\right). \tag{5.129}$$

As with discharging, the product RC is a time constant for the capacitor and resistor. When $t = RC$, $V = V_0(1 - \frac{1}{e}) \approx 0.63V_0$.

Figure 5.8(d) is a plot of a capacitor charging and then discharging. The capacitor voltage V is plotted against time. $R = 100\ \Omega$, $C = 0.123$ F; therefore, the time constant $RC = 12.3$ s. These parameters are those of the experimental setup depicted in Fig. 5.8(b). For the charging phase, $V_0 = 10$ V, and charging and discharging are both set to occur for $5RC$ s.

5.5.4 *Rules for combining capacitors in circuits*

Capacitors can be miniaturised, indeed to the extent that a modern integrated circuit may have *billions* on each chip. So, as with other fundamental components such as resistors and inductors, we have a fairly strong practical need to know the rules for how capacitances combine when wired in series and parallel.

Parallel wirings are fairly straightforward, as illustrated in Fig. 5.10(a). If each of the N capacitors (of capacitance C_i where $i = 1, 2, 3, \ldots, N$) is charged via a connection to a constant voltage source V_0, the total charge Q displaced is

$$Q = Q_1 + Q_2 + Q_3 + \cdots = C_1 V_0 + C_2 V_0 + C_3 V_0 + \cdots + C_N V_0. \tag{5.130}$$

Hence, if the total effective capacitance is $C = \frac{Q}{V_0}$,

$$C = C_1 + C_2 + C_3 + \cdots + C_N, \tag{5.131}$$

i.e. *capacitances add up in parallel* (whereas resistors add up in series).

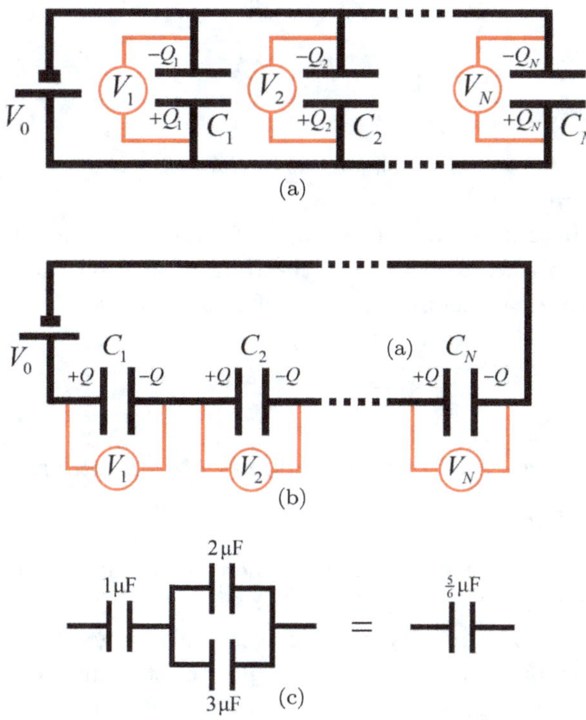

Fig. 5.10. (a) Capacitors wired in a parallel arrangement add up in an arithmetic sense, i.e. $C = C_1 + C_2 + \cdots + C_N$. This is because the total charge displaced is $Q = CV_0 = Q_1 + Q_2 + \cdots + Q_N$ and $Q_i = C_i V_0$, i.e. the potential differences V_i are all the same and equal the voltage source V_0. (b) Capacitors wired in series combine in reciprocals, i.e. $\frac{1}{C} = \frac{1}{C_1} + \frac{1}{C_2} + \cdots + \frac{1}{C_N}$. This is because the charge displaced Q for each capacitor must be the same, and the potential differences $V_i = \frac{Q}{C_i}$ must sum to the voltage source potential V_0. (c) Example of a capacitor combination in series and in parallel arrangements.

Alternatively, if N capacitors are wired in *series* to a constant voltage source V_0, as illustrated in Fig. 5.10(b), then the potential differences $V_1, V_2, \ldots, V_N$ across the capacitors must sum to V_0. Hence,

$$V_0 = V_1 + V_2 + V_3 + \cdots + V_N. \tag{5.132}$$

Now, if charge Q_1 is displaced across the plates of the first capacitor and Q_2 across the capacitor next to it, a plate with charge excess $-Q_1$ is connected (via conductive wires) to a plate with charge excess Q_2. Charge would, therefore, flow and reapportion (well, cancel out) between the plates until $Q_1 = Q_2$. This implies that the *same charge must be displaced across each capacitor, regardless of their individual capacitances.* The series-wired capacitors, from a charge separation perspective, is therefore Q on the first plate (connected to the positive terminal of the voltage source) and $-Q$ on the last plate (connected to the negative terminal of the voltage source). Therefore, using $Q = C_i V_i$ and the total effective capacitance C defined by $Q = CV_0$,

$$V_0 = \frac{Q}{C} = \frac{Q}{C_1} + \frac{Q}{C_2} + \frac{Q}{C_3} + \cdots + \frac{Q}{C_N}. \tag{5.133}$$

Hence,

$$\frac{1}{C} = \frac{1}{C_1} + \frac{1}{C_2} + \frac{1}{C_3} + \cdots + \frac{1}{C_N}, \tag{5.134}$$

i.e. *capacitances wired in series add up in reciprocals* (whereas resistors wired in parallel add up in reciprocals).

By way of an illustrative example in Fig. 5.10(c), we can combine both series and parallel combination rules to determine the total capacitance of a 1 μF capacitor wired in series with a parallel arrangement of 2 μF and 3 μF capacitors:

$$C = \frac{1}{\frac{1}{1 \ \mu\text{F}} + \frac{1}{(2+3) \ \mu\text{F}}} = \tfrac{5}{6} \ \mu\text{F} \approx 0.83 \ \mu\text{F}.$$

5.5.5   *The capacitance of charged spheres and cylinders*

5.5.5.1   *The capacitance of a charged sphere*

Consider a positively charged conductive sphere of radius a placed at the centre of a hollow conductive sphere of radius b, where $b \gg a$. Between the spheres is a vacuum with relative permittivity $\varepsilon_r = 1$. (Air is also a good enough approximation.) The inner sphere is connected to the positive terminal of a voltage source V_0, and the negative terminal is connected to the outer sphere. Assume that charge Q is displaced between the spheres, similar to the parallel plate capacitor situation. The electric field between the spheres is clearly radial by symmetry, and we can use Coulomb's law (or, indeed, Gauss' law) to express its magnitude E with distance r from the centre of the spheres:

$$E = \frac{Q}{4\pi\varepsilon_0 r^2}. \tag{5.135}$$

Since electric field strength is (negative of) the gradient of potential, i.e. $E = -\frac{dV}{dr}$ in our case, we can compute the potential difference $V_b - V_a$ between the spheres using

$$V_b - V_a = -\int_a^b \frac{Q}{4\pi\varepsilon_0 r^2}\,dr. \tag{5.136}$$

$$\therefore V_b - V_a = \frac{Q}{4\pi\varepsilon_0}\left[\frac{1}{r}\right]_a^b = \frac{Q}{4\pi\varepsilon_0}\left(\frac{1}{b} - \frac{1}{a}\right). \tag{5.137}$$

$V_a > V_b$, so let us define $V = V_a - V_b$. Hence,

$$V = \frac{Q}{4\pi\varepsilon_0}\left(\frac{1}{a} - \frac{1}{b}\right). \tag{5.138}$$

Therefore, the capacitance of the system is

$$C = \frac{Q}{V} = 4\pi\varepsilon_0 \left(\frac{1}{a} - \frac{1}{b}\right)^{-1}. \tag{5.139}$$

If $b \gg a$, we can ignore the $\frac{1}{b}$ compared to the $\frac{1}{a}$ term. This means

$$C \approx 4\pi\varepsilon_0 a, \tag{5.140}$$

which is the *capacitance of an isolated charged sphere*. This result helps to explain why a Van de Graaff generator[22] is both impressive (capable of generating small bolts of lightning on your laboratory bench) but also quite safe. For air to conduct, electric field strengths of more than about 3×10^6 Vm^{-1} are required. In my laboratory, I regularly achieve sparks that are about 0.1 m long between a charged Van de Graaf and a spherical conductor attached to an insulated rod. Therefore, to achieve three million Vm^{-1}, the potential of the charged spherical cap of the Van de Graaf must be about $V = 300,000$ V. This sounds impressive and excitingly dangerous, especially since 5,000 V is the highest setting on our power supplies, all of which carry warning signs saying 'Danger! High Voltage'. However, the radius of our spherical cap is about $a = 0.15$ m, so the charge on the cap is about

$$Q = 4\pi\varepsilon_0 a V$$

$$= 4\pi \times 8.85 \times 10^{-12} \times 0.15 \times 300,000$$

$$= 5.00 \times 10^{-6} \text{ C}.$$

The energy associated with a spark is therefore

$$E = \tfrac{1}{2}CV^2 = \tfrac{1}{2}QV$$

$$= \tfrac{1}{2} \times 5.00 \times 10^{-6} \times 300,000 \text{ J}$$

$$= 0.75 \text{ J},$$

---

[22]The inventor of perhaps *the* classic high-voltage source and spark-producing school laboratory demo was Robert Jemison Van de Graaff (1901–1967).

i.e. about a joule. The gravitational potential energy gain of a 75 kg physics teacher climbing a vertical height of 0.5 m onto a laboratory bench is about $75 \times 9.81 \times 0.5$ J ≈ 370 J, so the energy per spark is a fairly trivial fraction of the kinetic energy I would gain if I decided to jump off. (I don't tend to do this every day, so consider this a *gedankenexperiment*).

A nice illustrative capacitance problem, which often appears in different forms in the British Physics Olympiad[23] involves the joining of charged isolated spheres with a conductive wire and computing how the charge reapportions and energy is dissipated.

Let sphere #1 have capacitance $C_1 = 4\pi\varepsilon_0 a_1$ and sphere #2 have capacitance $C_2 = 4\pi\varepsilon_0 a_2$. Sphere #1 is charged and has a potential difference (relative to ground) of V_1. Sphere #2 is also charged and has potential V_2 at its surface. When the spheres are connected by a conductive wire, charge will flow until the potential difference between the spheres (i.e. across the wire) is zero. This means both spheres must now be at the same electric potential V. Now, if the system is isolated, the total amount of charge must be conserved. Therefore,

$$C_1 V_1 + C_2 V_2 = C_1 V + C_2 V, \tag{5.141}$$

which means

$$V = \frac{C_1 V_1 + C_2 V_2}{C_1 + C_2}, \tag{5.142}$$

i.e. a capacitance-weighted average of the voltages. The energy dissipated ΔE can now be computed from the difference in the potential energy associated with the spheres (which can be modelled as spherical capacitors):

$$\Delta E = \tfrac{1}{2}C_1 V_1^2 + \tfrac{1}{2}C_2 V_2^2 - \tfrac{1}{2}C_1 V^2 - \tfrac{1}{2}C_2 V^2 \tag{5.143}$$

$$= \tfrac{1}{2}C_1 V_1^2 + \tfrac{1}{2}C_2 V_2^2 - \tfrac{1}{2}V^2 (C_1 + C_2).$$

$$\therefore \Delta E = \tfrac{1}{2}C_1 V_1^2 + \tfrac{1}{2}C_2 V_2^2 - \tfrac{1}{2}\left(\frac{C_1 V_1 + C_2 V_2}{C_1 + C_2}\right)^2 (C_1 + C_2). \tag{5.144}$$

$$\therefore \Delta E = \tfrac{1}{2}C_1 V_1^2 + \tfrac{1}{2}C_2 V_2^2 - \tfrac{1}{2}\frac{(C_1 V_1 + C_2 V_2)^2}{C_1 + C_2}.$$

If the spheres have radii a_1 and a_2, respectively, then, since $C_{1,2} = 4\pi\varepsilon_0 a_{1,2}$,

$$\Delta E = 2\pi\varepsilon_0 \left\{ a_1 V_1^2 + a_2 V_2^2 - \frac{(a_1 V_1 + a_2 V_2)^2}{a_1 + a_2} \right\}. \tag{5.145}$$

---

[23] https://www.bpho.org.uk/.

5.5.5.2 *The capacitance of a coaxial cable (a charged cylinder inside a hollow conductive cylinder)*

In 1880, the British physicist Oliver Heaviside patented the *coaxial cable* and described how the design could enable high-frequency electrical signals to be transmitted over long distances (e.g. from Britain to America across the Atlantic Ocean) with low losses. The signals conveyed by the cable would also be protected from interference from external EM fields. A coaxial cable, illustrated in Fig. 5.8(f), is a long, cylindrical conductor surrounded by a dielectric (of relative permittivity ε_r), which is itself enclosed by another conductor (and then typically an insulating outer sheath too). The coaxial cable is one of the key technologies which has enabled our modern interconnected world[24] to be developed. Millions of domestic televisions will be connected by one to a roof-mounted aerial (or satellite dish).

Let us assume that the diameter of a coaxial cable is very much shorter than its length. Therefore, we can assume that the electric field $\mathbf{E}$ between a charged inner conductor acts purely radially within a cross-section of the cable. If the inner conductor carries charge Q in cable length l, we can use Gauss' law, $\int_S \mathbf{E} \cdot d\mathbf{S} = \frac{Q}{\varepsilon_r \varepsilon_0}$, to determine an expression for the electric field strength E a distance r away from the inner conductor, within range $a < r < b$, where a and b are the respective radii of the inner and outer conductors:

$$E \times 2\pi r l = \frac{Q}{\varepsilon_r \varepsilon_0}. \tag{5.146}$$

$$\therefore E = \frac{Q/l}{2\pi \varepsilon_r \varepsilon_0} \frac{1}{r}. \tag{5.147}$$

The potential difference $V = V_a - V_b$ between the inner and outer conductors is, using $E = -\frac{dV}{dr}$,

$$V = -(V_b - V_a) = -\left(-\int_a^b E\, dr \right) = \frac{Q/l}{2\pi \varepsilon_r \varepsilon_0} \int_a^b \frac{1}{r}\, dr.$$

$$\therefore V = \frac{Q/l}{2\pi \varepsilon_r \varepsilon_0} \ln\left(\frac{b}{a}\right), \tag{5.148}$$

which means the capacitance $\frac{Q}{V}$ of the cable is

$$C = \frac{2\pi \varepsilon_r \varepsilon_0 l}{\ln\left(\frac{b}{a}\right)}. \tag{5.149}$$

5.5.6 *The electric field between a conductive sphere placed between two charged parallel plates*

A conductive sphere of radius a placed in a uniform electric field (of strength E_0) will be polarised by it. In other words, charges will be separated on the surface of the

---

[24]These days a fibre-optic cable is possibly the mechanism of choice for a wired communications link rather than a coaxial cable. But this required the invention of the light emitting diode (LED), which was only in mass production from the late 1960s.

sphere to align with the field. Note that unless the sphere was initially charged, the total charge must still sum to zero. Either way, although the distribution of charge is modified by the external field, the total amount on the sphere must remain the same. To determine the electrical field outside the sphere (if we assume a hollow conductor, then the field must be exactly zero inside), we shall make a sensible guess that it comprises something which looks like a dipole superimposed upon a uniform field as well as (if the sphere was initially charged) a spherically symmetric field. We shall assemble this model by considering a *sum of electric potentials* associated with (i) a dipole, (ii) a uniform field between parallel capacitor plates and (iii) a point charge (which, by Gauss' theorem, is the same as a charged sphere, beyond the sphere).

Let's first define a coordinate system with origin being the centre of the sphere, as illustrated in Fig. 5.11(a). The uniform field between the charged plates shall act upwards, i.e. $E_0 \hat{\mathbf{y}}$. The distance from the centre of the sphere shall be r, and θ shall be the polar angle from the (upwards) y-axis. Hence, $x = r \sin\theta$ and $y = r \cos\theta$, and the polar unit vectors are $\hat{\mathbf{r}} = \hat{\mathbf{x}} \sin\theta + \hat{\mathbf{y}} \cos\theta$ and $\hat{\boldsymbol{\theta}} = \hat{\mathbf{x}} \cos\theta - \hat{\mathbf{y}} \sin\theta$. In terms of Cartesian x, y coordinates,

$$r = \sqrt{x^2 + y^2}, \tag{5.150}$$

$$\theta = \tan^{-1}\left(\frac{x}{y}\right).$$

In the earlier section on electric dipoles, the potential is derived to be

$$V_{\text{dipole}} \approx \frac{q}{4\pi\varepsilon_0} \frac{\mathbf{d} \cdot \hat{\mathbf{r}}}{r^2}. \tag{5.151}$$

Imagine a dipole corresponding to two opposing charges placed at the top and bottom of the sphere. The dipole moment is $q\mathbf{d} = 2qa\hat{\mathbf{y}}$, so $\mathbf{d} \cdot \hat{\mathbf{r}} = 2a\cos\theta$. So, if our polarised sphere behaved like a dipole,

$$V_{\text{dipole}} \approx \frac{1}{4\pi\varepsilon_0} \frac{2qa\cos\theta}{r^2}. \tag{5.152}$$

Let's assume this is reasonable model, but set $p = 2qa$ to be the 'polarisation', which we will determine later when we know more about the interaction of the uniform and point-charge potentials. The uniform field potential is

$$V_{\text{uniform}} = -E_0 y \tag{5.153}$$

since $E_0 = -\frac{\partial V_{\text{uniform}}}{\partial y}$. Therefore, in polar coordinates,

$$V_{\text{uniform}} = -E_0 r \cos\theta. \tag{5.154}$$

Finally, if the sphere has charge $Q = 4\pi\varepsilon_0 a V_0$, with potential V_0 being that of the sphere prior to being placed in the field between the conducting plates,

$$V_{\text{sphere}} = \frac{Q}{4\pi\varepsilon_0} \frac{1}{r} = \frac{a V_0}{r}. \tag{5.155}$$

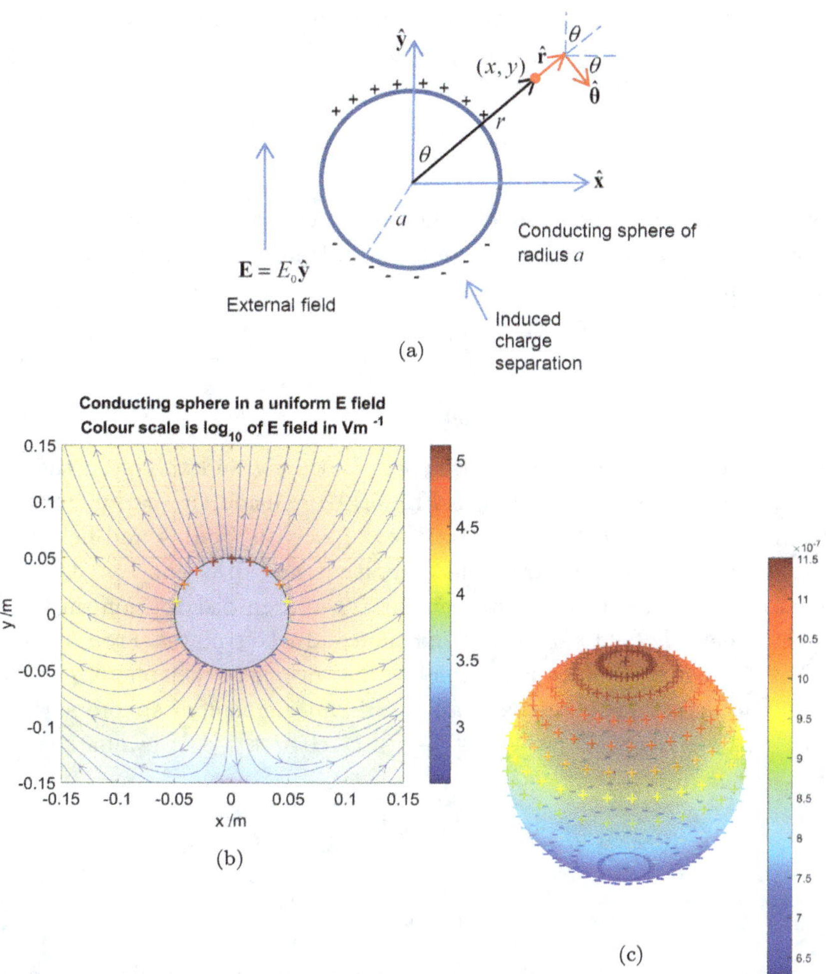

Fig. 5.11. (a) Coordinate system used to describe the electric field resulting from a charged conductive sphere of radius a placed between two charged parallel plates. (b) Field lines and field strength, assuming $E_0 = 10{,}000$ Vm^{-1}, $V_0 = 5000$ V and $a = 0.05$ m. (c) Surface charge density $\sigma = 3\varepsilon_0 E_0 \cos\theta + \frac{\varepsilon_0 V_0}{a}$ (in C/m^2) for the charged sphere, which is polarised by the uniform field (of strength E_0) in which it is placed.

The total potential is therefore

$$V = V_{\text{dipole}} + V_{\text{uniform}} + V_{\text{sphere}}. \tag{5.156}$$

$$\therefore V = \frac{1}{4\pi\varepsilon_0} \frac{p\cos\theta}{r^2} - E_0 r\cos\theta + \frac{aV_0}{r}. \tag{5.157}$$

The electric field strength (outside the sphere and between the charged plates) can be computed from $\mathbf{E} = -\nabla V$, which, in plane polar coordinates, is

$$\mathbf{E} = E_r\hat{\mathbf{r}} + E_\theta\hat{\boldsymbol{\theta}} = -\frac{\partial V}{\partial r}\hat{\mathbf{r}} - \frac{1}{r}\frac{\partial V}{\partial \theta}\hat{\boldsymbol{\theta}}. \tag{5.158}$$

Therefore, the radial and tangential components of the electric fields are

$$E_r = \frac{1}{4\pi\varepsilon_0}\frac{2p\cos\theta}{r^3} + E_0\cos\theta + \frac{aV_0}{r^2}, \tag{5.159}$$

$$E_\theta = \frac{1}{4\pi\varepsilon_0}\frac{p\sin\theta}{r^2} - E_0 r\sin\theta. \tag{5.160}$$

Now, there cannot be any tangential electric fields at the surface of the conductor. If there were, then the charges would move. Hence, when $r = a$, the tangential field component $E_\theta = 0$.

$$\therefore \frac{1}{4\pi\varepsilon_0}\frac{p\sin\theta}{a^2} - E_0 a\sin\theta = 0. \tag{5.161}$$

$$\therefore p = 4\pi\varepsilon_0 a^3 E_0. \tag{5.162}$$

We now have an expression for the polarisation p and therefore the problem is solved. An example is illustrated in Fig. 5.11(b), using the parameters of $E_0 = 10,000$ Vm^{-1}, $V_0 = 5,000$ V and $a = 0.05$ m. The colour coding and field lines help visualise, respectively, the electric field magnitude and direction.

It is also instructive to compute the surface charge distribution on the sphere as well as the electric field between it and the conductive plates. From Gauss's law, $\int_S \mathbf{E}\cdot d\mathbf{S} = \frac{Q}{\varepsilon_0}$; therefore, since $\mathbf{E}$ is radial at the surface of the sphere, $E_r dS = \frac{\sigma dS}{\varepsilon_0}$, where the *charge per unit area* is σ and dS is an element of sphere surface area. Hence, $\sigma = \varepsilon_0 E_r$, where $E_r = \frac{1}{4\pi\varepsilon_0}\frac{2p\cos\theta}{r^3} + E_0\cos\theta + \frac{aV_0}{r^2}$ is evaluated at $r = a$. Since $p = 4\pi\varepsilon_0 a^3 E_0$,

$$\sigma = \frac{1}{4\pi}\frac{2\times 4\pi\varepsilon_0 a^3 E_0\cos\theta}{a^3} + E_0\cos\theta + \frac{\varepsilon_0 V_0}{a}. \tag{5.163}$$

$$\therefore \sigma = 3\varepsilon_0 E_0\cos\theta + \frac{\varepsilon_0 V_0}{a}. \tag{5.164}$$

Colour-coded '+' and '−' signs on a sphere are used to visualise σ and its variation with polar angle θ in Fig. 5.11(c).

5.6   Curie, Ising and Ferromagnetism

5.6.1   *Diamagnetism, paramagnetism and ferromagnetism*

All atoms will exhibit *diamagnetism,* i.e. the effect of imposing a magnetic field on atomic orbital electrons is to produce a magnetic field which *opposes* the applied field. For certain special materials, diamagnetism can result in impressive repulsive forces. The graphene-based material *pyrolytic carbon* can be made to levitate above strong permanent magnets or a powerful solenoid. Many materials will also form magnetic dipoles that will instead align with an applied field, thereby strengthening it. This is called *paramagnetism* and would result in magnetic attraction between a paramagnetic material and the source of a magnetic field if the paramagnetic effect is stronger than the diamagnetic effect. As a general rule (there are some notable exceptions, such as copper), unpaired electrons in atomic orbitals result in paramagnetism, whereas if all atomic orbital electrons are paired, then a material is only diamagnetic. For a few *ferromagnetic* materials (involving compounds or alloys of iron, cobalt, nickel and certain rare-earth metals like neodymium), the paramagnetic strengthening effect can be significant. 'Hard' magnetic materials, such as steel, have *domain microstructures* of aligned atomic spins, which themselves will orient *en masse* to an applied field, intensifying it. Below their *Curie temperature*, magnetisation will persist once an applied field is removed. 'Soft' magnetic materials (such as annealed[25] iron) also have magnetic domains on a micrometer scale. However, when an applied field is removed, no alignment persists and the soft magnet ceases to retain its magnetism. This is why iron is so useful in the construction of an electromagnet, with a simple practical example being a loop of wire wound around an iron nail. When electric current flows through the wire, the iron intensifies the magnetic field within the loop, and the whole assembly now behaves like a macroscopic dipole electromagnet, i.e. a *solenoid*. But when the current is switched off, the nail ceases to be magnetic. Without the ability to 'switch off' magnetism, many vital devices of our modern world, including relays, transformers and scrapyard machines, would not function effectively.

To account for magnetisation effects, the net magnetic flux density $\mathbf{B}$ (in Tesla) is defined as the sum of a *magnetisation field* $\mathbf{M}$ and an *applied magnetic field* $\mathbf{H}$:

$$\mathbf{B} = \mu_0 \left(\mathbf{H} + \mathbf{M} \right), \tag{5.165}$$

where $\mu_0 = 1.257 \times 10^{-6}\,\mathrm{NA}^{-2}$ (to 4.s.f) is the *vacuum permeability*.

If a material is isotropic in a magnetic sense, we might sensibly assume magnetisation effects are parallel to the applied field. This means we can write

$$\mathbf{B} = \mu\mu_0 \mathbf{H}. \tag{5.166}$$

μ is a dimensionless parameter called the *relative permeability*. Air, wood, water and copper have a relative permeability close to unity. Nickel and carbon steel have a typical μ of about 100, whereas ferrite (iron oxide combined with nickel, zinc etc.) may have a μ in excess of 600. Pure iron may have $\mu > 5{,}000$. For relative permeabilities of various common materials, see https://www.engineeringtoolbox.com/permeability-d_1923.html.

5.6.2 *Curie's law: How paramagnetism varies with temperature*

Consider an isotropic magnetic medium with *magnetic dipole moments* $\mathbf{m}$ (see the section on electric and magnetic dipoles) aligned in random orientations. Let a magnetic field be applied of strength $\mathbf{H}$. If $mH\cos\theta = \mathbf{m} \cdot \mathbf{H}$, i.e. the angle between the vectors $\mathbf{m}$ and $\mathbf{H}$ is θ, the average magnetic dipole moment parallel

---

[25] In very simple terms, *annealing* means 'heat until glowing, then allow to cool slowly'. Annealing causes the number of *dislocations* in the crystal lattice of a material to decrease, which typically leads to a change in material hardness and *ductility* (i.e. ability to be stretched into wires).

to $\mathbf{H}$ is

$$\langle m_\parallel \rangle = \int_0^\pi m \cos\theta \times p(\theta) d\theta. \tag{5.167}$$

The probability of the dipole being within angular range θ to $\theta + d\theta$ is $p(\theta)d\theta$, where

$$p(\theta) = A \frac{2\pi r \sin\theta \times r d\theta}{4\pi r^2} e^{\frac{\mu_0 m H \cos\theta}{k_B T}}. \tag{5.168}$$

A is a constant to ensure $\int_0^\pi p(\theta)d\theta = 1$. The ratio $\frac{2\pi r \sin\theta \times r d\theta}{4\pi r^2} = \frac{1}{2}\sin\theta d\theta$ is the fraction of the surface area of a sphere of radius r that corresponds to a polar angle θ. The quantity $2\pi r \sin\theta \times r d\theta$ represents a strip of sphere surface area of radius $r \sin\theta$ and width $r d\theta$, i.e. the area of the sphere associated with polar angle θ. The quantity $\frac{2\pi r \sin\theta \times r d\theta}{4\pi r^2}$ must be a factor of $p(\theta)$ to account for the 3D nature of $\mathbf{m}$ and the fraction of total possible pointing directions that correspond to angle θ. The last term is a *Boltzmann factor* $e^{-\frac{\varepsilon}{k_B T}}$ (see Chapter 3), which relates a measure of thermal energy $k_B T$ to the energy $\varepsilon = -\mathbf{m} \cdot \mu_0 \mathbf{H} = \mu_0 m H \cos\theta$ associated with the dipole in the magnetic field.[26] Boltzmann's constant is $k_B = 1.38 \times 10^{-23} \mathrm{JK}^{-1}$ and T is the absolute temperature in K.

To find A and $\langle m_\parallel \rangle$ consider a substitution $u = \cos\theta$.

$$\therefore \quad du = -\sin\theta d\theta \tag{5.169}$$

$$\implies d\theta = -\frac{du}{\sin\theta}. \tag{5.170}$$

Also, define a dimensionless parameter

$$x = \frac{\mu_0 m H}{k_B T}. \tag{5.171}$$

Hence,

$$p(\theta)d\theta = A \left(\tfrac{1}{2}\sin\theta\right) e^{xu} \left(-\frac{du}{\sin\theta}\right) = -\tfrac{1}{2}A e^{xu} du. \tag{5.172}$$

Since $\cos(0) = 1$ and $\cos(\pi) = -1$, the probability distribution condition $\int_0^\pi p(\theta)d\theta = 1$ implies $\tfrac{1}{2}A \int_1^{-1} -e^{xu}du = 1 \Rightarrow \tfrac{1}{2}A \int_{-1}^1 e^{xu}du = 1$.

$$\therefore A = \frac{1}{\tfrac{1}{2}\int_{-1}^1 e^{xu}du} \tag{5.173}$$

$$= \frac{1}{\tfrac{1}{2}\left[\tfrac{1}{x}e^{xu}\right]_{-1}^1}. \tag{5.174}$$

$$\therefore A = \frac{x}{\sinh x}. \tag{5.175}$$

So, $p(\theta)d\theta$ becomes

$$p(\theta)d\theta = \frac{x}{\sinh x}\left(\tfrac{1}{2}\sin\theta d\theta\right) e^{x\cos\theta} = -\tfrac{1}{2}\frac{x}{\sinh x}e^{xu}du. \tag{5.176}$$

---

[26] See Jackson [18] or Bleaney & Bleaney [2].

Therefore, the average magnetic dipole moment parallel to **H** is

$$\langle m_\parallel \rangle = \int_0^\pi m \cos\theta \times p(\theta) d\theta \tag{5.177}$$

$$= \frac{1}{2}\frac{mx}{\sinh x} \int_{-1}^1 u e^{xu} du \tag{5.178}$$

$$= \frac{1}{2}\frac{mx}{\sinh x} \left\{ \left[\frac{1}{x} u e^{xu}\right]_{-1}^1 - \int_{-1}^1 \frac{1}{x} e^{xu}(1) du \right\} \tag{5.179}$$

$$= \frac{mx}{\sinh x}\frac{1}{x} \left\{ e^x + e^{-x} - \left[\frac{1}{x} e^{xu}\right]_{-1}^1 \right\} \tag{5.180}$$

$$= \frac{1}{2}\frac{m}{\sinh x} \left\{ 2\cosh x - \frac{2}{x}\sinh x \right\}. \tag{5.181}$$

$$\therefore \langle m_\parallel \rangle = m\left(\coth x - \frac{1}{x} \right). \tag{5.182}$$

$\coth x - \frac{1}{x}$ is called the *Langevin function*, named after Paul Langevin (1872–1946), who worked with Pierre Curie (1859–1906) in developing this model of paramagnetism. Let us define a *Curie temperature* when $x = 1$, i.e. since $x = \frac{\mu_0 mH}{k_B T}$,

$$T_C = \frac{m\mu_0 H}{k_B} \tag{5.183}$$

and therefore

$$x = \frac{T_C}{T}. \tag{5.184}$$

A sensible set of (dimensionless) plotting variables are

$$y = \frac{\langle m_\parallel \rangle}{m}, \tag{5.185}$$

$$z = \frac{T}{T_C} = \frac{1}{x}. \tag{5.186}$$

Hence, the generic Langevin function, as plotted in Fig. 5.12, is

$$y = \coth\left(\frac{1}{z}\right) - z. \tag{5.187}$$

As the graph of $y = \frac{\langle m_\parallel \rangle}{m}$ versus $z = \frac{T}{T_C}$ illustrates, the average magnetic dipole moment parallel to **H** tends to zero as $T \gg T_C$, which quantitatively explains the diminution of paramagnetic effects with increased temperature.

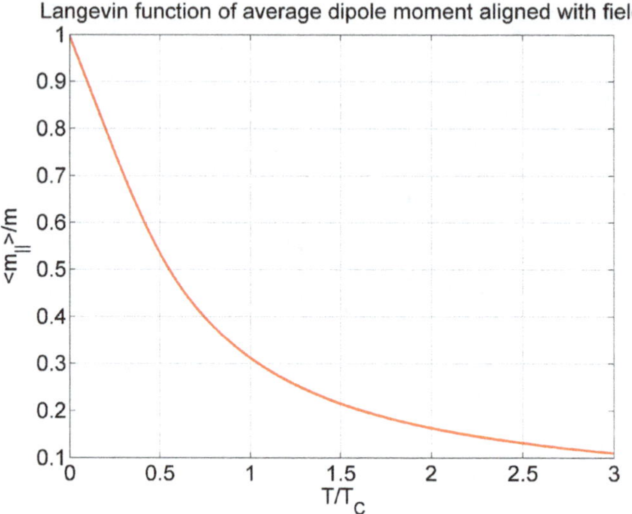

Fig. 5.12. Curie's model of *paramagnetism*. The average magnetic dipole moment parallel to the applied magnetic field $\mathbf{H}$ is given by $\langle m_\parallel \rangle = m \left(\coth x - \frac{1}{x} \right)$, where m is the maximum dipole moment and $x = \frac{m\mu_0 H}{k_B T}$. $\coth x - \frac{1}{x}$ is called the *Langevin function*. A measure of thermal energy is $k_B T$, where T is absolute temperature /K and k_B is Boltzmann's constant $k_B = 1.38 \times 10^{-23}\,\mathrm{JK^{-1}}$. μ_0 is the vacuum permeability $\mu_0 = 1.257 \times 10^{-6}\,\mathrm{NA^{-2}}$. $x = 1$ defines the *Curie temperature* $T_C = \frac{m\mu_0 H}{k_B}$. When $T \gg T_C$, random thermal motion causes a diminishment of any average alignment of magnetic dipoles with the applied field.

5.6.3 *The Ising model*

All atoms will respond in some fashion to magnetic fields. The angular momentum (and spin) properties of electrons imply a circulating charge[27] q of velocity $\mathbf{v}$, which means they will be subject to a *Lorentz* $q\mathbf{v} \times \mathbf{B}$ force in a magnetic field of flux density $\mathbf{B}$. However, the effects of *diamagnetism, paramagnetism* and *anti-ferromagnetism*[28] are typically very small for most materials. The exception are ferromagnetic materials (iron, cobalt, nickel and some rare-earth metal compounds) which respond strongly to applied magnetic fields and can intensify them by several

---

[27] Of course, the classical model of electrons orbiting a positively charged nucleus cannot be an entirely accurate representation. Charges in circular motion accelerate, and accelerating charges radiate. One can show (see the next volume of *Science by Simulation*) that an electron in an atom of Hydrogen should inspiral into the nucleus in about 10^{-10} s. The fact that electrons *are* indeed separated from nuclei (thank goodness—we would not exist otherwise) was one of the more fundamental challenges faced by physicists at the turn of the twentieth century. The solution given by Planck, Bohr, Einstein, de Broglie and many others gave rise to *quantum mechanics,* a theory with spectacular predictive power, but perhaps challenging, baffling even, from a conceptual standpoint. An electron is *both* some form of standing wave around the atom (which prevents it losing energy) *and* a particle which can be modelled (albeit incompletely) by the same laws of dynamics and forces that govern macroscopic objects.

[28] *Anti-ferromagnetism* is where magnetic dipoles align with an external field, but in alternating parallel and anti-parallel directions.

orders of magnitude, i.e. the relative permeability μ can be tens, hundreds or possibly even thousands. The *Ising*[29] *model* is a simplified description of a ferromagnet which exhibits a *phase transition* above the Curie temperature. Below the Curie temperature, magnetic dipoles will tend to cluster into *domains* of alignment, and it is these micro-scale groupings (i.e. involving thousands of atoms in each direction, which means billions of atoms in three dimensions) which give rise to ferromagnetic behaviour.

The Ising model, as illustrated in Figs. 5.13–5.15, can be used to demonstrate spontaneous mass alignment of magnetic dipoles and possibly a mechanism for domain formation. Perhaps the simplest model which yields characteristic behaviour is an $N \times N$ square grid, where each square is initially randomly assigned a $+1$ or -1 value, with equal probability. The $+/-1$ values correspond to a single direction of magnetic dipole moment in a rectangular lattice of ferromagnetic atoms, or, in the case of individual electrons, *spin* (i.e. its intrinsic angular momentum property, which in turn defines its intrinsic electric dipole moment). White squares represent $+1$, while black squares represent -1. In the Ising model, spin is an abstracted signed integer quantity, whereas in reality, it will have dimensions of mass $\times$ velocity $\times$ distance and indeed a vector direction.

The Ising model can be applied[30] using an iterative computational method attributed to Nicholas Metropolis (1915–1999).

5.6.3.1 *The Metropolis algorithm*

(1) Choose one square at random from the $N \times N$ grid. Let its spin be $s(n) = +1$ or -1.

(2) Find the spins of the nearest neighbours. Use *circular boundary conditions*. If $s(n)$ is at the *edge* of the grid, take a nearest neighbour to be that of the *opposite end* of the grid. This rather curious rule won't be needed for enormous grids, which would be more representative of real domains, but we want to see similar behaviour in a more computationally manageable 2D array of spins. Circular boundary conditions help us achieve this aspiration, and we might justify our reasoning on the basis that our mini-system could be thought of as a small subset of a domain.

(3) Compute a sum of spin-coupling energies for $s(n)$ and its (up, down, left, right) neighbours $s_n(k)$, and work out the energy change ΔE if $s(n)$ *were to flip sign from* $+1$ to -1 or from -1 to $+1$. The formula for the energy change ΔE is given by

$$\Delta E = 2 \left(F + J \sum_{k=1}^{4} s_n(k) \right) s(n). \tag{5.188}$$

---

[29] Ernst Ising (1900–1998).

[30] The Ising model is a popular addition to many university statistical thermodynamics and electromagnetism courses. A similar analysis to that presented here is at https://farside.ph.utexas.edu/teaching/329/lectures/node110.html (Accessed 25 August 23).

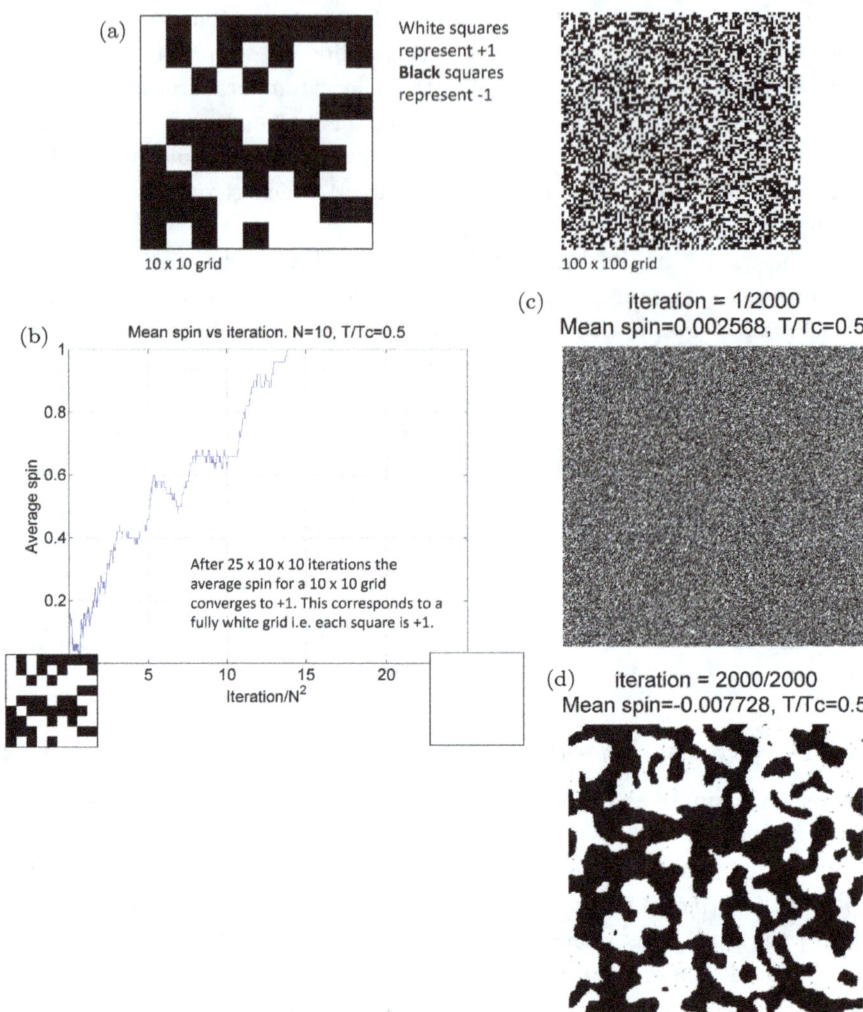

Fig. 5.13. (a) 10×10 or 100×100 randomised initial conditions for a simulation of the Ising model of ferromagnetism. White squares represent a magnetic dipole moment ('spin') in an 'up' orientation (numeric value of $+1$), whereas black squares represent a spin in a 'down' position (numeric value of -1). (b) After 2,500 iterations of the Metropolis algorithm, (with temperature ratio $\frac{T}{T_C} = 0.5$), the average spin $\langle s \rangle = \frac{1}{N^2} \sum_{n=1}^{N^2} s(n)$ tends to $+1$. (Note that it could have easily been -1. The point is a transition occurs from randomness to uniformity). (c) and (d) illustrate a 500×500 grid. In this case, spin uniformity is not yet reached, even after $I = 2000 \times 500 \times 500$ iterations. However, *domain*-like structures are clearly visible in this intermediate state.

J is the spin-coupling energy in eV and F is the energy in eV associated with the alignment of spin $s(n)$ with an applied external magnetic field. Let us ignore any energy contributions from non-nearest neighbours. We might be justified in making this assumption given that the magnetic field strength from magnetic dipoles diminishes with distance r as $\frac{1}{r^3}$. (See the previous section of this chapter on electric and magnetic dipoles.)

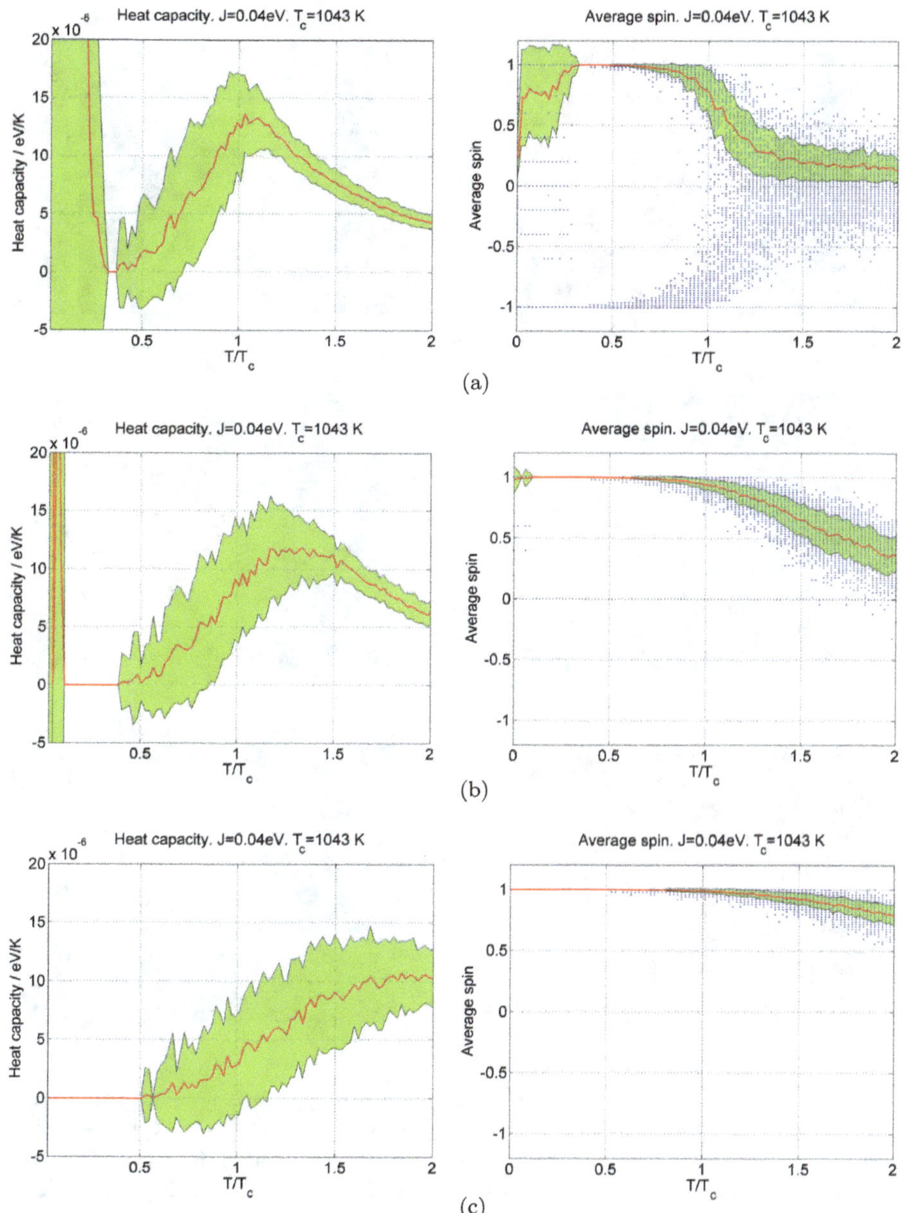

Fig. 5.14. Ising model of ferromagnetism. Heat capacity and average spin are plotted for (a) $F/J = 0$; (b) $F/J = 0.53$; (c) $F/J = 2$. Red lines are the mean values over $R = 100$ repeats, and the green patch corresponds to the standard deviation. Note that when $F/J = 0$, the heat capacity peaks at the Curie temperature. The Ising model is for a 10×10 grid and run for $I = 2{,}000 \times (10 \times 10)$ iterations of Metropolis algorithm for each temperature. 100 different linearly spaced temperatures are used with T/T_C in the range of 0.0 to 2.0.

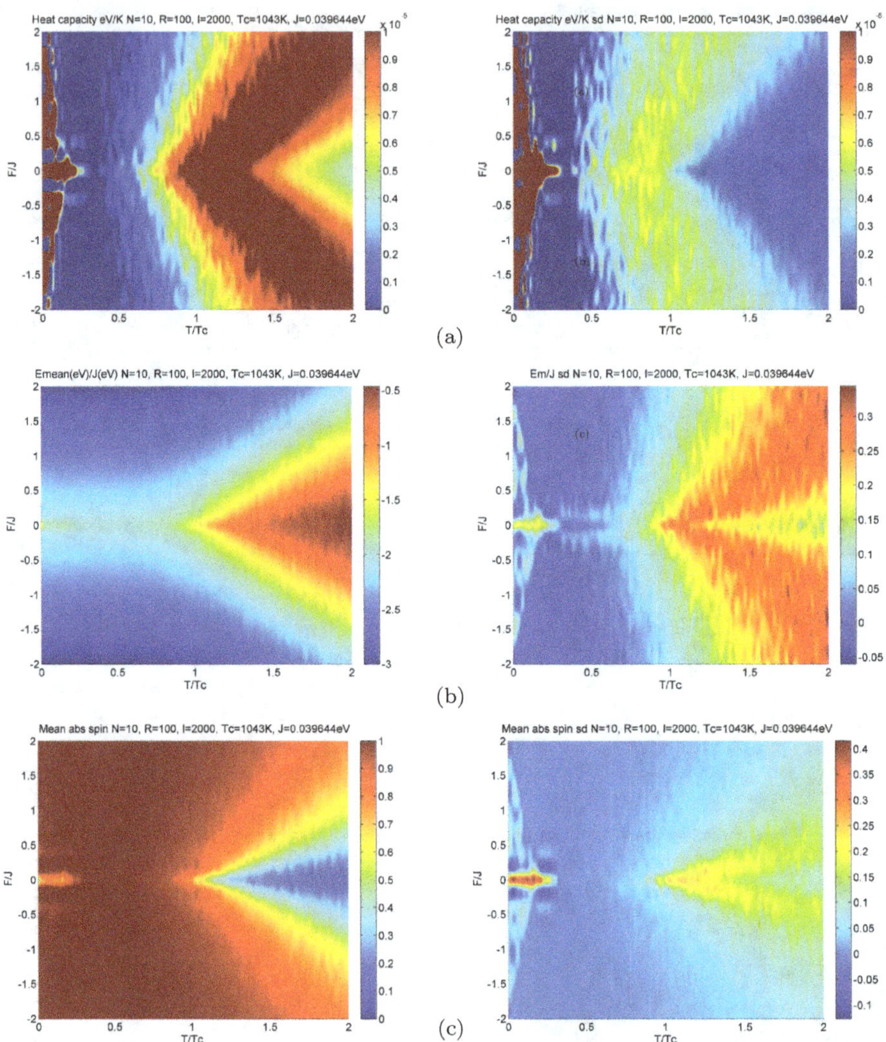

Fig. 5.15. Summary of Ising simulations run for a 10×10 grid for $2{,}000 \times 10 \times 10$ iterations. 100 repeats for every $\frac{T}{T_C}$ value in the range of 0 to 2 and $\frac{F}{J}$ values in the range of -2 to 2. Each plot is a coloured surface of $\frac{T}{T_C}$ and $\frac{F}{J}$, with the colour scale representing the height of the surface. (a) is the average heat capacity (left) and standard deviation (right) in eV/K, based on a Curie temperature of 1,043 K (i.e. iron). (b) is the mean energy ratio $\frac{\langle E \rangle}{J}$ and the standard deviation of $\frac{\langle E \rangle}{J}$, and (c) is the mean spin magnitude and the standard deviation of spin magnitude. The phase transition below the Curie temperature is most obviously revealed in (c). Below the Curie temperature, the spins will align with an applied field, regardless of the size of the alignment to coupling energy ratio $\frac{F}{J}$. Note that Onsager's relationship $J = \frac{1}{2} k_B T_C \ln(1 + \sqrt{2})$ is used to compute the coupling energy J from the Curie temperature.

(4) Now, change the sign of spin $s(n)$ according to the following rule, which is based on a uniformly distributed random number $0 \leq r \leq 1$: if $e^{-\frac{\Delta E}{k_B T}} \geq r$ **or** $\Delta E < 0$, *swap the sign* of $s(n)$. i.e. $s(n) \to -s(n)$. A negative energy change implies that the system moves from a higher energy to a lower energy state, which might be regarded as thermodynamically favourable.[31] Note that this is *deterministic*, given a prior system spin state. The $e^{-\frac{\Delta E}{k_B T}} \geq r$ option introduces an additional element of *chance*, and we might imagine the spin-flip process being driven by an energy (and entropy) exchange resulting from the *random* thermal motion of the spin system, which is why we use the *Boltzmann factor* $e^{-\frac{\Delta E}{k_B T}}$.

(5) Repeat from step 1 until $I \times N^2$ iterations have elapsed. Update the black-and-white grid after a suitable number of iterations.

Now, compute from the $N \times N$ grid the following parameters:
Average spin

$$\langle s \rangle = \frac{1}{N^2} \sum_{n=1}^{N^2} s(n). \tag{5.189}$$

Mean energy per spin

$$\langle E \rangle = \frac{1}{2} \frac{1}{N^2} \sum_{n=1}^{N^2} \left(F + J \sum_{k=1}^{4} s_n(k) \right) s(n). \tag{5.190}$$

Mean heat capacity $\langle C \rangle$, in eV per K, from the equation

$$k_B T^2 \langle C \rangle = \langle E^2 \rangle - \langle E \rangle^2, \tag{5.191}$$

where

$$\langle E^2 \rangle = \frac{1}{4} \frac{1}{N^2} \sum_{n=1}^{N^2} \left(s(n) F + J s(n) \sum_{k=1}^{4} s_n(k) \right)^2. \tag{5.192}$$

The relationship $k_B T^2 \langle C \rangle = \langle E^2 \rangle - \langle E \rangle^2$ assumes a *Boltzmann distribution* for a set of energies $\{E_i\}$, i.e the mean energy is

$$\langle E \rangle = \frac{\sum_i E_i e^{-\frac{E_i}{k_B T}}}{\sum_i e^{-\frac{E_i}{k_B T}}} \tag{5.193}$$

and also

$$\langle E^2 \rangle = \frac{\sum_i E_i^2 e^{-\frac{E_i}{k_B T}}}{\sum_i e^{-\frac{E_i}{k_B T}}}. \tag{5.194}$$

---

[31] More accurately, we should consider whether the total *entropy* increases during a sign swap for it to be thermodynamically favourable. But in order to do this, we must consider energy exchange with the surroundings as well as the spin system in isolation.

If we define heat capacity to be

$$C = \frac{\partial \langle E \rangle}{\partial T},$$ (5.195)

then using the quotient rule of differentiation,

$$\frac{\partial \langle E \rangle}{\partial T} = \frac{\left(\sum_i e^{-\frac{E_i}{k_B T}}\right) \sum_i E_i e^{-\frac{E_i}{k_B T}} \left(\frac{E_i}{k_B T^2}\right) - \left(\sum_i E_i e^{-\frac{E_i}{k_B T}}\right) \left(\sum_i e^{-\frac{E_i}{k_B T}} \left(\frac{E_i}{k_B T^2}\right)\right)}{\left(\sum_i e^{-\frac{E_i}{k_B T}}\right)^2}$$

$$= \frac{1}{k_B T^2} \frac{\sum_i E_i^2 e^{-\frac{E_i}{k_B T}}}{\sum_i e^{-\frac{E_i}{k_B T}}} - \frac{1}{k_B T^2} \left(\frac{\sum_i E_i e^{-\frac{E_i}{k_B T}}}{\sum_i e^{-\frac{E_i}{k_B T}}}\right)^2$$

$$= \frac{1}{k_B T^2} \left(\langle E^2 \rangle - \langle E \rangle^2\right).$$ (5.196)

Hence,

$$k_B T^2 \langle C \rangle = \langle E^2 \rangle - \langle E \rangle^2.$$ (5.197)

For a 2D Ising model, Lars Onsager[32] determined, in 1944, the relationship between the phase transition Curie temperature T_C and coupling energy J:

$$k_B T_C = \frac{2J}{\ln(1 + \sqrt{2})}.$$ (5.198)

In an implementation of the Ising model, define a scaled temperature

$$z = \frac{T}{T_C}.$$ (5.199)

For the simulations described in this chapter, $T_C = 1,043$ K is used, which is typical for iron. J (in eV) will be defined, using Onsager's formula, to be

$$\frac{J}{\text{eV}} = \tfrac{1}{2} \frac{k_B T_C}{q_e \times 1\text{V}} \ln(1 + \sqrt{2}),$$ (5.200)

where electron charge $q_e = 1.602 \times 10^{-19}$ C. A ratio $\frac{F}{J}$ will be defined rather than F separately.

5.6.3.2 Simulation 1

$N = 500$, $I = 2,000$, $F/J = 0$, $\frac{T}{T_C} = 0.5$. See Fig. 5.13.

For a 500 × 500 grid, a similar equilibrium is not yet reached, even after $I = 2000 \times 500 \times 500$ iterations. However, domain-like structures are clearly visible in this intermediate state.

---

[32] Lars Onsager (1903–1976).

5.6.3.3 *Simulation 2*

See Figs. 5.14 and 5.15. The specifications of this simulation are:

- 10×10 grid. $I = 2{,}000 \times (10 \times 10)$ iterations of the Metropolis algorithm.
- $R = 100$ repeats for each temperature.
- 100 different temperatures from $T/T_C = 0.0, \ldots, 2.0$.
- 21 different F/J values from -2 to 2.

That is, $2000 \times 10 \times 10 \times 100 \times 100 \times 21 = 42$ *billion* iterations of the Metropolis algorithm! The total running time using an Intel i5 PC was about five days—a huge opportunity for using distributed parallel processing here. Only the $2000 \times 10 \times 10$ iterations for a single Metropolis run follow a *serial* process, so this simulation could, in principle, be sped up by a factor of 210,000 if one could run a variant of the algorithm on 210,000 Intel i5 PCs simultaneously and then collate the results afterwards.[33]

An interesting result, illustrated in Fig. 5.14, is that heat capacity appears to peak around the Curie temperature and then diminish at higher temperatures. The formula $k_B T^2 \langle C \rangle = \langle E^2 \rangle - \langle E \rangle^2$ implies that the mean heat capacity $=$ the *variance* in energy per spin$/k_B T^2$, i.e. the greater the *fluctuations* in spin energy, the greater the heat capacity. My results broadly agree with this statement, although I do also observe a rise in fluctuations (and hence heat capacity) for lower temperatures when $\frac{F}{J}$ is small. I'm sceptical my somewhat crude simulation is revealing anything apart from numerical noise here, but in the spirit of scientific enquiry, I present what I did, anomalies and all. I encourage the reader to use the recipe described here to build their own Ising model and see if they observe similar results—and hopefully with a bit more parallel processing to speed up the simulation.

5.7 Ampère's Theorem and a Toroidal Inductor

5.7.1 *Ampère's theorem*

Ampère's theorem[34] is a foundational concept of electromagnetism and indeed forms the basis of one of Maxwell's four equations,[35] which, together, defines how the

---

[33] As of March 2023, the fastest supercomputer is the US\$600 million Hewlett Packard *Frontier*, hosted at the Oak Ridge Leadership Computing Facility in Tennessee, USA. It comprises an array of 9,472 sixty-four-core processors (CPUs) supplemented by 37,888 graphics processors (GPUs) with 220 cores each, resulting in a total of 8,941,568 cores. So, a parallel processing challenge equivalent to 210,000 standard desktop PCs would be pretty trivial for Frontier! My Intel i5 PC had a measly *dual*-core processor, so one might anticipate a speed boost of $\approx 105{,}000$, which means the five-day simulation would take about 4.1 s. If the cores of Frontier also had a higher clock rate, then this could be even shorter. Note that we can't exploit the full parallel processing capacity of Frontier because the *serial* $2000 \times 100 \times 100$ iterations of the Metropolis algorithm cannot be split up between processor cores (https://en.wikipedia.org/wiki/Frontier_(supercomputer)).

[34] André-Marie Ampère (1775–1836).

[35] James Clerk Maxwell (1831–1879).

distribution and dynamics of charges give rise to electric and magnetic fields. A brief presentation of Maxwell's equations, and some of the important theorems that can be deduced from them, comprises the final section of this chapter.

If a closed loop with path element vector $d\mathbf{l}$ encircles 'free' current I_{enclosed} (i.e. the sort of current which flows in conductors in a circuit, rather than any circulation of charge resulting from any magnetisation effects),

$$\oint_{\text{loop}} \mathbf{H} \cdot d\mathbf{l} = I_{\text{enclosed}}. \tag{5.201}$$

As discussed in the previous section, to account for magnetisation effects, the net magnetic flux density $\mathbf{B}$ is defined as the sum of a *magnetisation field* $\mathbf{M}$ and an *applied field* $\mathbf{H}$:

$$\mathbf{B} = \mu_0 \left(\mathbf{H} + \mathbf{M} \right). \tag{5.202}$$

Note that it is $\mathbf{B}$, not $\mathbf{H}$, that we often tend to refer to as 'magnetic field' (although officially it is called *magnetic flux density*), and indeed, it is $\mathbf{B}$ that we will measure (in Tesla) via a *Hall probe* for the air-gap toroidal inductor, which shall be the practical subject of this chapter section. If a material is isotropic in a magnetic sense, we might sensibly assume magnetisation effects are parallel to the applied field. Hence,

$$\mathbf{B} = \mu \mu_0 \mathbf{H}. \tag{5.203}$$

The quantity $\mu_0 \approx 4\pi \times 10^{-7}$ NA^{-2} is the *vacuum permeability*, and μ is the *relative permeability*, which could be a large (dimensionless) number, i.e. several hundreds or higher for a *ferromagnetic* material such as annealed iron.

5.7.2 *A toroidal inductor with a small air gap*

Consider a circular iron ring of radius r with N coils of wire wound around it, as illustrated in Fig. 5.16. The wire coils carry an electric current I. This type of solenoid is known as a *toroidal inductor* since the ring is a torus and a coil of wire has the potential to be an *inductor*. If a time-varying current were to pass through it, *Faraday's law*[36] states that a *back EMF* of $\varepsilon = L\frac{dI}{dt}$ would be induced, where L is the *inductance*, a property dependent on the physical characteristics of the inductor but independent of current I.

The ring has a small air gap of width $d \ll 2\pi r$, whereas the remaining $2\pi r - d$ of the ring is assumed to be magnetically isotropic iron of relative permeability μ. By Ampère's theorem,

$$\oint_{\text{loop}} \mathbf{H} \cdot d\mathbf{l} = NI, \tag{5.204}$$

i.e. a total current of NI *enters the plane of the iron ring* and is therefore 'enclosed' by it. The circular symmetry of the torus suggests a loop comprising the circle of the ring, which passes through the axial centre of the ring cross-section. Along this

---

[36]Michael Faraday (1791–1867).

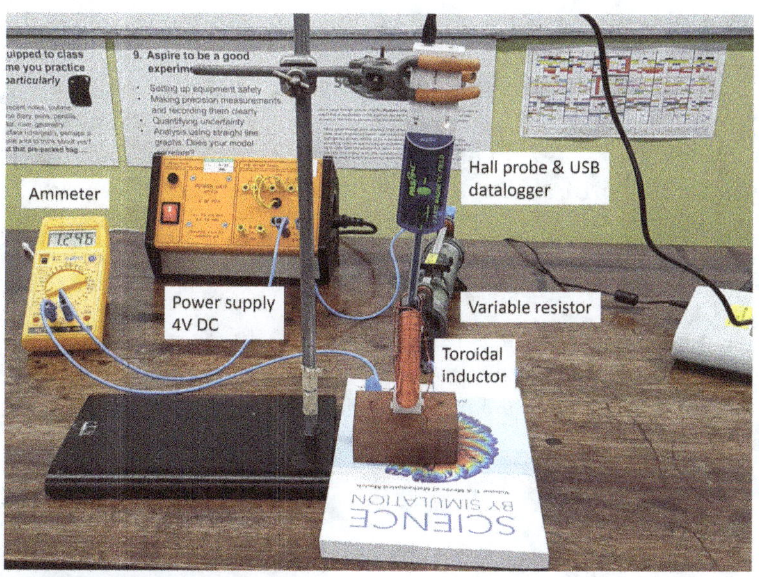

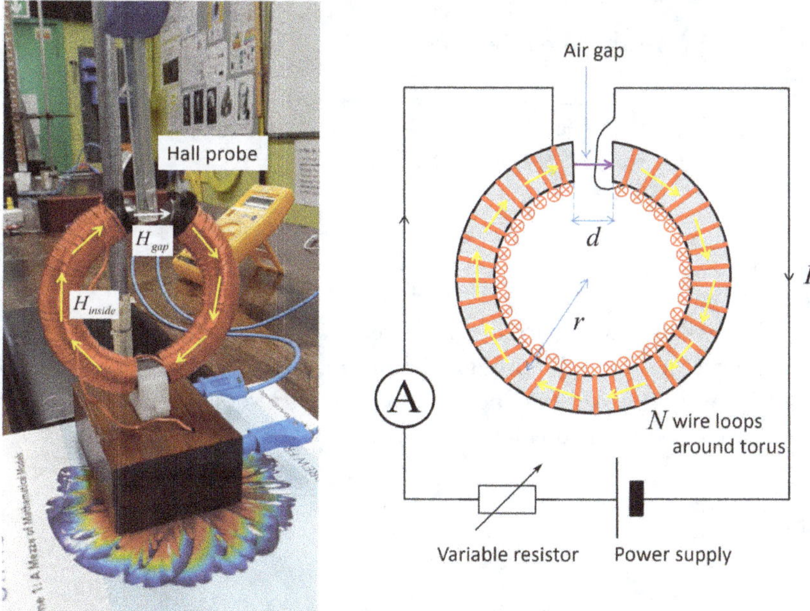

Fig. 5.16. A laboratory demonstration of a *toroidal inductor* with a small air gap. This consists of $N = 220$ turns of wire round a 80 mm radius iron metal ring that has a 18 mm air gap segment cut out from it. A multi-axis Hall probe can be mounted in the air gap to measure the magnetic flux density B in the ring as current I in the wires is varied. This is achieved by connecting the toroidal inductor to a 4 V rated DC power supply and a 8 Ω, 5 A rated variable resistor. Currents within a range of 0.4 to about 2.0 A should be expected. Current is measured using a digital ammeter wired in series to the circuit. Ampère's theorem $\oint_{\text{loop}} \mathbf{H} \cdot d\mathbf{l} = NI$ can be used to show that $B = \frac{\mu\mu_0 N}{2\pi r + (\mu-1)d} I$, i.e a direct proportion $B = kI$. We assume low currents and anticipate fairly negligible hysteresis and a lack of magnetic saturation of the ring core. From a line of best fit, we can find k and hence calculate the relative permeability μ of the ring.

path, we might assume that the magnetic field is tangential to the circle, i.e. $\mathbf{H}$ is parallel to $d\mathbf{l}$. Hence, we can express $\oint_{\text{loop}} \mathbf{H} \cdot d\mathbf{l} = NI$ as

$$H_{\text{inside}}(2\pi r - d) + H_{\text{gap}}d = NI \tag{5.205}$$

since I_{enclosed} by the loop is NI. (The crosses, indicating current entering into the plane of the page in Fig. 5.16, should hopefully make this clear.) If $\mathbf{B} = \mu\mu_0\mathbf{H}$, then

$$B_{\text{gap}} = \mu_0 H_{\text{gap}}, \tag{5.206}$$

$$B_{\text{inside}} = \mu\mu_0 H_{\text{inside}}. \tag{5.207}$$

The relative permeability of air is taken as unity. A fundamental Maxwell equation result (see Section 5.13.5) is that magnetic flux density $\mathbf{B}$ is *continuous perpendicular to the boundary between magnetic media*. This means $B_{\text{inside}} = B_{\text{gap}} = B$. Hence, we can write

$$H_{\text{gap}} = \frac{B}{\mu_0}, \tag{5.208}$$

$$H_{\text{inside}} = \frac{B}{\mu\mu_0}. \tag{5.209}$$

Therefore, Ampère's theorem becomes

$$\frac{B}{\mu\mu_0}(2\pi r - d) + \frac{B}{\mu_0}d = NI \tag{5.210}$$

$$B\left(2\pi r - d + \mu d\right) = \mu\mu_0 NI. \tag{5.211}$$

$$\therefore B = \frac{\mu\mu_0 N}{2\pi r + (\mu - 1)d}I. \tag{5.212}$$

So, if μ is constant for our range of currents,[37] we anticipate B versus I to be a straight-line graph, $B = kI$, through the origin, with constant of proportionality k. If B and I can be measured experimentally and B versus I plotted on a graph, the gradient k can be used to calculate μ:

$$k = \frac{\mu\mu_0 N}{2\pi r + (\mu - 1)d} \tag{5.213}$$

$$k\left(2\pi r - d + \mu d\right) = \mu\mu_0 N \tag{5.214}$$

$$k(2\pi r - d) = \mu\left(\mu_0 N - kd\right). \tag{5.215}$$

$$\therefore \mu = \frac{k(2\pi r - d)}{\mu_0 N - kd}. \tag{5.216}$$

An experiment involving a toroidal inductor of $N = 220$ turns wound around a (supposedly) iron ring of radius $r = 80$ mm, with an air gap of $d = 18$ mm is

---

[37] Most magnetic materials will undergo a process called *hysteresis*, i.e. a graph of B vs. H for an isotropic magnetic material will comprise a *loop* rather than a 1:1 graph. For a general model of our toroidal inductor, we might anticipate B vs. I to follow a slightly different relationship as I is increased and then decreased.

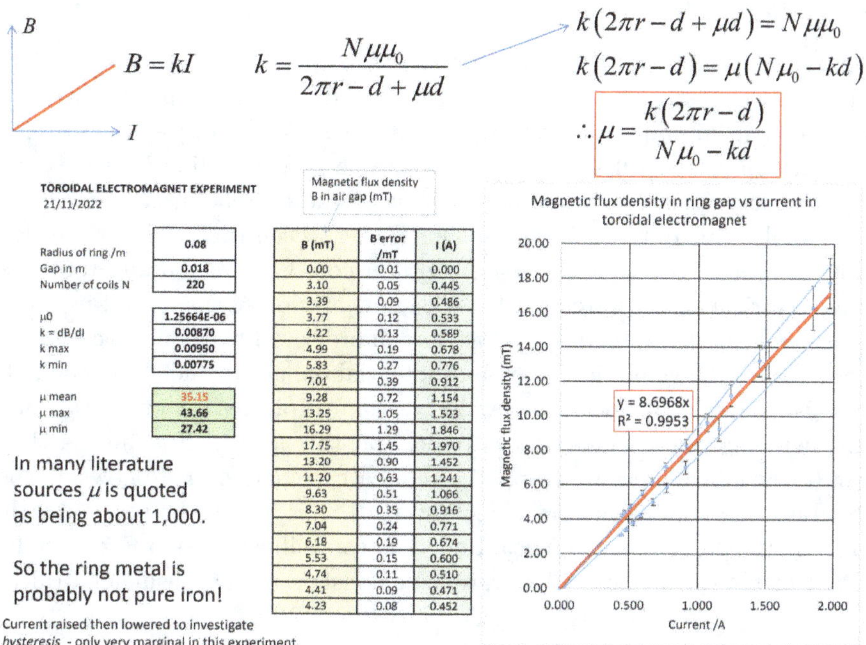

B (mT)	B error /mT	I (A)
0.00	0.01	0.000
3.10	0.05	0.445
3.39	0.09	0.486
3.77	0.12	0.533
4.22	0.13	0.589
4.99	0.19	0.678
5.83	0.27	0.776
7.01	0.39	0.912
9.28	0.72	1.154
13.25	1.05	1.523
16.29	1.29	1.846
17.75	1.45	1.970
13.20	0.90	1.452
11.20	0.63	1.241
9.63	0.51	1.066
8.30	0.35	0.916
7.04	0.24	0.771
6.18	0.19	0.674
5.53	0.15	0.600
4.74	0.11	0.510
4.41	0.09	0.471
4.23	0.08	0.452

Fig. 5.17. Analysis of an experiment with an iron-cored toroidal inductor with an air gap. A model of magnetic flux density B being directly proportional to wire current I is assumed, and a graph of the relationship $B = kI$ is used to calculate k. Ampère's theorem is used to calculate the relative permeability of the iron ring via the formula $\mu = \frac{k(2\pi r - d)}{\mu_0 N - kd}$. The ring radius $r = 80$ mm, the gap width $d = 18$ mm, the number of coils $N = 220$ and the vacuum permeability $\mu_0 \approx 4\pi \times 10^{-7}$ Hm^{-1}. In the experiment, the magnetic flux density was observed to oscillate, with an amplitude that increased with current. The mean and standard deviation of these oscillations were used to define the B values (and their errors) in the graph. I was increased from about 0.4 to 2.0 A and then reduced again to assess hysteresis. There is some evidence, but the overall effect is small. Using an envelope of lines of best fit, a range of k values are obtained, leading to an estimate for the ring core's relative permeability in the range $27.4 < \mu < 43.7$. This is rather low compared to that expected for annealed iron (i.e. several hundreds or more), so one assumes the ring is not pure iron, but some alloy which diminishes its ferromagnetic performance.

illustrated in Fig. 5.16. Experimental results are presented in Fig. 5.17. Current in the coils was varied from about 0.4 A to just under 2.0 A using a 4 V rated DC power supply and an 8 Ω, 5 A rated variable resistor. Only marginal hysteresis was evident as the current was raised and lowered, so a line of best fit to determine k from B versus I was deemed appropriate. $k = 8.70$ mT/A, with lower and upper estimated bounds of $k_{\mathrm{max}} = 9.50$ mT/A and $k_{\mathrm{min}} = 7.75$ mT/A, respectively, which means

$$27.4 < \mu < 43.7, \tag{5.217}$$

i.e. an average of about $\mu = 35$. This is rather low compared to the 'several hundreds to a few thousands' suggested for annealed iron, which implies that the ring (which I think bears a striking similarity to my doorknocker) is probably *not* pure annealed iron.

5.7.3 *A Rogowski coil*

The air-gap toroidal inductor has a more practical application than a school physics experiment to investigate relative permeability. If the coil is placed over a wire carrying AC (the air gap permits this without any surgery to wire or coil), then the effect of EM induction will result in a measurable potential difference between the ends of the coil. This can be used to calculate the current in the wire, without the need to place an ammeter in series. This *Rogowski coil* (named after the German physicist Walter Rogowski 1881–1947) can therefore be a very useful tool in spot-checking electrical power systems, where instrumentation such as wired-in ammeters may be sparse and 'turning off' a power line in order to fit one may be undesirable for reasons of cost, system integrity and potential danger to the electrician. Unlike our iron(ish)-cored coil, most Rogowski coils are typically air cored to maintain linearity with magnetic fields and current, and this lack of iron means that the coil can be flexible and easily wrapped around a live conductor. Rogowski coils can also be used as a *transformer* (see Section 5.10). The use of the toroidal inductor described in the previous section as a Rogowski coil is illustrated in Fig. 5.18. In this 'please don't try this at home' laboratory demonstration, an unshielded (although

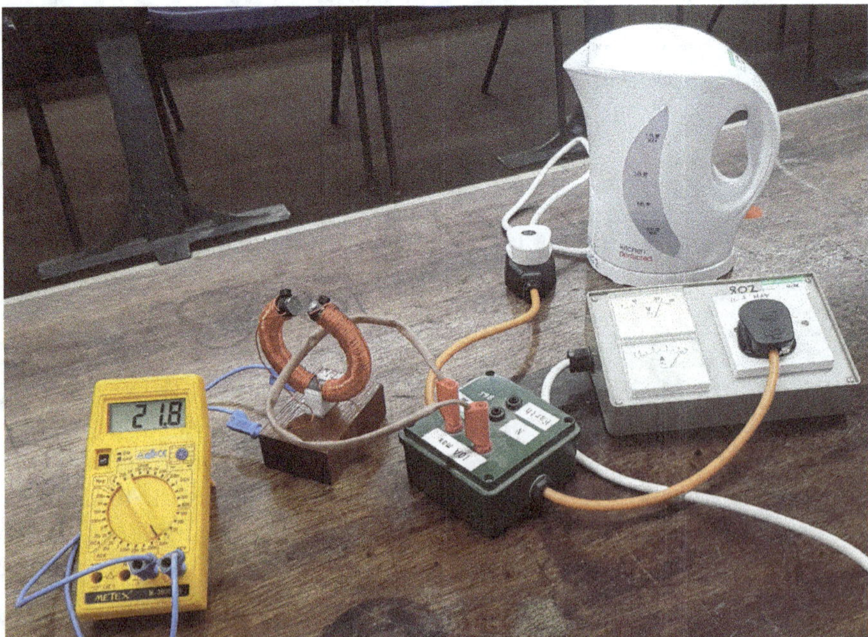

Fig. 5.18. A *Rogowski* coil of wire wrapped around a conductor can be used to determine the RMS voltage ε_{RMS} of an AC signal in an (unshielded) cable. In this 'please don't try this at home' laboratory demonstration, an unshielded (although insulated) short section of a mains power cable from an electric kettle is passed through the centre of the Rogowski coil. The mains supply voltage is 220 V RMS and 8.35 A RMS are drawn from the mains by the boiling kettle. An AC mode voltmeter connected across the Rogowski coil yields an induced RMS voltage of 21.8 mV.

insulated) short section of a mains power cable[38] from an electric kettle is passed through the centre of the Rogowski coil. The mains supply voltage is 220 V RMS and 8.35 A RMS are drawn from the mains by the boiling kettle. An AC-mode voltmeter connected across the Rogowski coil yields an induced RMS voltage of 21.8 mV.

Consider a Rogowski coil of radius r with N turns wrapped around a wire carrying AC with time variation $I = I_0 \cos \omega t$, i.e. of frequency $f = \frac{\omega}{2\pi}$. An application of Ampère's theorem gives the magnetic flux density at a radius r from the wire as

$$B = \frac{\mu_0 I}{2\pi r}. \tag{5.218}$$

If the coil has cross-sectional area A, then the magnetic flux linked by the coil is

$$\Phi = NBA = \frac{\mu_0 N A I_0}{2\pi r} \cos \omega t. \tag{5.219}$$

Faraday's law of electromagnetic induction implies that the EMF ε induced across the coil is

$$\varepsilon = -\frac{d\Phi}{dt}, \tag{5.220}$$

with the negative sign indicating that any current induced in the coil would 'oppose the change that produced it', i.e. *Lenz's law*. One assumes a Rogowski coil used to measure AC current would be connected in series to a very high resistance (AC-mode) voltmeter; therefore, one might expect very little current to actually flow in the coil. Using $\Phi = \frac{\mu_0 N A I_0}{2\pi r} \cos \omega t$, the EMF induced is

$$\varepsilon = \frac{\mu_0 \omega N A I_0}{2\pi r} \sin \omega t. \tag{5.221}$$

So, if we measure the root-mean-squared EMF $\varepsilon_{RMS} = \frac{\varepsilon_{\max}}{\sqrt{2}}$ with a voltmeter, the peak current in the wire is expected to be

$$I_0 = \frac{2\sqrt{2}\pi r \varepsilon_{RMS}}{\mu_0 \omega N A}. \tag{5.222}$$

For our split-iron toroidal inductor, we might assume that $B = \frac{\mu \mu_0 I}{2\pi r}$, so we can multiply the induced EMF by the core relative permeability μ, which means $I_0 = \frac{2\sqrt{2}\pi r \varepsilon_{RMS}}{\mu \mu_0 \omega N A}$. Is this effect likely to be measurable with the toroidal inductor described in the previous section? Let's assume that the mains AC in the United Kingdom (UK) is 50 Hz, and let's take the coil cross-section (in m^2) to be about $A = \pi \left(0.5 \times 10^{-2}\right)^2 = 7.9 \times 10^{-5}$. Hence, $\frac{\varepsilon_{RMS}}{I_0} = \frac{\mu \mu_0 \omega N A}{2\sqrt{2}\pi r} \approx \frac{35 \times 4\pi \times 10^{-7} \times 2\pi \times 50 \times 220 \times 7.9 \times 10^{-5}}{2\sqrt{2}\pi \times 80 \times 10^{-3}} = 3.36 \times 10^{-4}$ V/A. For AC of only a few amperes, this means a voltage close to the limits of (classroom-rated!) devices, so NA/r should, ideally, be about $100\times$ larger.

How does this compare to our actual measurements? Well, an RMS current of 8.35 A implies $I_0 = \sqrt{2} \times 8.35$ A $= 11.8$ A and, therefore, a predicted ε_{RMS}

---

[38]This was safely prepared for me by Winchester College's ever-amazing chief lab technician Andy Chesters.

of 3.36×10^{-4} V/A $\times$ 11.8 A $= 3.97$ mV. This is clearly much smaller than our measurement of 21.8 mV. The ratio $\frac{21.8}{0.51} = 5.50$, which perhaps implies that our actual relative permeability μ of the inductor is $\mu \approx 35 \times 5.5 \approx 193$, which is perhaps more believable for an iron(ish)-cored inductor than the $\mu \approx 35$ result of the previous experiment. I shall leave an investigation of the sixfold discrepancy between the steady current and AC versions of this experiment to a keen reader!

5.8   The Hall Probe

A cuboidal semiconductor, as shown in Fig. 5.19, of width w and height h, carrying current I, is placed in a magnetic field of flux density $\mathbf{B}$. The *Lorentz force* $q\mathbf{v} \times \mathbf{B}$ on charges q moving at drift velocity $\mathbf{v}$ through the semiconductor will result in a build-up of charge at the sides. This separation of charge, as in a *capacitor*, will manifest an electric field of strength $\mathbf{E}$ mutually perpendicular to both $\mathbf{v}$ and $\mathbf{B}$ since the charge separation is parallel to $\mathbf{v} \times \mathbf{B}$. This electric field will cause a force to act on the moving charges, which will eventually balance the Lorentz force. A steady-state charge separation will soon result, which means a steady voltage across the semiconductor. This phenomenon is called the *Hall effect*[39] and can be

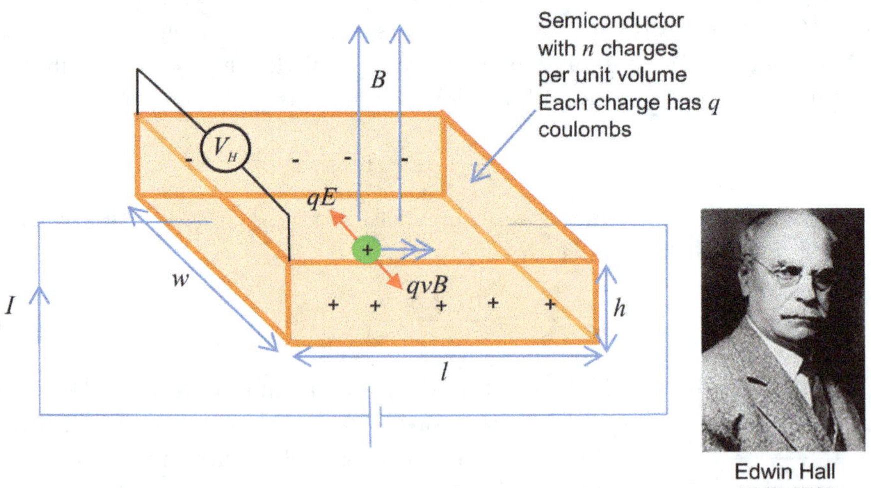

Fig. 5.19. Schematic diagram of a *Hall probe* used to determine magnetic flux density B from the measurement of a Hall voltage V_H across a slice of semiconductor material. By considering an equilibrium between the Lorentz force on charge carriers moving through the semiconductor (constituting current I) and the effect of the electric field resulting from the build-up of charge at the edges of the semiconductor slice, $B = qnh\frac{V_H}{I}$. The density of charge carriers is qn, and h is the width of the slice. Note that q is not well modelled using (just) electrons in a semiconductor—a weighted average involving 'holes', and their respective mobility is required. The semiconductor slice can be miniaturised, which means a Hall probe can be used to map out a spatially variable magnetic field if the probe is physically moved.

---

[39] The Hall effect is named after Edwin Hall (1855–1938).

utilised in miniature sensors that can measure local magnetic field strength. If we consider $\mathbf{v}$, $\mathbf{B}$ and $\mathbf{E}$ to be mutually perpendicular and v, B and E to be their respective magnitudes, the force balance at equilibrium implies

$$qE = qvB. \tag{5.223}$$

$$\therefore E = vB. \tag{5.224}$$

The electric field strength E across the semiconductor slice can be expressed as the *gradient of a potential* V_H. This is called the *Hall voltage* and is readily measurable across a semiconductor, as we will soon justify. Since we expect E to be uniform (as in a parallel plate capacitor),

$$E = \frac{V_H}{w}. \tag{5.225}$$

Now, since $E = vB$,

$$vB = \frac{V_H}{w}. \tag{5.226}$$

The current I in the semiconductor, i.e. the rate of flow of charge, can be expressed in terms of charge carrier *drift velocity* v:

$$I = qnwhv, \tag{5.227}$$

where n is the number density of charge carriers (i.e. charges per unit volume). Every second a volume of semiconductor of cross-section wh multiplied by v comprises $nwhv$ drifting charge carriers. If q is the charge on each of these charge carriers, then the rate of flow of charge, i.e. current I, is $I = qnwhv$. Hence, drift velocity v is

$$v = \frac{I}{qnwh}. \tag{5.228}$$

By substituting $v = \frac{I}{qnwh}$ into $vB = \frac{V_H}{w}$, we arrive at an expression for B in terms of the easily measurable quantities I and V_H and the semiconductor parameters q, h and n:

$$\frac{I}{qnwh}B = \frac{V_H}{w}. \tag{5.229}$$

$$\therefore B = qnh\frac{V_H}{I}. \tag{5.230}$$

Example calculation: Let $n = 7.0 \times 10^{21}$ m^{-3}, $q = 1.60 \times 10^{-19}$ C (i.e. charge carriers are electrons) and $h = 1.0 \times 10^{-4}$ m. $\therefore qnh = 0.11$ Cm^{-2} and $\therefore B = 0.11$ T $\times \frac{V_H/\mathrm{V}}{I/\mathrm{A}}$. Note that, with semiconductors (and several metals), a simple model of a single charge carrier of charge q is not accurate.[40] In a *p-type* semiconductor, for example, it is better to think of positively charged 'holes' to move rather than negatively charged electrons. So, for a practical piece of equipment, such as the PASCO Hall Probe described in the previous section about the toroidal inductor,

---

[40] See the discussion by Kittel [20] on pp. 164–166 as a starting point.

the constant of proportionality between B and $\frac{V_H}{I}$ will probably be determined empirically using a known magnetic field strength.

It is possible to measure V_H in a small semiconductor, so the Hall effect can be used to determine how a non-uniform magnetic field varies spatially, by physically moving the probe and exploring the variation of the field. Some Hall probes have semiconductors oriented at 90° to each other, which means they can record perpendicular $\mathbf{B}$ field components simultaneously, avoiding the need to rotate the probe to determine the full vector nature of the magnetic field.

5.9 The Mass Spectrometer, the Velocity Selector and the Cyclotron

A balance, and indeed *imbalance*, between the Lorentz forces of magnetism and those resulting from electric fields forms the basis of machines that can allow us to quantify the physical properties of charged particles. And for other applied science purposes, such as cathode ray tubes in old televisions and oscilloscopes, the interplay of these forces of electromagnetism provides a mechanism for precision control over the direction of beams of charged particles via changes in voltage and current in an electrical circuit.

5.9.1 *The mass spectrometer*

The ratio of charge q to mass m of the molecular constituents of a material can be determined using a *mass spectrometer*. There are many different designs, but in essence the material being analysed is first incinerated and ionised (i.e. electrons transferred by a vigorous heating process), and then a beam of the resulting charged particulates is accelerated in a vacuum through an electric potential V, resulting in particle velocity v. At this point, the particle enters a vacuum chamber free of electric fields. The accelerated beam is then bent into a circle of radius r via a Lorentz force, which results from the mutually perpendicular movement of the charges in a uniform magnetic field of flux density B, as illustrated in Fig. 5.21. This could be achieved by placing the vacuum chamber between a pair of *Helmholtz coils*.[41] These are a pair of identical current carrying coils of radius a and N windings, i.e. electromagnets that are positioned one above (or beside) the other. The net field between the coils can be surprisingly uniform. In fact, if the Helmholtz coils are separated by distance $2h$, the on-axis field strength can be calculated via the *Biot–Savart law*,[42] to be

$$B = \frac{\mu_0 N I a^2}{(a^2 + h^2)^{\frac{3}{2}}}. \tag{5.231}$$

---

[41] Hermann von Helmholtz (1821–1894).

[42] For an illustration of the Biot–Savart law, see Fig. 5.20.

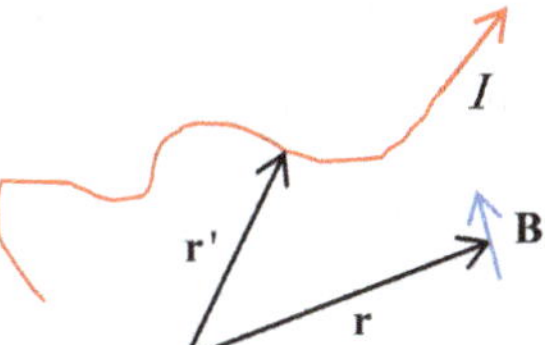

Fig. 5.20. The *Biot–Savart law* $\mathbf{B} = \frac{\mu_0 I}{4\pi} \int \frac{d\mathbf{l}\times(\mathbf{r}-\mathbf{r}')}{|\mathbf{r}-\mathbf{r}'|^3}$ can be used to compute magnetic flux density $\mathbf{B}$ at position $\mathbf{r}$ due to a conductor carrying current I within a line element vector $d\mathbf{l}$ along its length. The position vector $\mathbf{r}'$ is from the coordinate origin to the location on the conductor of $d\mathbf{l}$. The magnetic flux density on axis from a circular current loop of N turns and radius a is $B = \frac{1}{2} \frac{\mu_0 N I a^2}{\left(a^2+z^2\right)^{\frac{3}{2}}}$, where z is the on-axis distance from the centre of the loop. This result is derived from the Biot–Savart law in http://www.eclecticon.info/index_htm_files/Magnetism.pdf.

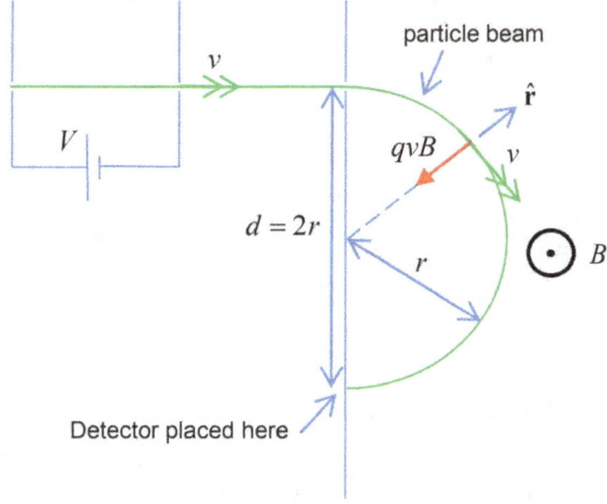

Fig. 5.21. A *mass spectrometer* is an electromagnetic device that can be used to determine the *charge-to-mass ratio* $\frac{q}{m}$ of a particle, such as an ionised molecule. Ionised particles are accelerated by potential V and injected into a vacuum chamber, where they are exposed to a uniform magnetic flux density B. This causes the particle trajectories to become circles, whose diameter d depends on $\frac{q}{m}$ via $d = \sqrt{\frac{8V}{B^2 \frac{q}{m}}}$. An array of detectors can measure d and hence determine the charge-to-mass ratio via $\frac{q}{m} = \frac{8V}{B^2 d^2}$.

The charge-to-mass ratio $\frac{q}{m}$ can then be determined from the diameter $d = 2r$ of the circle, where an array of detectors is placed. i.e. each detector corresponds to a suitably narrow range of d.

By Newton's second law and noting that acceleration is centripetal since the motion of the particle beam is uniform and circular (i.e. there are no electric fields to deviate particle motion from a circular trajectory, and we can neglect gravitational effects compared to the strength of the Lorentz force),

$$\frac{mv^2}{r} = qvB. \tag{5.232}$$

$$\therefore r = \frac{mv}{qB}. \tag{5.233}$$

By conservation of energy during the particle acceleration process (i.e. prior to entering the space where the magnetic field acts),

$$\tfrac{1}{2}mv^2 = qV. \tag{5.234}$$

$$\therefore v = \sqrt{\frac{2qV}{m}}. \tag{5.235}$$

Here, we assume that the particle velocity v is much less than the speed of light, so we can use the non-relativistic (i.e. classical) expression for kinetic energy and indeed the non-relativistic expression of Newton's second law. In the voltage range $V \gg V_{\text{max}}$, when this is *not* the case, different formulæ must be used, such as $(\gamma - 1)\,mc^2$ for kinetic energy, where $\gamma = \left(1 - \frac{v^2}{c^2}\right)^{-\frac{1}{2}}$. These modifications will be explored in the next volume of *Science by Simulation* in a chapter on *special relativity*. A practical limit might be $v = \frac{1}{100}c = 3.00 \times 10^6 \text{m/s}$. In this case, $qV_{\text{max}} = \tfrac{1}{2}m\left(\frac{1}{100}c\right)^2$.

$$\therefore V_{\text{max}} = \frac{\tfrac{1}{2}m\left(\frac{1}{100}c\right)^2}{q} = \frac{1}{20{,}000}\frac{mc^2}{q} \tag{5.236}$$

will give an estimate of the maximum permissible accelerating potential. I say approximation because $V_{\text{max}} = (\gamma - 1)\frac{mc^2}{q}$ would perhaps be a more appropriate limit, i.e. starting from the relativistic expression rather than the classical. For $v = \frac{1}{100}c$, $\gamma - 1 = \left(1 - \frac{1}{10{,}000}\right)^{-\frac{1}{2}} - 1 = 5.0004 \times 10^{-5} \approx 3.75 \times 10^{-9} + \frac{1}{20{,}000}$. So, to a very good approximation, the classical and relativistic limits converge.

By combining $d = 2r = \frac{2mv}{qB}$ and $v = \sqrt{\frac{2qV}{m}}$, we can derive an expression for the diameter d of the circular trajectory of the charged particle in terms of $\frac{q}{m}$ and the 'system parameters' B and V, and hence enable $\frac{q}{m}$ to be calculated in terms of what is readily measurable:

$$d = \frac{2m}{qB}\sqrt{\frac{2qV}{m}} \tag{5.237}$$

$$d^2 q^2 B^2 = \frac{8m^2 qV}{m}. \tag{5.238}$$

$$\therefore \frac{q}{m} = \frac{8V}{B^2 d^2}. \tag{5.239}$$

5.9.2   *The fine beam tube*

A *fine beam tube* (see Figs. 5.22 and 5.23) is a beautiful laboratory demonstration of the experimental verification of the charge-to-mass ratio $\frac{e}{m_e}$ of the electron. The principle is exactly as described for the mass spectrometer, except that, rather than incinerated charged particulates, an electron beam is bent into a complete circle within a hollow glass sphere containing low-pressure hydrogen gas. The flow of

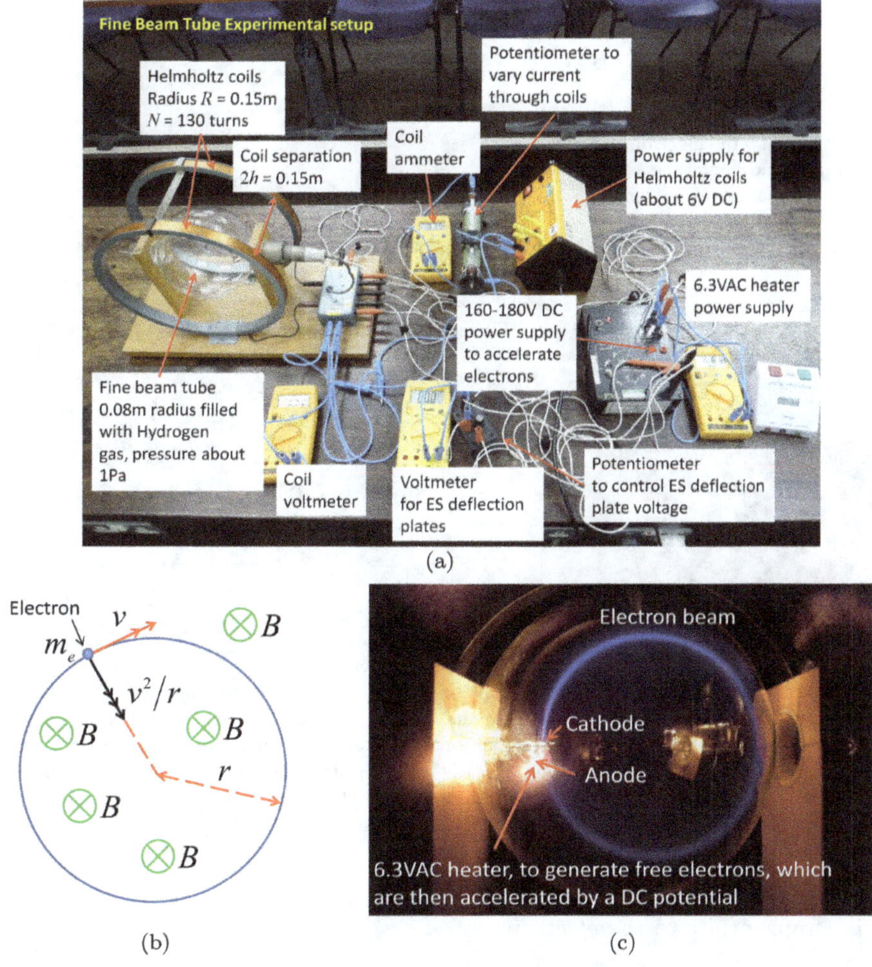

Fig. 5.22. (a) A fine beam tube is a hollow glass sphere of radius 8 cm that is filled with low-pressure hydrogen gas. Electrons emitted from a low voltage 6.3 VAC heating element are accelerated by a $V = 180$ V potential and ejected vertically upwards into the tube space. A pair of Helmholtz coils provides a fairly uniform magnetic flux density B in the vicinity of the sphere, which causes the electrons to follow circular trajectories of radius r. The electrons cause the hydrogen gas to produce a blue-purple glow, which enables r to be measured and hence the charge-to-mass ratio of the electron $\frac{e}{m_e} = 1.759 \times 10^{11}$ Ckg^{-1} from $\frac{8V}{d^2} = B^2 \frac{e}{m_e}$, where beam diameter $d = 2r$. (b) Geometric relationship between the electron velocity and the magnetic field (which acts into the page). (c) illustrates the circular path of electrons in the fine beam tube, as indicated by the blue-purple glow of the low-pressure hydrogen gas.

electrons causes the gas to produce a faint blue-purple light, indicating the trajectory of the electrons. The radius of this circle can therefore be measured quite easily. Personally, I like to do this by taking a long-exposure digital photograph of the equipment and then taking measurements from the resulting images. The vintage Leybold hydrogen sphere I have used has a radius of about 0.08 m, with Helmholtz coils of radius $a = 0.15$ m, $N = 130$ turns and a separation of $2h = 0.15$ m. Rearranging the charge-to-mass ratio equation $\frac{q}{m} = \frac{8V}{B^2 d^2}$ and using charge e and

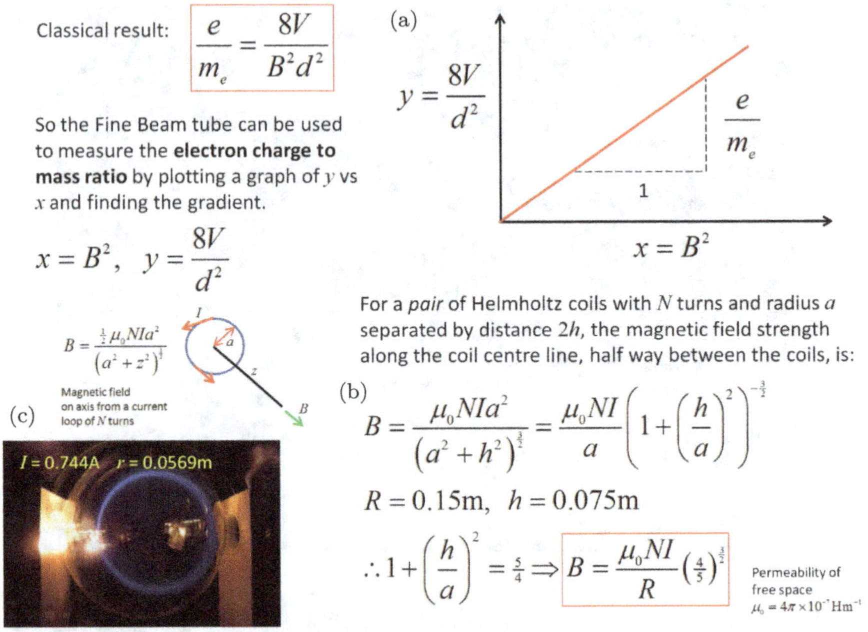

Classical result:
$$\frac{e}{m_e} = \frac{8V}{B^2 d^2}$$

(a)

$$y = \frac{8V}{d^2}$$

$$\frac{e}{m_e}$$

$$x = B^2$$

So the Fine Beam tube can be used to measure the **electron charge to mass ratio** by plotting a graph of y vs x and finding the gradient.

$$x = B^2, \quad y = \frac{8V}{d^2}$$

(c)

$$B = \frac{\frac{1}{2}\mu_0 N I a^2}{\left(a^2 + z^2\right)^{\frac{3}{2}}}$$

Magnetic field on axis from a current loop of N turns

$I = 0.744\text{A} \quad r = 0.0569\text{m}$

For a *pair* of Helmholtz coils with N turns and radius a separated by distance $2h$, the magnetic field strength along the coil centre line, half way between the coils, is:

(b)
$$B = \frac{\mu_0 N I a^2}{\left(a^2 + h^2\right)^{\frac{3}{2}}} = \frac{\mu_0 N I}{a}\left(1 + \left(\frac{h}{a}\right)^2\right)^{-\frac{3}{2}}$$

$$R = 0.15\text{m}, \quad h = 0.075\text{m}$$

$$\therefore 1 + \left(\frac{h}{a}\right)^2 = \tfrac{5}{4} \Rightarrow B = \frac{\mu_0 N I}{R}\left(\tfrac{4}{5}\right)^{\frac{3}{2}}$$

Permeability of free space $\mu_0 = 4\pi \times 10^{-7}\,\text{Hm}^{-1}$

Fig. 5.23. (a) describes the predicted direct proportion between $\frac{8V}{d^2}$ and B^2 for a fine beam tube. V is the electron-accelerating potential, d is the electron beam diameter $d = 2r$ and B is the magnetic flux density experienced by the electrons as they execute a circular trajectory. (b) describes the calculation of B from a pair of Helmholtz coils of radius a, N turns, separated by distance $2h$ and each carrying current I. (c) is an example long-exposure digital photograph of a particular I, r pairing. For a more precise calculation of the charge-to-mass ratio of the electron, I is varied, which allows $\frac{e}{m_e}$ to be determined from a line of best fit of $\frac{8V}{d^2}$ vs B^2.

mass m_e appropriate to the electron,

$$d = \sqrt{\frac{8V}{B^2 \frac{e}{m_e}}}. \tag{5.240}$$

The charge-to-mass ratio for the electron is (to 4.s.f)

$$\frac{e}{m_e} = \frac{1.602 \times 10^{-19}\text{C}}{9.109 \times 10^{-31}\text{kg}} = 1.759 \times 10^{11}\ \text{Ckg}^{-1}. \tag{5.241}$$

For a Helmholtz coil current of 1.00 A, we expect a magnetic flux density of $B = \frac{\mu_0 N I a^2}{(a^2 + h^2)^{\frac{3}{2}}}$. Using the parameters of the Leybold equipment,

$$B = \frac{4\pi \times 10^{-7} \times 130 \times 1.00 \times 0.15^2}{(0.15^2 + 0.075^2)^{\frac{3}{2}}}\ \text{(T)} \tag{5.242}$$

$$= 7.79 \times 10^{-4}\ \text{T}. \tag{5.243}$$

If the accelerating voltage is $V = 180$ V, then we expect an electron beam diameter of

$$d = \sqrt{\frac{8 \times 180}{(7.79 \times 10^{-4})^2 \times 1.759 \times 10^{11}}} \text{ (m)} \tag{5.244}$$

$$= 0.12 \text{ m}, \tag{5.245}$$

i.e. a 6 cm radius, which should fit nicely into the 8 cm radius sphere. To obtain a value for $\frac{e}{m_e}$, the idea is to vary the current I in the Helmholtz coils and record this versus the electron beam diameter d. Since $\frac{8V}{d^2} = B^2 \frac{e}{m_e}$, one expects a graph of $\frac{8V}{d^2}$ versus B^2 to be a straight line through the origin with gradient $\frac{e}{m_e}$. The magnetic flux density is proportional to current I using $B = \frac{\mu_0 N I a^2}{(a^2+h^2)^{\frac{3}{2}}} = 7.79 \times 10^{-4} \text{ TA}^{-1} \times I$. So, by plotting $\frac{8V}{d^2}$ versus B^2 and measuring the gradient (e.g. manually with a pencil and ruler or via a line-of-best-fit function in a computer program such as Microsoft Excel), one can determine $\frac{e}{m_e}$.

Let's check whether non-relativistic physics is appropriate. For the fine beam tube, our accelerating voltage limit V_{max} for a maximum electron speed $= \frac{1}{100}c$ is $V_{\text{max}} \approx \frac{1}{20,000} \frac{mc^2}{q} = \frac{1}{20,000} \times \frac{(2.998\times10^8)^2}{1.759\times10^{11}} \approx 25.6$ V. An accelerating potential of 180 V is about seven times this. So, is it still valid to use classical physics? Using the relativistic expression for kinetic energy $eV = (\gamma - 1)\, m_e c^2$, the Lorentz factor γ is

$$\gamma = 1 + \frac{e}{m_e} \frac{V}{c^2}, \tag{5.246}$$

which for the fine beam tube experiment is

$$\gamma = 1 + 1.759 \times 10^{11} \times \frac{180}{(2.998 \times 10^8)^2} \tag{5.247}$$

$$= 1 + 3.523 \times 10^{-4}. \tag{5.248}$$

So, although we might expect a very small effect on the electron beam radius, it is probably unmeasurable in a school-laboratory.

5.9.3 *The velocity selector*

An alternative system for determining the ratio of charge q to mass m of the molecular constituents of a material is a *velocity selector*, as illustrated in Fig. 5.24. In this case, a second electric potential is varied until the Lorentz force balances an electric field between two charged plates, which are placed in a uniform magnetic field that acts in a perpendicular direction to the electric field. When these forces balance, a particle beam travelling with velocity v will *not* be deflected into one of the plates and can therefore be detected once it emerges from a small aperture.

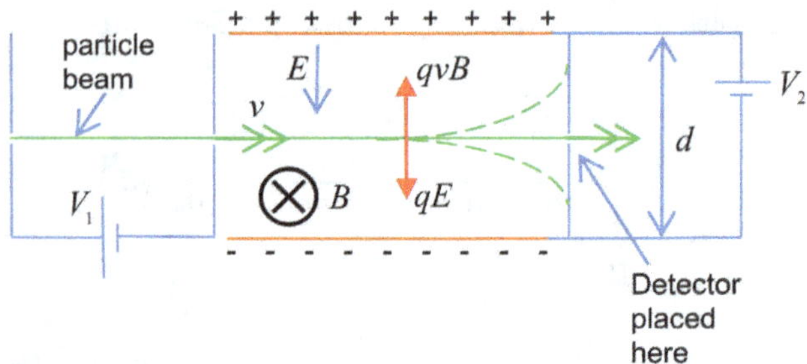

Fig. 5.24. A *velocity selector* is an electromagnetic device which permits charged particles to leave a vacuum chamber containing (perpendicular) electric and magnetic fields only when the particle speed v is such that the electric field $E = vB$. The magnetic flux density is B.

If there is a force balance between the magnetic and electric fields,

$$qE = qvB. \tag{5.249}$$

$$\therefore E = vB. \tag{5.250}$$

Let the accelerating potential be V_1. For modest voltages such that classical physics can be assumed,

$$\tfrac{1}{2}mv^2 = qV_1. \tag{5.251}$$

$$\therefore v = \sqrt{\frac{2qV_1}{m}}. \tag{5.252}$$

Now, let the distance between the charged plates which manifest the uniform electric field be d and the potential difference which produces the field be V_2. Therefore, since the magnitude of electric field strength is the potential gradient[43]

$$E = \frac{V_2}{d}. \tag{5.253}$$

Combining $E = \frac{V_2}{d}$, $v = \sqrt{\frac{2qV_1}{m}}$ and $E = vB$,

$$\frac{V_2}{d} = \sqrt{\frac{2qV_1}{m}}B, \tag{5.254}$$

$$\frac{V_2^2}{d^2} = \frac{2qV_1}{m}B^2. \tag{5.255}$$

$$\therefore \frac{q}{m} = \frac{V_2^2}{2V_1 B^2 d^2}. \tag{5.256}$$

Let's perform an example calculation using similar parameters to those of the *fine beam tube*, i.e. we will consider electrons accelerated by potential $V_1 = 180$ V, and let the magnetic flux density be 7.79×10^{-4} T, i.e. corresponding to the fine

---

[43]In more general terms, electric field $\mathbf{E} = -\nabla V$ where ∇ is the vector gradient operator and $V(x,y,z)$ is a *scalar electric potential*. In Cartesian, $\nabla = \hat{\mathbf{x}}\frac{\partial}{\partial x} + \hat{\mathbf{y}}\frac{\partial}{\partial y} + \hat{\mathbf{z}}\frac{\partial}{\partial z}$ where $\hat{\mathbf{x}}, \hat{\mathbf{y}}, \hat{\mathbf{z}}$ are *orthonormal* unit vectors which constitute the Cartesian basis set.

beam tube Helmholtz coils with 1.00 A of current. In this case, the electric field strength E required to offset the Lorentz force on the electron is

$$E = B\sqrt{\frac{2qV_1}{m}} \tag{5.257}$$

$$= 7.79 \times 10^{-4} \times \sqrt{\frac{2 \times 1.602 \times 10^{-19} \times 180}{9.109 \times 10^{-31}}} \; (\mathrm{Vm}^{-1}) \tag{5.258}$$

$$= 6.20 \times 10^3 \; \mathrm{Vm}^{-1}, \tag{5.259}$$

which means, if the charged plate gap is $d = 0.100$ m, this means a voltage of $V_2 = 620$ V, which is easily achievable in a school laboratory.

5.9.4 *The cyclotron*

A *cyclotron*, as depicted in Fig. 5.25, is a space-efficient mechanism for accelerating charges to high speeds and was invented by Ernest Lawrence (1901–1958) around 1930 at the University of California, Berkeley. Much like a mass spectrometer, a beam of particles of mass m and charge q are bent into a circle within a vacuum chamber, which is subject to a uniform magnetic flux density B. However, in this case, the charges receive a velocity boost *twice per cycle* from a sinusoidally time-varying potential difference between two halves ('Dees') of the cyclotron.

After a half-circular path of radius r_n, a particle beam enters a small gap (compared to the radius of the circular motion) between the Dees with velocity v_n and is accelerated to velocity v_{n+1} by potential difference V_0. If non-relativistic dynamics can be assumed, conservation of energy implies

$$\tfrac{1}{2}mv_{n+1}^2 = \tfrac{1}{2}mv_n^2 + qV_0. \tag{5.260}$$

$$\therefore v_{n+1} = \sqrt{v_n^2 + \frac{2qV_0}{m}}. \tag{5.261}$$

As for the mass spectrometer and the fine beam tube, Newton's second law implies

$$\frac{mv_n^2}{r_n} = qv_n B. \tag{5.262}$$

$$\therefore r_n = \frac{mv_n}{qB}. \tag{5.263}$$

Now, since motion is circular with constant speed v_n between boosts,

$$v_n = \frac{2\pi r_n}{T_n}, \tag{5.264}$$

where $\tfrac{1}{2}T_n$ is the time to complete a half-circular orbit between boosts. Since $r_n = \frac{mv_n}{qB}$, this means

$$T_n v_n = \frac{2\pi m v_n}{qB}. \tag{5.265}$$

$$\therefore T_n = \frac{2\pi m}{qB}. \tag{5.266}$$

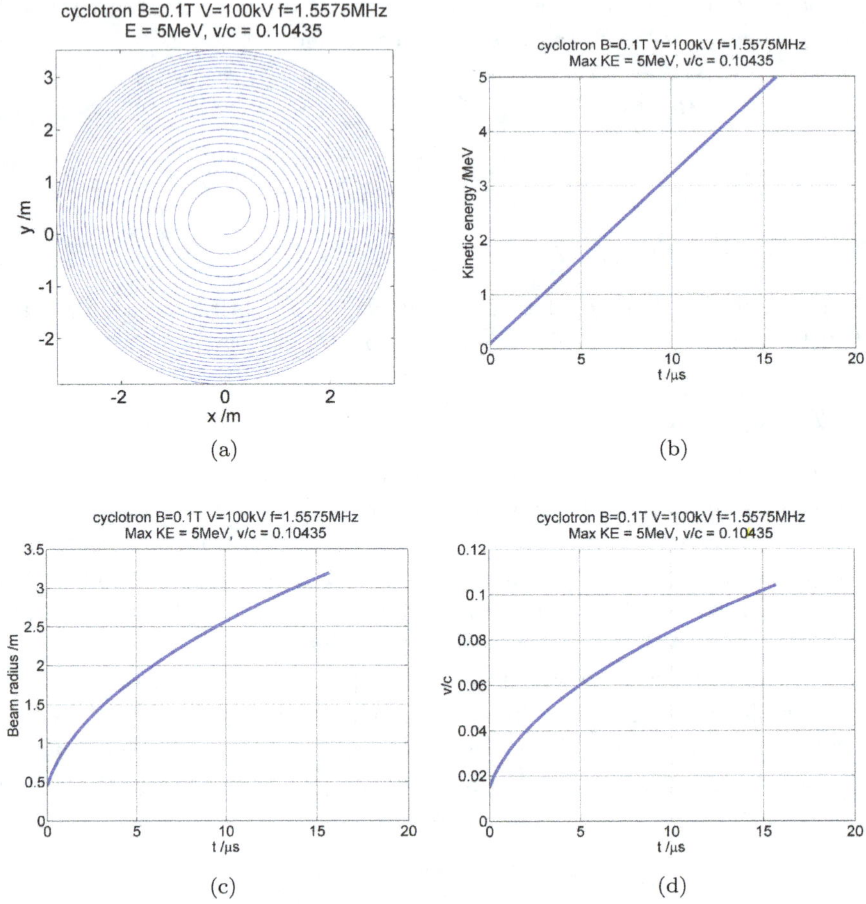

Fig. 5.25. A *cyclotron* is a space-efficient mechanism for accelerating charges to high speeds. Much like a mass spectrometer, a beam of particles of mass m and charge q are bent into a circle of radius r within a vacuum chamber, which is subject to a uniform magnetic flux density B. However, in this case, the charges receive a velocity boost *twice per cycle* from a sinusoidally time-varying potential difference between two halves ('Dees') of the cyclotron. This works because the accelerating potential oscillates at the *cyclotron frequency* $f_c = \frac{1}{2\pi}\frac{qB}{m}$, which is independent of the speed of the charged particle. (a) describes the path of a proton over 50 boosts of 100 kV. (b) tracks the (linear) increase in kinetic energy between each boost, whereas (c) tracks the beam radius and (d) tracks the proton speed as a fraction of the speed of light c. As the proton energy increases, the fractional increase in r and v decreases in a cyclotron. In this simple model, radiative and relativistic effects are ignored. Both must be considered at higher energies, rendering a cyclotron design less preferable than a *synchrotron* concept.

Therefore, T_n is a *constant*, independent of the speed of the particle. It is common to define $T_n = \frac{1}{f_c}$, where

$$f_c = \frac{1}{2\pi}\frac{qB}{m} \tag{5.267}$$

is known as the *cyclotron frequency*. f_c, being independent of particle speed, can be used as the basis of a simple mechanism for applying the accelerating voltage V_0 twice per cycle. The solution is an *alternating voltage $V(t)$ at the cyclotron frequency* applied across the Dees:

$$V(t) = V_0 \cos\left(2\pi f_c t\right) = V_0 \cos\left(\frac{qBt}{m}\right). \tag{5.268}$$

Every half-circle, i.e. every $\frac{1}{2}T_n = \frac{\pi m}{qB}$ seconds, $V(t)$ has magnitude V_0 but with an alternating $+/-$ sign. This means the electron acceleration between the Dees causes successively faster circular motion in the same clockwise (or anticlockwise) sense. In the pioneering years of particle physics in the middle part of the 20th century, cyclotrons were a useful laboratory system. Let's consider a proton beam device with parameters $V_0 = 100$ kV and $B = 0.1$ T. For a proton, $q = 1.602 \times 10^{-19}$ C and $m = 1.637 \times 10^{-27}$ kg.

The cyclotron frequency $f_c = \frac{qB}{2\pi m}$ is

$$f_c = \frac{1.602 \times 10^{-19} \times 0.1}{2\pi \times 1.637 \times 10^{-27}} \text{ (Hz)} \tag{5.269}$$

$$= 1.56 \times 10^6 \text{ Hz}, \tag{5.270}$$

i.e. about 1.6 MHz, which is at the lower end of commercial radio frequencies[44] and, therefore, easily achievable in a laboratory (although perhaps not a school laboratory!) The goal of the cyclotron is to accelerate the charges to a high enough energy that makes them useful for particle physics experiments. Once the particle reaches the required energy, it can leave the cyclotron. The equation for the circular path radius $r_n = \frac{mv_n}{qB}$ shows how this might be achieved. As the particle gets faster, the path radius increases in direct proportion. In the example shown in Fig. 5.25, a 5 MeV proton beam[45] is established after 50 boosts using a magnetic flux density of $B = 0.1$ T. The proton is assumed to start from rest before the initial boost. In Fig. 5.25, the circular paths are plotted, which get progressively larger in radius as the particle energy increases. However, $v_{n+1} = \sqrt{v_n^2 + \frac{2qV_0}{m}}$ means that the increase in r_n between each boost reduces since the fractional change in velocity $\frac{v_{n+1}-v_n}{v_n}$ diminishes as v_n increases:

$$v_{n+1} = v_n\sqrt{1 + \frac{2qV_0}{mv_n^2}}. \tag{5.271}$$

$$\therefore \frac{v_{n+1}-v_n}{v_n} = 1 - \sqrt{1 + \frac{2qV_0}{mv_n^2}}, \tag{5.272}$$

which tends to zero as $v_n \to \infty$ and $\frac{2qV_0}{mv_n^2} \to 0$.

Assuming classical physics and no radiative losses, the gain in KE is qV_0 per boost, so a graph of KE versus boost (and hence time, given the constant cyclotron frequency) will be linear.

---

[44] For example, the *frequency modulated* (FM) broadcast of BBC Radio 4 in the UK operates between 92 and 95 MHz and also from 103 to 105 MHz.

[45] The proton kinetic energy in MeV ('Mega electron-volts') is the energy in J divided by 1.602×10^{-13}.

Since the particles are executing circular motion, they must be accelerating. Accelerating charges *radiate*; therefore, the assumption that the particles will maintain their kinetic energy between boosts will become increasingly flawed as they move faster. As discussed previously, higher particle energies will also eventually violate the assumption of non-relativistic dynamics. Modern particle accelerators, which in the case of the Large Hadron Collider (LHC) at CERN operated up to 6.5 TeV (i.e. 6,500,000 MeV) per beam in 2015, tend to work on different principles to those of cyclotrons. The LHC is essentially a *synchrotron*, which uses *time-varying magnetic fields* to *maintain a fixed circular beam* as the charges are accelerated. In the LHC, over 10,000 superconducting magnets are used to either maintain the circular trajectory or focus the beam. In the UK, the *Diamond Light Source*[46] is the national (electron) synchrotron facility, located at the Harwell Science and Innovation Campus in Oxfordshire. This source of high-energy electron beams has a myriad of scientific uses, particularly biological applications such as the characterisation of viruses and the development of vaccines. The circular beam lines at Diamond have a circumference of 561.571 m. According to https://www.diamond.ac.uk/Science/Machine.html (Accessed January 2023), the beam energy is 3 GeV.

5.10 The Frequency Response of a Transformer

5.10.1 *Introducing the transformer*

A transformer is a very useful electrical device for changing the amplitude of an oscillating voltage signal. In the power cable problem at the start of this chapter, we alluded to the very high voltage of transmission lines in an electrical distribution system, such as a national grid. Since transmitted power is $P = VI$, if the voltage V can be raised, then the current I in a transmission line must be reduced in inverse proportion in order for P to be preserved. Reducing I during transmission is useful since the power loss is $P_{\text{loss}} = I^2 R$, where R is the resistance of the power cable. Given that P_{loss} is proportional to the square of current, a reduction in I by a factor of 10 has ten times the impact of the same fractional reduction in R. Unless a cost-effective mechanism for resistivity reduction is discovered (e.g. the electrical holy grail of the room-temperature superconductor, which has *no resistance* at room temperature), the reduction of current (and raising of transmission voltage) is a practical solution. But in order to raise the voltage of transmission and then reduce it down to sensible values (e.g. 230V RMS in the UK), a mechanism is required to achieve this without excessive energy loss. A design that achieves this requirement is a pair of coils wound around opposing sides of a physical loop of magnetically 'soft' material, such as annealed, laminated iron. These coils are often labelled *primary* and *secondary*. A model of such a *transformer* is illustrated in Fig. 5.26, and a practical example, common in school laboratories, is the assembly of a transformer from an inductor ($N_p = 120$ primary turns in the example shown),

---

[46] https://www.diamond.ac.uk/.

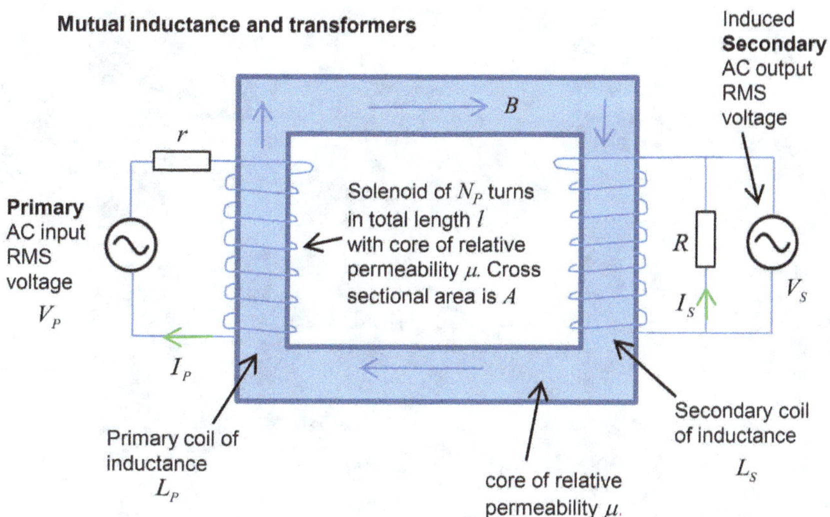

Fig. 5.26. Schematic diagram of an electrical transformer. An oscillating primary voltage, modelled by $V_p = V_0 \cos \omega t$, drives current through a coil of N_p turns of resistance r. The coil is wound around the left side of a ring of magnetically soft material (e.g. laminated annealed iron) with relative permeability μ. A circulating magnetic flux density B is set up in the iron ring, which oscillates in alternate clockwise and anticlockwise directions at frequency $\frac{\omega}{2\pi}$ Hz. A secondary coil of N_s windings and resistance R therefore experiences a rate of change of magnetic flux linked, and an EMF V_s is induced. If R is not large compared to r and the mutual inductance between the coils is inefficient, then back EMFs resulting from the coil currents will cause a deviation from the *transformer equation* $\frac{V_s}{V_p} = \frac{N_s}{N_p}$ and the power transfer equation $V_p I_p = V_s I_s$.

a pair of iron C-cores and a secondary coil of wire connected to a terminal block. The C-shaped cores are laminated (i.e. constructed from a multi-layer sandwich of thin iron sheets) to minimise *eddy currents*, which may be induced in them due to the changing magnetic fields. The eddy currents would result in a heating of the cores and hence cause a loss of energy transferred from the primary to the secondary coils.

The essential idea is that current in the primary coil sets up a time-varying magnetic field in the 'magnetic circuit' established by the loop of iron. The field internal to the loop is intensified by the ferromagnetic iron and enhances the field experienced by the secondary coil wound around the opposing side of the iron loop. By *Faraday's law of electromagnetic induction*, '*a rate of change of magnetic flux linked by the secondary coil will result in an EMF induced*'.

If magnetic flux density B is established in the iron loop, the flux linked by the secondary coil is $\Phi_s = N_s B A$, where N_s is the number of secondary coil turns and A is the cross-sectional area of the coils. The EMF induced in the secondary coil is, by Faraday's law,

$$V_s = -\frac{d\Phi_s}{dt} = -N_s A \frac{dB}{dt}. \tag{5.273}$$

If the primary coil has N_p turns and also has a cross-section A, we might suggest the similar-looking relationship

$$V_p = -N_p A \frac{dB}{dt} \tag{5.274}$$

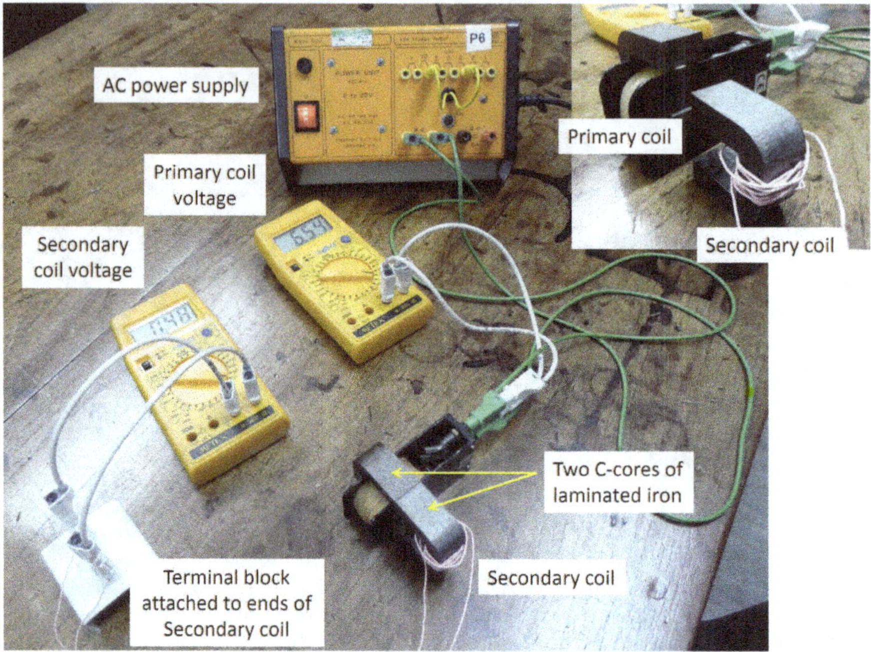

Fig. 5.27. Experimental setup to explore the *transformer equation* $\frac{V_s}{V_p} = \frac{N_s}{N_p}$. An AC power supply provides a 6.54 V RMS primary voltage V_p to a $N_p = 120$ turn coil. This is connected via a pair of laminated iron C-cores to a secondary coil of N_s windings. The latter is connected in series to a voltmeter via a terminal block. The series connection and the high resistance of the voltmeter means very little current will flow in the secondary circuit. $\frac{V_s}{V_p} = \frac{N_s}{N_p}$ can be assessed by plotting a graph of V_s vs. a variable number of secondary windings N_s.

between the oscillating primary voltage applied to the circuit V_p and the rate of change of magnetic flux density that it establishes. Therefore, if $\frac{dB}{dt} = -\frac{V_p}{N_p A}$, we can substitute this result into $V_s = -N_s A \frac{dB}{dt}$ to arrive at the *transformer equation*:

$$V_s = -N_s A \left(-\frac{V_p}{N_p A} \right). \tag{5.275}$$

$$\therefore \frac{V_s}{V_p} = \frac{N_s}{N_p}. \tag{5.276}$$

The validity of this result is nicely demonstrated in Figs. 5.27 and 5.28, which confirms by experiment the linearity of V_s with respect to N_s, given a fixed primary voltage V_p and the number of primary turns $N_p = 120$. The 'wind your own transformer' equipment illustrated in Fig. 5.27 was used to produce the measurements in the table. The idea is to record the primary and secondary voltages using multimeters, holding the cores together tightly while you turn on the power to the primary circuit. The tight grip (which can more tediously be achieved with a metal bracket) helps maintain contact between both sides of the C-cores, which have a tendency to buzz dramatically due to the *magnetostriction* effect of magnetic material undergoing strain during the process of magnetisation. In between measurements, extra turns of wire are added to the secondary coil total N_s, which allows V_s to vary, and hence a graph of V_s versus N_s can be plotted. Since $\frac{V_s}{V_p} = \frac{N_s}{N_p}$, voltage can be increased or reduced depending on the ratio $\frac{N_s}{N_p}$.

In many physics courses appropriate for 16-year-olds, the concept of power $=$ voltage $\times$ current is used to justify the idea of power transfer between primary and secondary coils. In other words, if I_p is the primary current and I_s is the current induced in the secondary coil, we might consider writing down something

Number of secondary coil turns N2	Secondary coil voltage V2 /volts	Number of primary coil turns N1	Primary coil voltage /volts V1	Predicted secondary voltage /volts V2 = (N2/N1)*V1
5	0.27	120	6.5	0.27
10	0.54	120	6.5	0.54
15	0.81	120	6.5	0.81
20	1.08	120	6.5	1.08
25	1.35	120	6.5	1.35
30	1.63	120	6.5	1.63
35	1.90	120	6.5	1.90
40	2.17	120	6.5	2.17
45	2.44	120	6.5	2.44
50	2.71	120	6.5	2.71

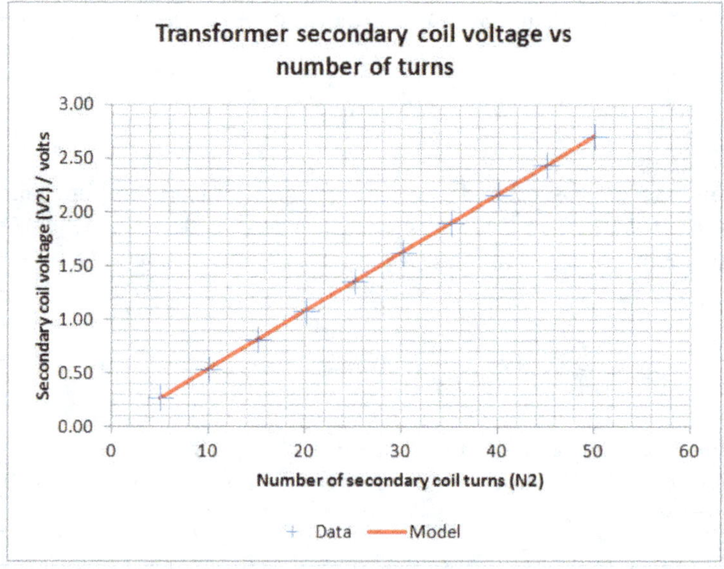

Fig. 5.28. Experimental results for the equipment described in Fig. 5.27. The highly linear graph of induced secondary coil voltage V_s vs. secondary coil windings N_s can be used to confirm the efficacy of the transformer equation model $\frac{V_s}{V_p} = \frac{N_s}{N_p}$. Note that in this Microsoft Excel sheet screenshot, $N_p = $ N1, $N_s = $ N2 etc.

like

$$I_p V_p = I_s V_s \tag{5.277}$$

and hence predict the induced secondary current to be

$$I_s = \frac{I_p V_p}{V_s} = I_p \frac{N_p}{N_s}. \tag{5.278}$$

The problem is, if you actually try to do this with the same school-laboratory equipment, you may discover that power transfer between primary and secondary is far from efficient. You will also, in a step-down system, probably blow the fuse in your secondary multimeter that is configured as an AC ammeter, causing your long-suffering lab technician to roll their eyes and shake their head. $\frac{V_s}{V_p} = \frac{N_s}{N_p}$ will appear to work very nicely, but only if a voltmeter is connected in series with the secondary coil. Since a voltmeter has a very high resistance, very little current will flow in the secondary circuit. To model power transfer from primary to secondary coils, we will need to consider the resistances of both primary and secondary circuits, as well as the (i) *back EMF* resulting from current in the secondary coil and (ii) the *mutual inductance* between the coils. Before we assemble this model, let us first explore the concept of inductance using the example of a solenoid.

5.10.2 *Calculating the inductance of a (long) solenoid*

The magnetic flux density B inside the centre of a *long* solenoid of N turns in l metres, carrying current I and wound around a core of relative permeability μ, is given by[47]

$$B = \frac{\mu \mu_0 N I}{l}. \tag{5.279}$$

If the coil has cross-sectional area A, the flux linked by the coil is NBA. By Faraday's law, if current I varies with time, then a *back EMF* will be induced, which will oppose the voltage driving current through the coil. This back EMF is

$$\varepsilon_{\text{back}} = \frac{d}{dt}(NBA) = \frac{d}{dt}\left(N\frac{\mu \mu_0 N I}{l} A\right) = \frac{\mu \mu_0 N^2 A}{l}\frac{dI}{dt}. \tag{5.280}$$

If we define *inductance* L by the equation $\varepsilon_{\text{back}} = L\frac{dI}{dt}$, i.e. the constant of proportionality between EMF induced and the rate of change of current, the inductance

---

[47]This result is easily proven using *Ampère's theorem* $\oint_{\text{loop}} \mathbf{H} \cdot d\mathbf{l} = I_{\text{enclosed}}$. If one assumes the magnetic field $\mathbf{H}$ to be uniform, $H = \frac{B}{\mu \mu_0}$ within the long solenoid and effectively zero outside it. Therefore, a loop through the solenoid, 'well beyond' the solenoid, and then back to complete the loop, will imply $\frac{Bl}{\mu \mu_0} = NI$. Note that if current I flows through each coil and there are N coils in length l, then NI is the current enclosed by the loop. Hence, flux density $B = \frac{\mu \mu_0 N I}{l}$.

of a long solenoid is

$$L = \frac{\mu\mu_0 N^2 A}{l}. \tag{5.281}$$

The unit of inductance is *Henry* (H), named after the American electrical pioneer Joseph Henry (1797–1878), which explains why the permeability of free space $\mu_0 = 4\pi \times 10^{-7}$ Hm$^{-1} = \frac{Ll}{\mu N^2 A}$ is often expressed in units of Henry per metre since $\frac{1}{\mu N^2}$ is dimensionless and the (SI) dimensions of A/l are metres.

5.10.3 *Modelling the transformer*

Let us consider a transformer model as illustrated in Fig. 5.26 and apply Kirchhoff's second law to both primary and secondary circuits. The resistive load in the primary circuit is r, and the resistive load in the secondary circuit is R. For the primary circuit,

$$V_p - L_p \frac{dI_p}{dt} + M \frac{dI_s}{dt} = I_p r, \tag{5.282}$$

i.e. the applied voltage V_p (e.g. from an AC power supply) is reduced by the back EMF $L_p \frac{dI_p}{dt}$ from the primary coil and (possibly) enhanced by the *mutual induction* $M \frac{dI_s}{dt}$ between the coils. M is defined as the *mutual inductance* between the coils. Let us express M in terms of the primary and secondary inductances and an empirical *coupling constant* k:

$$M = k\sqrt{L_p L_s}. \tag{5.283}$$

If back EMF $\varepsilon_{\text{back}} = L \frac{dI}{dt}$, then $I\varepsilon_{\text{back}} = LI \frac{dI}{dt}$ is an an expression of power transfer to (or from) an inductor. Now, $\frac{d}{dt}\left(\frac{1}{2}LI^2\right) = LI \frac{dI}{dt}$; therefore, we can associate $\frac{1}{2}LI^2$ with an *energy stored in the magnetic field of the inductor*. Incorporating all the terms of the form $LI \frac{dI}{dt}$, the rate of energy flow $\frac{dE}{dt}$ into the primary and secondary coils can be written as

$$\frac{dE}{dt} = L_p I_p \frac{dI_p}{dt} + L_s I_s \frac{dI_s}{dt} + M I_s \frac{dI_s}{dt} + M I_p \frac{dI_p}{dt}, \tag{5.284}$$

assuming that the mutual inductance from primary to secondary M is the same as that from secondary to primary.

$$\therefore \frac{dE}{dt} = \frac{d}{dt}\left(\tfrac{1}{2}L_p I_p^2 + \tfrac{1}{2}L_s I_p^2 + M I_s I_p\right). \tag{5.285}$$

If there is net energy input into the system (e.g. from a power supply), then $\frac{dE}{dt} \geq 0$, which means $\frac{1}{2}L_p I_p^2 + \frac{1}{2}L_s I_s^2 + M I_s I_p \geq 0$. This equation can be written in matrix form:

$$\tfrac{1}{2}\begin{pmatrix} I_p & I_s \end{pmatrix} \begin{pmatrix} L_p & M \\ M & L_s \end{pmatrix} \begin{pmatrix} I_p \\ I_s \end{pmatrix} \geq 0. \tag{5.286}$$

For this to be true for any (signed) values of I_p and I_s, the determinant $\begin{vmatrix} L_p & M \\ M & L_s \end{vmatrix} > 0$, which means $L_p L_s - M^2 \geq 0$, which in turn implies $M \leq \sqrt{L_p L_s}$. So, if $M = k\sqrt{L_p L_s}$, then this means the coupling constant k can only be in the range $0 \leq k \leq 1$. We still have to determine its value empirically, but at least we know it must be between zero and unity.

Let us now apply Kirchhoff's second law to the secondary coil circuit:

$$M\frac{dI_p}{dt} - L_s\frac{dI_s}{dt} = I_s R = V_s, \tag{5.287}$$

i.e. the induced secondary EMF $V_s = I_s R$ is the EMF resulting from mutual induction between the primary and secondary coils, minus the back EMF due to the secondary coil. Let us assume that a steady state is established in the transformer system, i.e. the frequency of the current and voltage associated with the secondary coil is the same as that of the 'driving' primary voltage V_p. As we will use in the following section on resonance in L, C, R circuits, it will be convenient to model the voltages and currents as the real parts of a complex number representation:

$$V_p = V_0 e^{i\omega t}, \tag{5.288}$$

i.e. the actual primary voltage is $V_0 \cos \omega t$ since $\text{Re}\left(e^{i\omega t}\right) = \text{Re}\left(\cos \omega t + i \sin \omega t\right)$. Let us define (using the '$V = IR$' dimensional relationship between voltage, current and resistance)

$$I_p = \frac{V_0}{r}\alpha e^{i\omega t}, \tag{5.289}$$

$$I_s = \frac{V_0}{R}\beta e^{i\omega t}, \tag{5.290}$$

where α and β are *dimensionless* complex numbers that will depend on the parameters of the system, which include frequency $f = \frac{\omega}{2\pi}$. Note that, although the frequency of the secondary coil current and voltage should be of the same frequency as the primary, there is no reason to justify that the amplitude and phase should be the same, as in the driven SHM mass–spring system described at the conclusion of the previous chapter. The substitution of $V_p = V_0 e^{i\omega t}$, $I_p = \frac{V_0}{r}\alpha e^{i\omega t}$ and $I_s = \frac{V_0}{R}\beta e^{i\omega t}$ into our Kirchhoff's second law equations allows us to solve for α and β.

Primary coil:

$$V_p - L_p\frac{dI_p}{dt} + M\frac{dI_s}{dt} = I_p r \tag{5.291}$$

$$V_0 e^{i\omega t} - i\omega L_p\frac{V_0}{r}\alpha e^{i\omega t} + i\omega M\frac{V_0}{R}\beta e^{i\omega t} = V_0\alpha e^{i\omega t} \tag{5.292}$$

$$1 - \frac{i\omega L_p\alpha}{r} + \frac{i\omega M\beta}{R} = \alpha. \tag{5.293}$$

Secondary coil:

$$M\frac{dI_p}{dt} - L_s\frac{dI_s}{dt} = I_s R = V_s \tag{5.294}$$

$$Mi\omega\frac{V_0}{r}\alpha e^{i\omega t} - L_s i\omega\frac{V_0}{R}\beta e^{i\omega t} = \frac{V_0}{R}\beta e^{i\omega t} R \tag{5.295}$$

$$\frac{Mi\omega\alpha}{r} - \frac{L_s i\omega}{R}\beta = \beta \tag{5.296}$$

$$\frac{Mi\omega\alpha}{r} = \beta\left(1 + \frac{L_s i\omega}{R}\right). \tag{5.297}$$

$$\therefore \beta = \frac{\frac{Mi\omega\alpha}{r}}{1 + \frac{L_s i\omega}{R}} = \frac{Mi\omega}{r + \frac{L_s i\omega r}{R}}\alpha. \tag{5.298}$$

The substitution of $\beta = \frac{Mi\omega}{r + \frac{L_s i\omega r}{R}}\alpha$ into $1 - \frac{i\omega L_p\alpha}{r} + \frac{i\omega M\beta}{R} = \alpha$ yields

$$1 - \frac{i\omega L_p\alpha}{r} + \frac{i\omega M}{R}\left(\frac{Mi\omega\alpha}{r + \frac{L_s i\omega r}{R}}\right) = \alpha. \tag{5.299}$$

$$\therefore 1 = \alpha\left(1 - \frac{M^2 i^2\omega^2}{rR + L_s i\omega r} + \frac{i\omega L_p}{r}\right). \tag{5.300}$$

$$\therefore A = \left(1 + \frac{M^2\omega^2\alpha}{rR + L_s i\omega r} + \frac{i\omega L_p}{r}\right)^{-1}, \tag{5.301}$$

noting that $i^2 = -1$. Hence, the complete model is

$$V_p = V_0 e^{i\omega t}, \tag{5.302}$$

$$I_p = \frac{V_0}{r}\alpha e^{i\omega t}, \tag{5.303}$$

$$I_s = \frac{V_0}{R}\beta e^{i\omega t}, \tag{5.304}$$

$$\alpha = \left(1 + \frac{M^2\omega^2}{rR + L_s i\omega r} + \frac{i\omega L_p}{r}\right)^{-1}, \tag{5.305}$$

$$\beta = \frac{Mi\omega}{r + \frac{L_s i\omega r}{R}}\alpha, \tag{5.306}$$

with (approximate) inductances (i.e. modelling both coils as long solenoids of cross-section A and length l)

$$L_p = \frac{\mu\mu_0 N_p^2 A}{l}, \tag{5.307}$$

$$L_s = \frac{\mu\mu_0 N_s^2 A}{l} \tag{5.308}$$

and mutual inductance

$$M = k\sqrt{L_p L_s}. \tag{5.309}$$

To visually explore this, quite literally, 'complex' transformer model (!), the equations above were coded into MATLAB, with a graphical user interface (GUI) comprising sliders which vary the parameters r, R, N_p, N_s, k and with the other parameters V_0, A, l, μ as user-editable quantities in little text boxes. In Figs. 5.29 and 5.30, the graphs of $\left|\frac{V_s}{V_0}\right|, |I_p|$ and $|I_s|$ versus frequency are plotted. In Fig. 5.29, the GUI screenshots represent a step-up transformer of turn ratio $\frac{N_s}{N_p} = 2$, with a modest coupling constant $k = 0.7$ and a low secondary load $R = 30\ \Omega$, only three times the primary load $r = 10\ \Omega$. Interestingly, the graph of $\left|\frac{V_s}{V_0}\right|$ versus frequency in subplot (a) shows a peak at around 17 Hz and then a decay as frequency increases. Both primary and secondary currents decrease after this 'resonance' frequency. The same transformer is modelled in Fig. 5.30, but this time, the maximum coupling constant of $k = 1$ is set. Above about 50 Hz, the ratio $\left|\frac{V_s}{V_0}\right| \to$ constant, although this is somewhat less than what the trans-former equation would predict, i.e. $\left|\frac{V_s}{V_0}\right|$ should ideally equal $\frac{N_s}{N_p} = 2$. Both primary and secondary currents tend to a constant, roughly in a ratio of 2:1, which is what the 'possibly dubious power transfer equation' $I_s = I_p \frac{N_p}{N_s}$ would predict.

The school-physics equations of $\frac{V_s}{V_p} = \frac{N_s}{N_p}$ and $I_s = I_p \frac{N_p}{N_s}$ only become valid when $R \gg r$, $k \to 1$ and ω becomes large (i.e. much greater than the 'resonance' frequency when $k < 1$).

5.11 Resonance in an LCR Circuit

Consider a series circuit comprising a coil (an inductor) of inductance L, a capacitor of capacitance C and a total resistance R, as shown in Fig. 5.31(a). R could be a variable (but otherwise ohmic) resistor, plus the resistance of the wires that form the inductor. Let an AC voltage source drive current through the circuit. The sum of the potential differences across the capacitor V_C and resistor V_R must equate to the source voltage V, minus a 'back EMF' V_L resulting from electromagnetic induction due to a time-varying magnetic field being produced by the coil. Applying Kirchhoff's second law around the LCR loop (or simply by conservation of energy),

$$V - V_L = V_C + V_R. \tag{5.310}$$

Now, let $V(t)$ be a sinusoidally varying voltage source at frequency $f = \frac{\omega}{2\pi}$.

$$\therefore V = V_0 e^{i\omega t}. \tag{5.311}$$

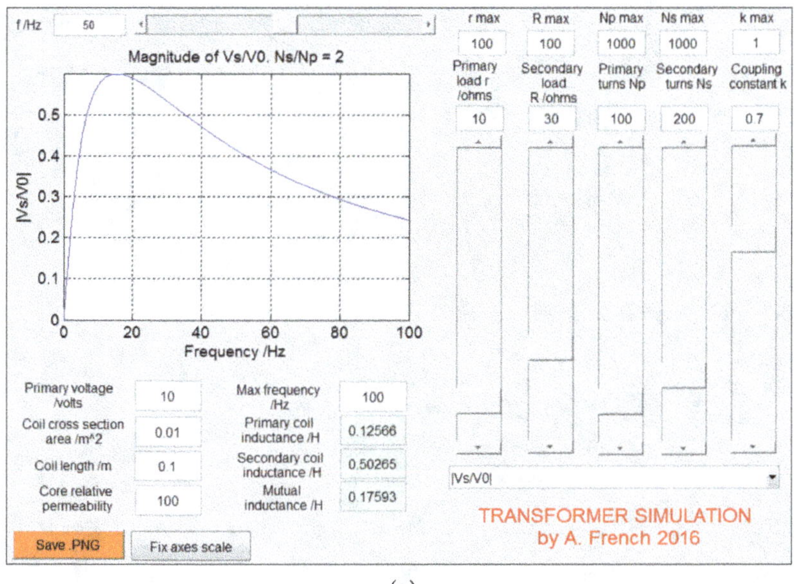

(a)

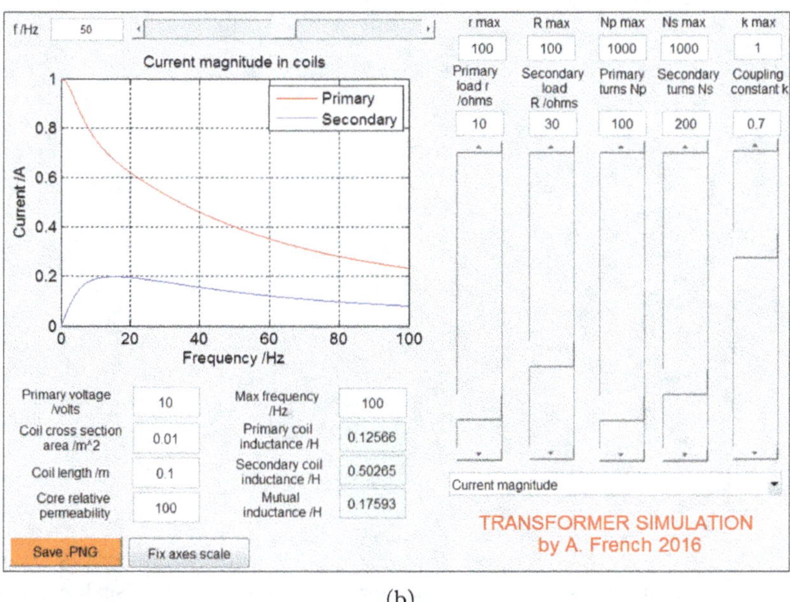

(b)

Fig. 5.29. A MATLAB app with a graphical user interface is used to visualise a model of a transformer consisting of N_p primary coils and N_s secondary coils. In (a), a graph of $\left|\frac{V_s}{V_0}\right|$ vs. frequency is plotted, where $\left|\frac{V_s}{V_0}\right|$ is the magnitude of the ratio of secondary to primary coil voltages. A modest secondary load to primary load ratio of $\frac{R}{r} = \frac{30\Omega}{10\Omega}$ and a mutual inductance of $M = 0.7\sqrt{L_p L_s}$ (with L being the coil inductances) result in a significant departure from the transformer equation, which would predict $\left|\frac{V_s}{V_0}\right| = \frac{N_s}{N_p} = \frac{200}{100} = 2$, independent of frequency. (b) plots the current magnitude in both primary and secondary coils. There is also significant variation with frequency and indeed a departure from an assumption of efficient power transfer between the primary and secondary coils.

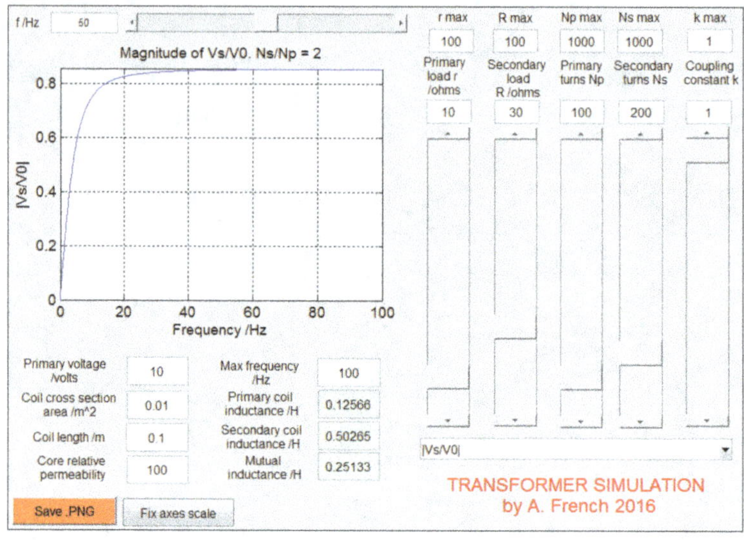

(a)

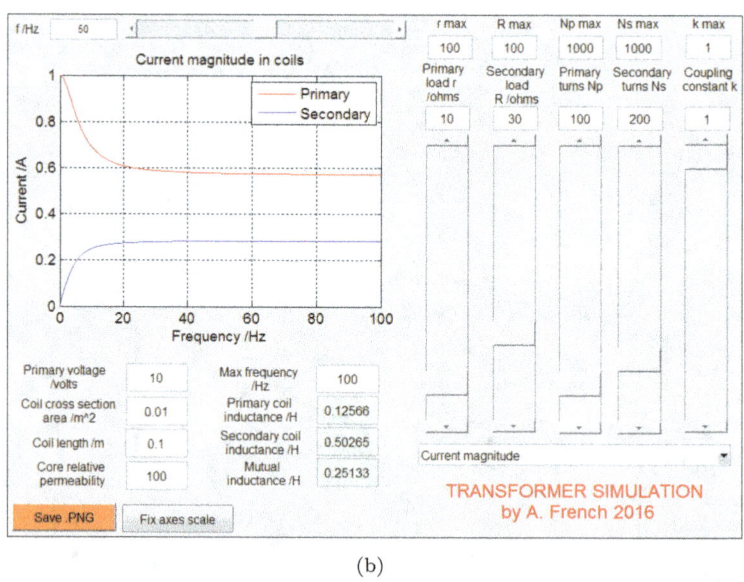

(b)

Fig. 5.30. A MATLAB app with a graphical user interface is used to visualise a model of a transformer consisting of N_p primary coils and N_s secondary coils. In (a), a graph of $\left|\frac{V_s}{V_0}\right|$ vs. frequency is plotted, where $\left|\frac{V_s}{V_0}\right|$ is the magnitude of the ratio of secondary to primary coil voltages. A maximum mutual inductance of $M = \sqrt{L_p L_s}$ (with L being the coil inductances), i.e. coupling constant $k = 1$, results in an asymptotic $\left|\frac{V_s}{V_0}\right| \to$ constant. However, this constant is less than half of what the *transformer equation* would predict, i.e. $\left|\frac{V_s}{V_0}\right| = \frac{N_s}{N_p} = \frac{200}{100} = 2$. The ratio of $\left|\frac{V_s}{V_0}\right|$ approaches this ideal limit as secondary to primary load ratio $\frac{R}{r} \to \infty$. This can be observed by shifting the secondary load slider to its maximum, and setting the Rmax box to be something like 2,000 Ω. In (b), the asymptotic ratio of primary to secondary currents is actually relatively close to the 2:1 ratio predicted by the 'school physics' power transfer equation $I_s = I_p \frac{N_p}{N_s}$.

If current I flows through the circuit,[48]

$$V_R = IR, \tag{5.312}$$

$$CV_C = \int I\,dt, \tag{5.313}$$

$$V_L = L\frac{dI}{dt}. \tag{5.314}$$

Putting this together, using $V_S - V_L = V_C + V_R$,

$$V_0 e^{i\omega t} = \frac{1}{C}\int I\,dt + IR + L\frac{dI}{dt}. \tag{5.315}$$

In the steady state, assume a steady AC current I is established at the same frequency as the source:

$$I = I_0 e^{i\omega t - i\phi}. \tag{5.316}$$

Hence,

$$V_0 e^{i\omega t} = I_0 e^{i\omega t - i\phi}\left(\frac{1}{i\omega C} + R + i\omega L\right). \tag{5.317}$$

$$\therefore V = IZ, \tag{5.318}$$

where *impedance* $Z = \frac{1}{i\omega C} + R + i\omega L$. We can use the analogy between $V = IR$ for DC circuits to define a *complex impedance* (i.e. a generalised resistance for time-varying 'active' circuits) for the capacitor and inductor components. This means we can perform circuit calculations for active circuits in exactly the same 'potential divider' way as for DC circuits. No more differential equations are required! Let Z_C be the impedance of a capacitor and Z_L be the impedance of an inductor:

$$Z_C = \frac{1}{i\omega C}, \tag{5.319}$$

$$Z_L = i\omega L, \tag{5.320}$$

$$Z_R = R. \tag{5.321}$$

For a series LCR circuit, the voltage across the capacitor is, using the idea of the potential divider,

$$V_C = \frac{Z_C}{Z_L + Z_C + Z_R}V_0 e^{i\omega t} \tag{5.322}$$

$$= \frac{\frac{1}{i\omega C}}{i\omega L + \frac{1}{i\omega C} + R}V_0 e^{i\omega t} \tag{5.323}$$

$$= \frac{1}{-\omega^2 LC + 1 + i\omega RC}V_0 e^{i\omega t}. \tag{5.324}$$

---

[48]The charge separated on the capacitor plates is $Q = CV_C$, which must equate to the sum of charges $dQ = I\,dt$ deposited in time interval dt. Hence, $CV_C = \int dQ = \int I\,dt$.

Let

$$V_C = \left| \frac{V_C}{V_0} \right| V_0 e^{i\omega t} e^{-\Phi}. \tag{5.325}$$

Now, using the complex number result, $\frac{1}{a+ib} = \frac{a-ib}{a^2+b^2} = \frac{e^{-i\tan^{-1}\left(\frac{b}{a}\right)}}{\sqrt{a^2+b^2}}$,

$$\frac{1}{-\omega^2 LC + 1 + i\omega RC} = \frac{e^{-i\tan^{-1}\left(\frac{\omega RC}{1-\omega^2 LC}\right)}}{\sqrt{(1-\omega^2 LC)^2 + \omega^2 R^2 C^2}}. \tag{5.326}$$

Hence,

$$V_C = \left| \frac{V_C}{V_0} \right| V_0 e^{i\omega t} e^{-\Phi} = \frac{e^{-i\tan^{-1}\left(\frac{\omega RC}{1-\omega^2 LC}\right)} V_0 e^{i\omega t}}{\sqrt{(1-\omega^2 LC)^2 + \omega^2 R^2 C^2}}. \tag{5.327}$$

Therefore, we define *gain* as

$$\left| \frac{V_C}{V_0} \right| = \frac{1}{\sqrt{(1-\omega^2 LC)^2 + \omega^2 R^2 C^2}} \tag{5.328}$$

and *phase* as

$$\Phi = \tan^{-1}\left(\frac{\omega RC}{1-\omega^2 LC} \right). \tag{5.329}$$

The LCR circuit has a *resonance peak* of a gain versus ω curve when

$$\frac{d}{d\omega} \left| \frac{V_C}{V_0} \right| = 0. \tag{5.330}$$

Hence, using $\left| \frac{V_C}{V_0} \right| = \frac{1}{\sqrt{(1-\omega^2 LC)^2 + \omega^2 R^2 C^2}}$,

$$\frac{-\frac{1}{2}\left\{ 2\left(1-\omega^2 LC\right)\left(-2\omega LC\right) + 2\omega R^2 C^2 \right\}}{\left((1-\omega^2 LC)^2 + \omega^2 R^2 C^2 \right)^{\frac{3}{2}}} = 0. \tag{5.331}$$

$$\therefore \left(1-\omega^2 LC\right)\left(-2\omega LC\right) + \omega R^2 C^2 = 0. \tag{5.332}$$

$$\therefore R^2 C^2 = 2\left(LC - \omega^2 L^2 C^2 \right). \tag{5.333}$$

$$\therefore \omega = \frac{\sqrt{LC - \frac{1}{2} R^2 C^2}}{LC}. \tag{5.334}$$

$$\therefore \omega = \frac{1}{\sqrt{LC}} \sqrt{1 - \frac{1}{2}\frac{R^2 C}{L}}. \tag{5.335}$$

Using frequency $f = \frac{\omega}{2\pi}$, the resonance frequency which corresponds to the maximum gain $\left|\frac{V_C}{V_0}\right|$ is

$$f = \frac{1}{2\pi\sqrt{LC}}\sqrt{1 - \frac{1}{2}\frac{R^2 C}{L}}. \tag{5.336}$$

For a typical LCR circuit that could be constructed in a school laboratory, $L \approx 1.00$ H, $R = 100\ \Omega$ and $C = 0.1\ \mu$F. This means

$$\frac{1}{2\pi\sqrt{LC}} \approx 503\ \text{Hz}, \tag{5.337}$$

$$\frac{1}{2}\frac{R^2 C}{L} \approx \frac{1}{2,000}. \tag{5.338}$$

We can therefore neglect the $\frac{1}{2}\frac{R^2 C}{L}$ term and set the LCR resonance frequency to be

$$f \approx \frac{1}{2\pi}\frac{1}{\sqrt{LC}}. \tag{5.339}$$

Radio receivers use tunable LCR circuits to resonate with electromagnetic waves with particular carrier frequencies. Typically, it is the capacitance C which is varied rather than inductance L or resistance R. In the UK, BBC Radio 4 is broadcast at the analogue FM frequency of about 104 MHz. Assuming a 1 pF capacitor, this means an inductance of

$$L \approx \frac{1}{(2\pi f)^2 C} = \frac{1}{(2\pi \times 104 \times 10^6)^2 \times 10^{-12}}\ \text{(H)}, \tag{5.340}$$

$$L \approx 2.3\ \mu\text{H}. \tag{5.341}$$

A different LCR circuit is illustrated in Fig. 5.31(e), which results in a *notch filter*. In this case, the capacitor and inductor are wired in parallel rather than in series. Using the complex impedance concept, we can immediately write down an expression for the voltage across the resistor:

$$V_R = \frac{Z_R}{\left(\frac{1}{Z_C} + \frac{1}{Z_L}\right)^{-1} + Z_R}V_0 e^{i\omega t}. \tag{5.342}$$

As plotted in Fig. 5.31(f), the gain $\left|\frac{V_R}{V_0}\right|$ across the resistor is unity, except close to the resonant frequency $\approx \frac{1}{2\pi}\sqrt{\frac{1}{LC}}$, where the gain tends to zero. This particular circuit could be used to filter UK 50 Hz mains emissions if L and C are suitable values, such that $\frac{1}{2\pi}\sqrt{\frac{1}{LC}} \approx 50$ Hz.

5.12　The Fresnel Equations

If an EM wave experiences a change in *refractive index* n, i.e. the ratio between the speed of light in a vacuum and the speed of light in the medium it is travelling, the general response will be for both *reflected* and *transmitted* (i.e. refracted) waves to be created. The power of the incident wave will be shared between these, and

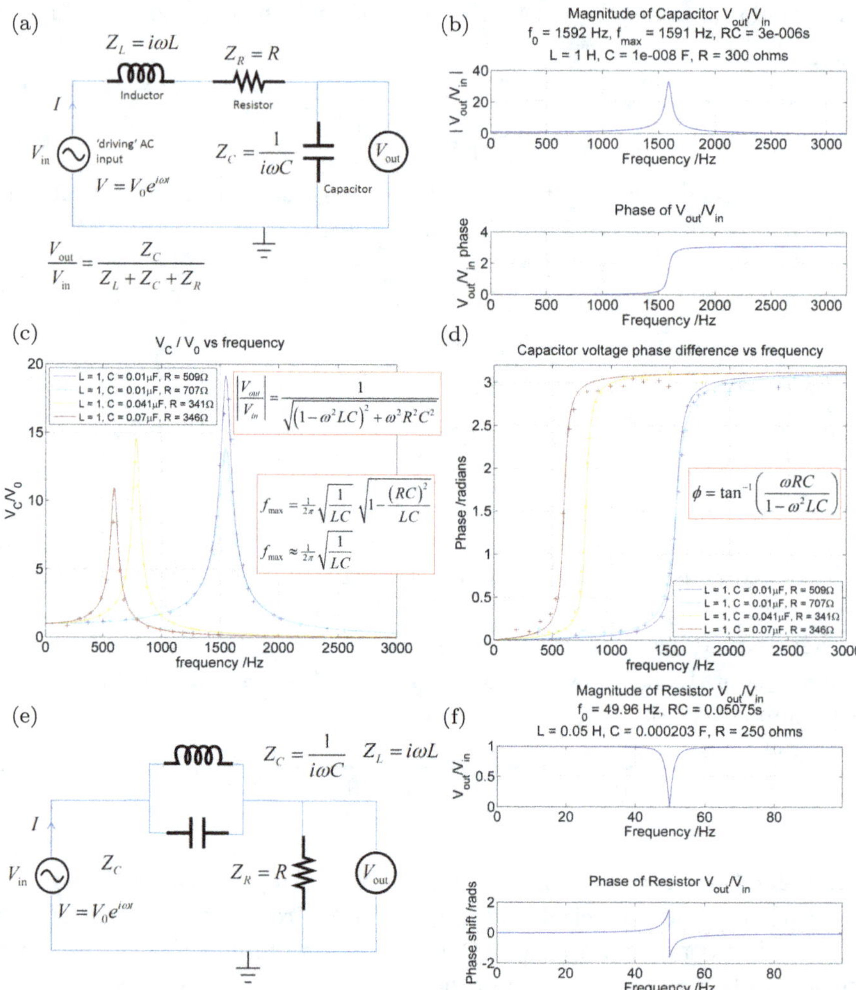

Fig. 5.31. (a) shows an LCR circuit consisting of an inductor (a coil of inductance L), a resistor (of resistance R) and a capacitor (of capacitance C). This is driven by a sinusoidal voltage source of frequency $f = \frac{\omega}{2\pi}$ and amplitude V_0. (b) describes the gain and phase of the voltage response across the capacitor. There is resonance peak (and a π phase change) at 1,591 Hz, which is $\approx \frac{1}{2\pi}\sqrt{\frac{1}{LC}}$. (c) and (d) show similar graphs, with theoretical curves (solid lines) overlaid with experimental data. In every case, $L = 1.00$H, but C and R are varied. Lowering C increases the resonance peak frequency, whereas lowering R increases the peak gain. A different circuit is illustrated in (e) and (f), which results in a *notch filter*. The gain across the resistor is unity, except close to the resonant frequency $\approx \frac{1}{2\pi}\sqrt{\frac{1}{LC}}$, where the gain tends to zero. This particular circuit could be used to filter UK 50 Hz mains emissions.

the balance of power depends on a number of factors, which are modelled using the *Fresnel equations*. At a refractive index boundary n_1 to n_2, the electric and magnetic fields that constitute an EM wave are defined as follows:

- E_i is the magnitude of the incident electric field; E_r is the magnitude of the reflected electric field; and E_t is the magnitude of the transmitted electric field.

- Assuming that all media are isotropic, magnetic flux density $\mathbf{B} = \mu\mu_0\mathbf{H}$. For EM waves in isotropic media,[49] $|\mathbf{B}| = \frac{n}{c}|\mathbf{E}|$, so $H_i = \frac{n_1}{\mu_1\mu_0 c}E_i, H_r = \frac{n_1}{\mu_1\mu_0 c}E_r, H_t = \frac{n_2}{\mu_2\mu_0 c}E_t$.

Maxwell's equations (see Section 5.13) tell us: 'At an interface between media of differing refractive index, vector components of total magnetic flux density $\mathbf{B}$ perpendicular to the interface surface must be *continuous* across the boundary. Also, components of the $\mathbf{E}$ and $\mathbf{H}$ fields which are *parallel* to the surface, must be continuous across the boundary, as long as there are no surface charges or currents.' (These conditions will be proven from Maxwell's equations in Section 5.13.5.)

Let us consider two scenarios (see Fig. 5.32) for the polarisation of the $\mathbf{E}$ field of the EM wave relative to the plane containing incident, reflected and transmitted EM wavevectors $\mathbf{k}_i$, $\mathbf{k}_r$ and $\mathbf{k}_t$, respectively. In both cases, assume that the incident EM wavevector is at an angle of incidence θ_i from the boundary normal and that there is a reflected wave, also at angle of reflection θ_i from the boundary normal. If $n_1 < n_2$, a transmitted wave is always refracted by angle θ_t from the boundary normal. By *Snell's law* of refraction,

$$n_2 \sin\theta_t = n_1 \sin\theta_i. \tag{5.343}$$

$$\therefore \theta_t = \sin^{-1}\left(\frac{n_1}{n_2}\sin\theta_{\cdot i}\right). \tag{5.344}$$

If $n_1 > n_2$, then a transmitted wave is produced at angle $\theta_t = \sin^{-1}\left(\frac{n_1}{n_2}\sin\theta_i\right)$ from the normal, but only for angles of incidence within the range of $0 \leq \theta_i < \theta_c$. The *critical angle* θ_c is when $\theta_t = 90°$.

$$\therefore n_2 \sin 90° = n_1 \sin\theta_c.$$

$$\therefore \theta_c = \sin^{-1}\left(\frac{n_2}{n_1}\right).$$

If $n_1 > n_2$ and $\theta_i \geq \theta_c$, then the incident EM wave will undergo *total internal reflection*, which means there will be *no transmitted wave* from medium 1 to medium 2. All incident wave power will be reflected. To explore these basic wave behaviours through an easy-to-access online simulation, I highly recommend *Bending Light* from PhET (https://phet.colorado.edu/sims/html/bending-light/latest/bending-light_all.html). However, it stops short of modelling the $\mathbf{E}$ and $\mathbf{H}$ fields. We need our own simulation!

5.12.1 *Case 1: S (perpendicular) polarisation*

In S polarisation, 'S' denotes *senkrecht* in German, which means *perpendicular* in English. In this scenario (see (a) in Fig. 5.32), the $\mathbf{H}$ vector (and hence $\mathbf{B}$ in the diagram) is in the *same plane* as the wavevectors (but necessarily perpendicular to

---

[49]See the concluding section of this chapter on *Maxwell's equations* and the derivations associated with EM waves, which propagate in a vacuum at a constant speed $c = \frac{1}{\sqrt{\mu_0\varepsilon_0}}$.

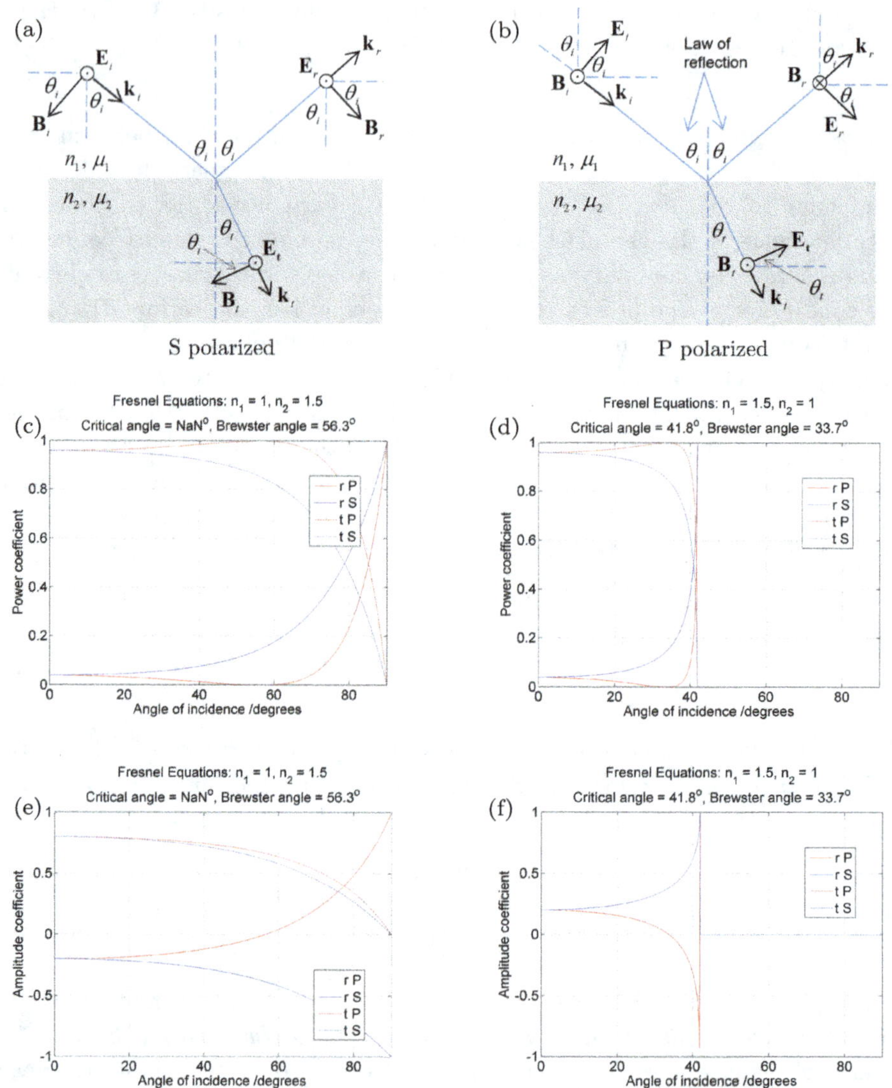

Fig. 5.32. (a) describes the arrangement of electric field $\mathbf{E}$, magnetic flux density $\mathbf{B}$ and wavevector $\mathbf{k}$ of electromagnetic waves that may be reflected and transmitted through a surface between isotropic media of differing refractive indices $n_{1,2}$ and relative permeabilities $\mu_{1,2}$. (a) depicts S-polarised waves, i.e. $\mathbf{E}$ is *perpendicular* to the plane of the wavevectors. (b) depicts P-polarised waves, i.e. $\mathbf{E}$ is in the *same plane* as the wavevectors but still mutually perpendicular to them, as all (transverse) EM waves are. (c) plots the reflection and transmission power coefficients given by the *Fresnel equations* for S and P polarisations, given $n_1 = 1.0$ and $n_2 = 1.5$. Note a minimum of the P-polarised reflection at the *Brewster angle* of $\tan^{-1}(1.5) = 56.3°$. In this simulation, $\mu_1 = \mu_2 = 1$. (d) is a similar plot but with the refractive indices reversed. Note that this means no transmission (i.e. 100% reflection) beyond a *critical angle* of $\sin^{-1}\frac{1}{1.5} = 41.8°$. (e) and (f) give the corresponding amplitude coefficients, which incorporate a change in sign, implying a π phase shift.

each corresponding wavevector), whereas the **E** field is perpendicular to the plane containing **H** and **k**.

$E_{||}$ continuity across the boundary means

$$E_i + E_r = E_t. \tag{5.345}$$

$H_{||}$ continuity across the boundary implies (recalling $H_i = \frac{n_1}{\mu_1 \mu_0 c} E_i$, $H_r = \frac{n_1}{\mu_1 \mu_0 c} E_r$, $H_t = \frac{n_2}{\mu_2 \mu_0 c} E_t$)

$$\frac{n_1}{\mu_1 \mu_0 c} E_i \cos\theta_i - \frac{n_1}{\mu_1 \mu_0 c} E_r \cos\theta_i = \frac{n_2}{\mu_2 \mu_0 c} E_t \cos\theta_t. \tag{5.346}$$

Define, for brevity

$$\alpha = \frac{n_1 \mu_2}{n_2 \mu_1}. \tag{5.347}$$

Hence,

$$\alpha E_i \cos\theta_i - \alpha E_r \cos\theta_i = (E_i - E_r) \cos\theta_t. \tag{5.348}$$

$$\therefore E_i \left(\alpha \cos\theta_i + \cos\theta_t \right) = E_r \left(\alpha \cos\theta_i - \cos\theta_t \right).$$

Define a reflection coefficient $r_\perp = \frac{E_r}{E_i}$ and transmission coefficient $t_\perp = \frac{E_t}{E_i}$.

$$\therefore r_\perp = \frac{\alpha \cos\theta_i - \cos\theta_t}{\alpha \cos\theta_i + \cos\theta_t}. \tag{5.349}$$

Since $E_i + E_r = E_t \Rightarrow 1 + \frac{E_r}{E_i} = \frac{E_t}{E_i} = t_\perp$.

$$\therefore t_\perp = 1 + r_\perp. \tag{5.350}$$

$$\therefore t_\perp = \left(\frac{\alpha \cos\theta_i + \cos\theta_t + \alpha \cos\theta_i - \cos\theta_t}{\alpha \cos\theta_i + \cos\theta_t} \right). \tag{5.351}$$

$$\therefore t_\perp = \frac{2\alpha \cos\theta_i}{\alpha \cos\theta_i + \cos\theta_t}. \tag{5.352}$$

The corresponding wave power coefficients are

$$R_\perp = |r_\perp|^2, \tag{5.353}$$

$$T_\perp = 1 - |r_\perp|^2. \tag{5.354}$$

The plots of $r_\perp$, $t_\perp$, $R_\perp$ and $T_\perp$ versus the angle of incidence θ_i are shown in Figs. 5.32(c)–(f). In (c) and (e), $n_1 = 2$ and $n_2 = 1.5$, which means transmission at all angles of incidence.

5.12.2 *Case 2: P (parallel) polarisation*

In this scenario (see (b) in Fig. 5.32), the **E** vector is in the *same plane as the wavevectors* (but necessarily perpendicular to each corresponding wavevector since

EM waves are transverse). The $\mathbf{H}$ field is perpendicular to this plane (and hence $\mathbf{B}$ in the diagram) since $\mathbf{k}$, $\mathbf{E}$, $\mathbf{H}$ for an EM wave must all be mutually perpendicular.[50]

$\mathbf{E}_\parallel$ continuity across the boundary means

$$E_i \cos\theta_i + E_r \cos\theta_i = E_t \cos\theta_t. \tag{5.355}$$

$\mathbf{H}_\parallel$ continuity across the boundary (noting that $\mathbf{H}$ is perpendicular to the plane) implies (recalling $H_i = \frac{n_1}{\mu_1 \mu_0 c} E_i$, $H_r = \frac{n_1}{\mu_1 \mu_0 c} E_r$, $H_t = \frac{n_2}{\mu_2 \mu_0 c} E_t$)

$$\frac{n_1}{\mu_1 \mu_0 c} E_i + \frac{n_1}{\mu_1 \mu_0 c} E_r = \frac{n_2}{\mu_2 \mu_0 c} E_t. \tag{5.356}$$

$$\therefore E_t = \frac{n_1 \mu_2}{n_2 \mu_1} \left(E_i - E_r \right). \tag{5.357}$$

Again, for brevity define

$$\alpha = \frac{n_1 \mu_2}{n_2 \mu_1}. \tag{5.358}$$

Therefore, substituting $E_t = \frac{n_1 \mu_2}{n_2 \mu_1} \left(E_i - E_r \right)$ into $E_i \cos\theta_i + E_r \cos\theta_i = E_t \cos\theta_t$ yields

$$E_i \cos\theta_i + E_r \cos\theta_i = \alpha \left(E_i - E_r \right) \cos\theta_t. \tag{5.359}$$

$$\therefore E_r \left(\alpha \cos\theta_t + \cos\theta_i \right) = E_i \left(\alpha \cos\theta_t - \cos\theta_i \right).$$

Define a *reflection coefficient* $r_\parallel = \frac{E_r}{E_i}$ and a *transmission coefficient* $t_\parallel = \frac{E_t}{E_i}$.

$$\therefore r_\parallel = \frac{\alpha \cos\theta_t - \cos\theta_i}{\alpha \cos\theta_t + \cos\theta_i}. \tag{5.360}$$

Hence, since $E_t = \frac{n_1 \mu_2}{n_2 \mu_1} \left(E_i - E_r \right)$ and therefore $\frac{E_t}{E_i} = \alpha \left(1 - \frac{E_r}{E_i} \right)$,

$$t_\parallel = \alpha \left(1 - r_\parallel \right) \tag{5.361}$$

$$= \alpha \left(\frac{\alpha \cos\theta_t + \cos\theta_i - \alpha \cos\theta_t + \cos\theta_i}{\alpha \cos\theta_t + \cos\theta_i} \right). \tag{5.362}$$

$$\therefore t_\parallel = \frac{2\alpha \cos\theta_i}{\alpha \cos\theta_t + \cos\theta_i}. \tag{5.363}$$

The corresponding wave power coefficients[51] are

$$R_\parallel = \left| r_\parallel \right|^2, \tag{5.364}$$

$$T_\parallel = 1 - \left| r_\parallel \right|^2. \tag{5.365}$$

Note that we can avoid errors by calculating power coefficients in this way, rather than simply (and erroneously) *squaring*. If $\theta_i > \sin^{-1} \frac{n_2}{n_1}$, then $\cos\theta_t$ will be a complex number with (potentially) real and imaginary parts. So, squaring will yield

---

[50]See the final section on Maxwell's equations.

[51]i.e. the *fraction* of incident wave power transmitted is T, and the fraction of wave power reflected is R. Hence, by conservation of energy, $R + T = 1$.

a complex number, whereas $|r_{\shortparallel}|^2$ will always be real (and positive). Alternatively, we can consider $n_1 > n_2$ as a special case and set $T_{\shortparallel} = 0$ and $R_{\shortparallel} = 1$ if $\theta_i \geq \theta_c$. The plots of $r_{\shortparallel}$, $t_{\shortparallel}$, $R_{\shortparallel}$ and $T_{\shortparallel}$ versus the angle of incidence θ_i are shown in Fig. 5.32(c)–(f). Note that in (d) and (f), $n_1 = 1.5$ and $n_2 = 1$, which means a critical angle of $\sin^{-1}\left(\frac{1}{1.5}\right) = 41.8°$.

5.12.3 *Brewster's angle*

The Fresnel equations yields an interesting and very useful special case for P-polarised EM radiation: the possibility of $r_{\shortparallel} = 0$ at a particular angle of incidence θ_i. This is called the *Brewster angle*[52] and is a useful concept to adopt in the design of optical systems, from solar panels to sunglasses. If $r_{\shortparallel} = 0$, then from $r_{\shortparallel} = \frac{\alpha\cos\theta_t - \cos\theta_i}{\alpha\cos\theta_t + \cos\theta_i}$,

$$\alpha\cos\theta_t - \cos\theta_i = 0 \tag{5.366}$$

$$\alpha^2\cos^2\theta_t = \cos^2\theta_i \tag{5.367}$$

$$\alpha^2\left(1 - \sin^2\theta_t\right) = 1 - \sin^2\theta_i. \tag{5.368}$$

Now, from Snell's law, $\sin\theta_t = \frac{n_1}{n_2}\sin\theta_i$. Therefore,

$$\alpha^2\left(1 - \frac{n_1^2}{n_2^2}\sin^2\theta_i\right) = 1 - \sin^2\theta_i \tag{5.369}$$

$$\left(1 - \alpha^2\frac{n_1^2}{n_2^2}\right)\sin^2\theta_i = 1 - \alpha^2 \tag{5.370}$$

$$\sin^2\theta_i = \frac{1 - \alpha^2}{1 - \alpha^2\frac{n_1^2}{n_2^2}}. \tag{5.371}$$

If relative permeability $\mu_1 = \mu_2$, then

$$\alpha = \frac{n_1\mu_2}{n_2\mu_1} = \frac{n_1}{n_2}, \tag{5.372}$$

which means

$$\sin^2\theta_i = \frac{1 - \alpha^2}{1 - \alpha^4} = \frac{1 - \alpha^2}{\left(1 - \alpha^2\right)\left(1 + \alpha^2\right)} = \frac{1}{1 + \alpha^2}. \tag{5.373}$$

Now,

$$\sin^2\theta_i + \cos^2\theta_i = 1 \tag{5.374}$$

$$\implies 1 + \frac{1}{\tan^2\theta_i} = \frac{1}{\sin^2\theta_i}. \tag{5.375}$$

---

[52]Sir David Brewster (1781–1868).

Therefore,

$$1 + \frac{1}{\tan^2 \theta_i} = 1 + \alpha^2. \qquad (5.376)$$

$$\therefore \tan \theta_i = \frac{1}{\alpha} = \frac{n_2}{n_1}. \qquad (5.377)$$

So, at the Brewster angle when the angle of incidence θ_i is

$$\theta_B = \tan^{-1}\left(\frac{n_2}{n_1}\right), \qquad (5.378)$$

you will *only* get reflection from S-polarised EM waves since $r_\parallel = 0$. Note the similarity of expression to the inverse-trigonometric formula for the critical angle under total internal reflection:

$$\theta_C = \sin^{-1}\left(\frac{n_2}{n_1}\right). \qquad (5.379)$$

An intuitive (and algebraically more efficient) method for deriving the expression for the Brewster angle, which doesn't explicitly require the Fresnel equations, is to consider P-polarised radiation when the *transmitted and reflected wavevectors are perpendicular to each other*, as illustrated in Fig. 5.33. P-polarised reflected radiation would have an electric field parallel to the transmitted wavevector in this scenario, and since **E**, **H** and **k** must all be mutually perpendicular for EM waves, *there can't be any reflected P-polarised EM waves*.

For this geometrical argument,

$$\theta_i + 90° + \theta_t = 180°. \qquad (5.380)$$

$$\therefore \theta_t = 90° - \theta_i. \qquad (5.381)$$

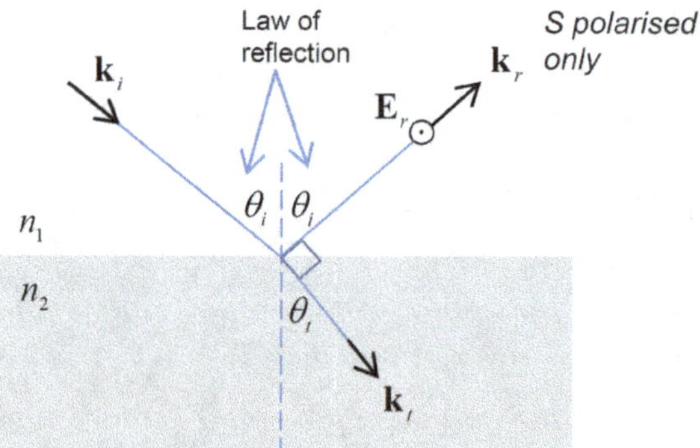

Fig. 5.33. *Brewster angle* $\theta_B = \tan^{-1}\left(\frac{n_2}{n_1}\right)$ between two media of refractive indices n_1 and n_2 can be calculated by considering when the reflected wavevector of EM waves is perpendicular to the transmitted wavevector. P-polarised reflected radiation would have an electric field parallel to the transmitted wavevector in this scenario, and since **E**, **H** and **k** must all be mutually perpendicular for EM waves, there can't be any reflected P-polarised EM waves in this situation.

From Snell's law,

$$n_1 \sin \theta_i = n_2 \sin \theta_t \tag{5.382}$$

$$n_1 \sin \theta_i = n_2 \sin (90^\circ - \theta_i) \tag{5.383}$$

$$n_1 \sin \theta_i = n_2 \cos \theta_i. \tag{5.384}$$

$$\therefore \tan \theta_i = \frac{n_2}{n_1}. \tag{5.385}$$

$$\therefore \theta_B = \tan^{-1} \left(\frac{n_2}{n_1} \right). \tag{5.386}$$

5.13 Maxwell's Equations of Electromagnetism

This final section is a very brief summary of ideas adapted from *A Student's Guide to Maxwell's Equations* [8], *Electricity & Magnetism* [2], *Classical Electrodynamics* [18] and *Theoretical Concepts in Physics* [22]. The aim is to set out a fairly complete theoretical structure which relates charges (and their motion) to electric and magnetic fields. Three important theorems (the emergence of EM waves, continuity conditions for $\mathbf{E}$ and $\mathbf{B}$ fields and Poynting's equation for energy flow) are then derived from Maxwell's equations.

5.13.1 *Electric and magnetic fields*

The study of *electromagnetism* is fundamentally an understanding of the relationships between *electric fields*, which are sourced from *charges*, and *magnetic fields*, which are sourced from moving charges, i.e. *electric currents*. If a charge of q coulombs moving with velocity $\mathbf{v}$ experiences an electric field $\mathbf{E}$ and magnetic flux density $\mathbf{B}$, there will be a force $\mathbf{f}$ acting on the charge:

$$\mathbf{f} = q \left(\mathbf{E} + \mathbf{v} \times \mathbf{B} \right). \tag{5.387}$$

In a vacuum, we might possibly have an isolated charge, but in most matter, which comprises atoms containing protons and electrons, the presence of external electric and magnetic fields may cause an internal drift of charge, which itself modifies the electric and magnetic fields that a given charge may experience. In Maxwell's equations, $\mathbf{D}$ is the 'net' electric field, which comprises a sum of an imposed electric field $\mathbf{E}$ and a *polarisation field* $\mathbf{P}$ set up by the drift of charge in a medium that experiences $\mathbf{E}$:

$$\mathbf{D} = \varepsilon_0 \mathbf{E} + \mathbf{P}. \tag{5.388}$$

By convention, there is a scaling factor ε_0 between the $\mathbf{D}$ and $\mathbf{E}$ fields. $\varepsilon_0 = 8.854 \times 10^{-12}$ F m^{-1} (to 4.s.f), the *permittivity of free space*, is the constant in Coulomb's inverse-square law, which relates the force F between two charges q and Q that are separated by distance r:

$$F = \frac{1}{4\pi\varepsilon_0} \frac{qQ}{r^2}. \tag{5.389}$$

Note that this means the dimensions of $\mathbf{D}$ and $\mathbf{E}$ fields are *not* the same. Similarly, an imposed magnetic field $\mathbf{H}$ may cause a material (e.g. iron, steel) to be magnetised and, therefore, contribute to the net magnetic field. The magnetic flux density $\mathbf{B}$ is related to magnetic field $\mathbf{H}$ and magnetisation field $\mathbf{M}$ by

$$\mathbf{B} = \mu_0 \left(\mathbf{H} + \mathbf{M} \right). \tag{5.390}$$

Again, we have a convention of a (non-dimensionless) scale factor, which, in this case, is the *permeability of free space* $\mu_0 = 1.257 \times 10^{-6}$ m kg s^{-2}A^{-2} (to 4.s.f). This constant relates the magnetic flux density B (in Tesla) at a perpendicular distance r from a wire carrying current I:

$$B = \frac{\mu_0 I}{2\pi r}. \tag{5.391}$$

Charged and magnetisable media will respond in different ways to applied fields, and the polarisation and magnetisation fields can be expressed in a generic fashion using a *susceptibility matrix* χ. Electric and magnetic fields are three-dimensional, so χ will be a 3×3 matrix:

$$\mathbf{P} = \varepsilon_0 \chi \mathbf{E}, \tag{5.392}$$

$$\mathbf{M} = \chi_M \mathbf{H}. \tag{5.393}$$

Isotropic media don't have any particular directions that are more (or less) susceptible to applied fields. In this case, the *polarisation and magnetisation fields are assumed to be parallel*, so we can write

$$\mathbf{D} = \varepsilon_r \varepsilon_0 \mathbf{E}, \tag{5.394}$$

$$\mathbf{B} = \mu \mu_0 \mathbf{H}. \tag{5.395}$$

ε_r is a dimensionless number called the *relative permittivity*, or 'dielectric constant'. For free space or a completely non-polarisable material, $\varepsilon_r = 1$, whereas for water at 298 K, $\varepsilon_r \approx 78$. Barium strontium titanate has a relative permittivity of about 500, which makes it a useful material for constructing a *dielectric* between capacitor plates.

μ is a dimensionless number called the *relative permeability*. For a non-magnetic material, $\mu = 1$. Carbon steel has a relative permeability of about 100; for iron oxide (ferrite), it may be in excess of 500; and for pure iron, it could be over 5,000.

5.13.2 *Maxwell's equations in integral form*

There are *four equations* involving electric and magnetic fields (and their sources), which are known as *Maxwell's equations*.

M1: The *flux* of $\mathbf{D}$ through a surface S enclosing volume V which contains charge per unity volume ρ is equal to the total charge within V. Vector element $d\mathbf{S}$ defines an outward normal to surface S:

$$\int_S \mathbf{D} \cdot d\mathbf{S} = \int_V \rho dV. \tag{5.396}$$

M2: There are *no magnetic monopoles*,[53] so the net flux of $\mathbf{B}$ through a closed surface S is zero:

$$\int_S \mathbf{B} \cdot d\mathbf{S} = 0. \tag{5.397}$$

M3: *Faraday's law.* The rate of change of total magnetic flux linked through an open surface S, bounded by loop C, is proportional to the EMF induced. The latter is the line integral of the electric field $\mathbf{E}$ around the loop. $d\mathbf{l}$ is a vector line element which, when added up, describes the loop spatially:

$$\oint_C \mathbf{E} \cdot d\mathbf{l} = -\frac{\partial}{\partial t} \int_S \mathbf{B} \cdot d\mathbf{S}. \tag{5.398}$$

M4: *Ampère's law* plus *Maxwell's displacement current.* Magnetic fields are generated from electric currents *and* a rate of change of electric fields:

$$\oint_C \mathbf{H} \cdot d\mathbf{l} = \int_S \mathbf{J} \cdot d\mathbf{S} + \frac{\partial}{\partial t} \int_S \mathbf{D} \cdot d\mathbf{S}. \tag{5.399}$$

Current density (i.e. current per unit volume) $\mathbf{J} = \rho\mathbf{v}$, where $\mathbf{v}$ is the velocity of a charged volume element with charge density ρ. The current I flowing through surface S is therefore

$$I = \int_S \mathbf{J} \cdot d\mathbf{S}. \tag{5.400}$$

M4 and the concepts of loop and surface integrals are illustrated in Fig. 5.34. In this figure, we assume $\mathbf{H} = \frac{1}{\mu_0}\mathbf{B}$, $\mathbf{D} = \varepsilon_0\mathbf{E}$ and current $I = \int_S \mathbf{J} \cdot d\mathbf{S}$ is flowing through surface S. i.e. a 'free-space' solution for electric fields and magnetic flux density.

5.13.3 *Maxwell's equations in differential form*

Now, let us consider two theorems from the branch of mathematics known as *vector calculus*.

---

[53]i.e. magnetic poles always come in north (N) and south (S) pairs. Magnetic field lines are *always closed loops*. The N and S designations are useful to consider, but in reality there are no point sources (or sinks) of magnetic fields.

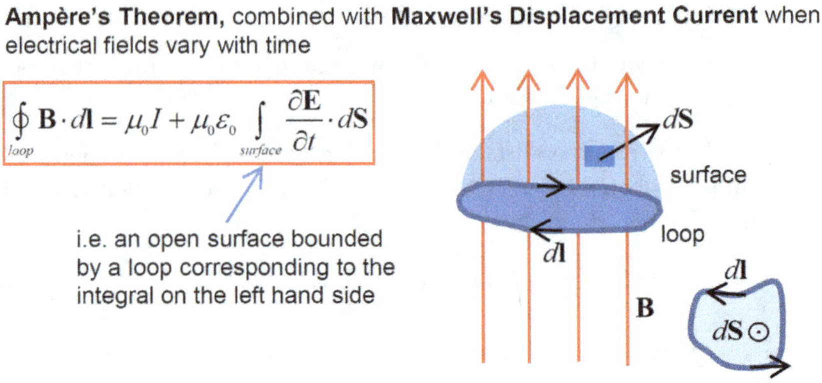

Ampère's Theorem, combined with **Maxwell's Displacement Current** when electrical fields vary with time

$$\oint_{loop} \mathbf{B} \cdot d\mathbf{l} = \mu_0 I + \mu_0 \varepsilon_0 \int_{surface} \frac{\partial \mathbf{E}}{\partial t} \cdot d\mathbf{S}$$

i.e. an open surface bounded by a loop corresponding to the integral on the left hand side

Fig. 5.34. A visualisation of Maxwell's equation **M4**, $\oint_C \mathbf{H} \cdot d\mathbf{l} = \int_S \mathbf{J} \cdot d\mathbf{S} + \frac{\partial}{\partial t} \int_S \mathbf{D} \cdot d\mathbf{S}$, and the concept of loop C and surface S integrals. We assume $\mathbf{H} = \frac{1}{\mu_0}\mathbf{B}$, $\mathbf{D} = \varepsilon_0 \mathbf{E}$, and current $I = \int_S \mathbf{J} \cdot d\mathbf{S}$ is flowing through surface S, i.e. a 'free-space' solution for electric fields $\mathbf{E}$ and magnetic flux density $\mathbf{B}$. Note only the magnetic flux density $\mathbf{B}$ is shown in the figure. Electric field $\mathbf{E}$ is also present in the general case, but it only contributes to $\mathbf{B}$ if $\mathbf{E}$ varies with time.

The *Divergence theorem* for a vector field $\mathbf{f}$ passing through volume V bounded by surface S is

$$\int_V (\nabla \cdot \mathbf{f}) \, dV = \int_S \mathbf{f} \cdot d\mathbf{S}. \tag{5.401}$$

In Cartesian (x, y, z) coordinates with unit vectors $\{\hat{\mathbf{x}}, \hat{\mathbf{y}}, \hat{\mathbf{z}}\}$, the *divergence* of $\mathbf{f}$ is

$$\nabla \cdot \mathbf{f} = \frac{\partial f_x}{\partial x} + \frac{\partial f_y}{\partial y} + \frac{\partial f_z}{\partial z}, \tag{5.402}$$

where

$$f_x = \mathbf{f} \cdot \hat{\mathbf{x}},$$
$$f_y = \mathbf{f} \cdot \hat{\mathbf{y}},$$
$$f_z = \mathbf{f} \cdot \hat{\mathbf{z}}.$$

There is also *Stokes' theorem,*[54] which relates to a loop C from which surface S emerges, like a bubble blown from a ring before it detaches:

$$\oint_C \mathbf{f} \cdot d\mathbf{l} = \int_S (\nabla \times \mathbf{f}) \cdot d\mathbf{S}. \tag{5.403}$$

In Cartesian (x, y, z) coordinates with unit vectors $\{\hat{\mathbf{x}}, \hat{\mathbf{y}}, \hat{\mathbf{z}}\}$, the *curl* of $\mathbf{f}$ is

$$\nabla \times \mathbf{f} = \hat{\mathbf{x}} \left(\frac{\partial f_z}{\partial y} - \frac{\partial f_y}{\partial z} \right) + \hat{\mathbf{y}} \left(\frac{\partial f_x}{\partial z} - \frac{\partial f_z}{\partial x} \right) + \hat{\mathbf{z}} \left(\frac{\partial f_y}{\partial x} - \frac{\partial f_x}{\partial y} \right).$$

---

[54]Sir George Stokes (1819–1903).

Note that the gradient, divergence and curl of a vector field $\mathbf{f}$ take slightly different algebraic forms in spherical polar or cylindrical coordinate systems.[55]

We can use the divergence theorem $\int_V (\nabla \cdot \mathbf{f})\, dV = \int_S \mathbf{f} \cdot d\mathbf{S}$ and Stokes' theorem $\oint_C \mathbf{f} \cdot d\mathbf{l} = \int_S (\nabla \times \mathbf{f}) \cdot d\mathbf{S}$ to express Maxwell's four equations in a simpler, differential (rather than integral) form:

$$\nabla \cdot \mathbf{D} = \rho, \tag{5.404}$$

$$\nabla \cdot \mathbf{B} = 0, \tag{5.405}$$

$$\nabla \times \mathbf{E} = -\frac{\partial \mathbf{B}}{\partial t}, \tag{5.406}$$

$$\nabla \times \mathbf{H} = \mathbf{J} + \frac{\partial \mathbf{D}}{\partial t}. \tag{5.407}$$

5.13.4 *Electromagnetic waves in free space*

In free space (i.e. where there is no charge density, polarisation, magnetisation or currents),

$$\mathbf{D} = \varepsilon_0 \mathbf{E}, \tag{5.408}$$

$$\mathbf{B} = \mu_0 \mathbf{H}, \tag{5.409}$$

$$\rho = 0, \tag{5.410}$$

$$\mathbf{J} = 0. \tag{5.411}$$

Therefore, Maxwell's equations reduce to

$$\nabla \cdot \mathbf{E} = 0, \tag{5.412}$$

$$\nabla \cdot \mathbf{B} = 0, \tag{5.413}$$

$$\nabla \times \mathbf{E} = -\frac{\partial \mathbf{B}}{\partial t}, \tag{5.414}$$

$$\nabla \times \mathbf{B} = \varepsilon_0 \mu_0 \frac{\partial \mathbf{E}}{\partial t}. \tag{5.415}$$

We can make good use of the cross-product[56] identity

$$\mathbf{a} \times (\mathbf{b} \times \mathbf{c}) = \mathbf{b}\,(\mathbf{a} \cdot \mathbf{c}) - (\mathbf{a} \cdot \mathbf{b})\,\mathbf{c} \tag{5.422}$$

---

[55] For a fairly complete 'one page print out and keep' summary of vector calculus theorems, see http://www.eclecticon.info/index_htm_files/Calculus%20-%20Calculus%20toolbox%20summary.pdf.

[56] Cross-product identities:

$$\mathbf{a} \times \mathbf{b} = -\mathbf{b} \times \mathbf{a}, \tag{5.416}$$

$$\mathbf{a} \times \mathbf{a} = 0, \tag{5.417}$$

$$(\mathbf{a} \times \mathbf{b}) \cdot (\mathbf{c} \times \mathbf{d}) = (\mathbf{a} \cdot \mathbf{c})(\mathbf{b} \cdot \mathbf{d}) - (\mathbf{a} \cdot \mathbf{d})(\mathbf{b} \cdot \mathbf{c}), \tag{5.418}$$

$$(\mathbf{a} \times \mathbf{b}) \times \mathbf{c} = (\mathbf{a} \cdot \mathbf{c})\,\mathbf{b} - (\mathbf{b} \cdot \mathbf{c})\,\mathbf{a}, \tag{5.419}$$

$$\mathbf{a} \times (\mathbf{b} \times \mathbf{c}) = (\mathbf{a} \cdot \mathbf{c})\,\mathbf{b} - (\mathbf{a} \cdot \mathbf{b})\,\mathbf{c}, \tag{5.420}$$

$$\mathbf{a} \cdot (\mathbf{b} \times \mathbf{c}) = \mathbf{b} \cdot (\mathbf{c} \times \mathbf{a}) = \mathbf{c} \cdot (\mathbf{a} \times \mathbf{b}). \tag{5.421}$$

to write

$$\nabla \times (\nabla \times \mathbf{E}) = \nabla (\nabla \cdot \mathbf{E}) - (\nabla \cdot \nabla) \mathbf{E} = \nabla (\nabla \cdot \mathbf{E}) - \nabla^2 \mathbf{E} \qquad (5.423)$$

and therefore

$$\nabla \times (\nabla \times \mathbf{B}) = \nabla (\nabla \cdot \mathbf{B}) - \nabla^2 \mathbf{B}. \qquad (5.424)$$

Since both $\nabla \cdot \mathbf{E} = 0$ and $\nabla \cdot \mathbf{B} = 0$ in free space,

$$\nabla \times (\nabla \times \mathbf{E}) = -\nabla^2 \mathbf{E}, \qquad (5.425)$$

$$\nabla \times (\nabla \times \mathbf{B}) = -\nabla^2 \mathbf{B}. \qquad (5.426)$$

From Maxwell's equation $\nabla \times \mathbf{E} = -\frac{\partial \mathbf{B}}{\partial t}$ and noting that $\nabla \times \frac{\partial \mathbf{B}}{\partial t} = \frac{\partial}{\partial t} (\nabla \times \mathbf{B})$,

$$\nabla \times (\nabla \times \mathbf{E}) = -\frac{\partial}{\partial t} (\nabla \times \mathbf{B}).$$

Since $\nabla \times \mathbf{B} = \varepsilon_0 \mu_0 \frac{\partial \mathbf{E}}{\partial t}$,

$$\nabla \times (\nabla \times \mathbf{E}) = -\varepsilon_0 \mu_0 \frac{\partial}{\partial t} \left(\frac{\partial \mathbf{E}}{\partial t} \right) \qquad (5.427)$$

$$= -\varepsilon_0 \mu_0 \frac{\partial^2 \mathbf{E}}{\partial t^2}. \qquad (5.428)$$

From Maxwell's equation $\nabla \times \mathbf{B} = \varepsilon_0 \mu_0 \frac{\partial \mathbf{E}}{\partial t}$ and noting that $\nabla \times \frac{\partial \mathbf{E}}{\partial t} = \frac{\partial}{\partial t} (\nabla \times \mathbf{E})$,

$$\nabla \times (\nabla \times \mathbf{B}) = \varepsilon_0 \mu_0 \frac{\partial}{\partial t} (\nabla \times \mathbf{E}).$$

Since $\nabla \times \mathbf{E} = -\frac{\partial \mathbf{B}}{\partial t}$,

$$\nabla \times (\nabla \times \mathbf{B}) = \varepsilon_0 \mu_0 \frac{\partial}{\partial t} \left(-\frac{\partial \mathbf{B}}{\partial t} \right) \qquad (5.429)$$

$$= -\varepsilon_0 \mu_0 \frac{\partial^2 \mathbf{B}}{\partial t^2}. \qquad (5.430)$$

Hence, since $\nabla \times (\nabla \times \mathbf{E}) = -\nabla^2 \mathbf{E}$ and $\nabla \times (\nabla \times \mathbf{B}) = -\nabla^2 \mathbf{B}$, electric fields and magnetic flux density in free space obey the *wave equations*

$$\nabla^2 \mathbf{E} = \frac{1}{c^2} \frac{\partial^2 \mathbf{E}}{\partial t^2}, \qquad (5.431)$$

$$\nabla^2 \mathbf{B} = \frac{1}{c^2} \frac{\partial^2 \mathbf{B}}{\partial t^2} \qquad (5.432)$$

and can therefore be expressed as *electromagnetic waves* of the form

$$\mathbf{E} = \mathbf{E}_0 e^{i(\mathbf{k} \cdot \mathbf{r} - \omega t)}, \qquad (5.433)$$

$$\mathbf{B} = \mathbf{B}_0 e^{i(\mathbf{k} \cdot \mathbf{r} - \omega t)}, \qquad (5.434)$$

which both travel at speed c, where $\frac{1}{c^2} = \varepsilon_0 \mu_0$. This means the speed of EM waves in free space is

$$c = \frac{1}{\sqrt{\mu_0 \varepsilon_0}}. \tag{5.435}$$

Amazingly, the wave speed c doesn't depend on the velocity of the source of the electric and/or magnetic fields relative to the velocity of an observer, only the fundamental constants ε_0 and μ_0. *Therefore, we expect EM waves in free space to travel at the same speed, regardless of our frame of reference.* This important fact underpins the necessary transformations of space and time between observers in relative motion that form the theory of *special relativity*. Since the speed of light c is such a fundamental quantity, it may often make good sense to replace μ_0 with $\mu_0 = \frac{1}{\varepsilon_0 c^2}$ in Maxwell's equations in isotropic media when $\varepsilon = 1$ and $\mu = 1$ (i.e. not polarisable or magnetisable).

M4 states that

$$\oint_C \mathbf{H} \cdot d\mathbf{l} = \int_S \mathbf{J} \cdot d\mathbf{S} + \frac{\partial}{\partial t} \int_S \mathbf{D} \cdot d\mathbf{S}. \tag{5.436}$$

So, if $\mu_0 = \frac{1}{\varepsilon_0 c^2}$, $\mathbf{B} = \mu_0 \mathbf{H}$ and $\mathbf{D} = \varepsilon_0 \mathbf{E}$,

$$\oint_C \frac{1}{\mu_0} \mathbf{B} \cdot d\mathbf{l} = \int_S \mathbf{J} \cdot d\mathbf{S} + \frac{\partial}{\partial t} \int_S \varepsilon_0 \mathbf{E} \cdot d\mathbf{S}. \tag{5.437}$$

$$\therefore \oint_C \mathbf{B} \cdot d\mathbf{l} = \mu_0 \int_S \mathbf{J} \cdot d\mathbf{S} + \mu_0 \varepsilon_0 \frac{\partial}{\partial t} \int_S \mathbf{E} \cdot d\mathbf{S}. \tag{5.438}$$

$$\therefore \oint_C \mathbf{B} \cdot d\mathbf{l} = \mu_0 \int_S \mathbf{J} \cdot d\mathbf{S} + \frac{1}{c^2} \frac{\partial}{\partial t} \int_S \mathbf{E} \cdot d\mathbf{S}. \tag{5.439}$$

Returning to the wave formulations of $\mathbf{E}$ and $\mathbf{B}$, we can easily calculate their *divergence* given their $e^{i(\mathbf{k} \cdot \mathbf{r} - \omega t)}$ dependency and the result $\nabla \left(e^{i(\mathbf{k} \cdot \mathbf{r} - \omega t)} \right) = i\mathbf{k} e^{i(\mathbf{k} \cdot \mathbf{r} - \omega t)}$:

$$\nabla \cdot \mathbf{E} = i\mathbf{k} \cdot \mathbf{E}, \tag{5.440}$$

$$\nabla \cdot \mathbf{B} = i\mathbf{k} \cdot \mathbf{B}. \tag{5.441}$$

Since $\nabla \cdot \mathbf{E} = 0$ and $\nabla \cdot \mathbf{B} = 0$ in free space, this implies wavevector $\mathbf{k}$ is perpendicular to *both* $\mathbf{E}$ and $\mathbf{B}$, so $\mathbf{E}$ and $\mathbf{B}$ must be mutually perpendicular *transverse waves*.

Now, using the identity

$$\nabla \times (\phi \mathbf{F}) = \nabla \phi \times \mathbf{F} + \phi \nabla \times \mathbf{F},$$

where ϕ is a scalar-valued function (e.g. $\phi = e^{i(\mathbf{k} \cdot \mathbf{r} - \omega t)}$) and $\mathbf{F}$ is a vector field (e.g. $\mathbf{E}$ or $\mathbf{B}$), the plane wave formulations $\mathbf{E} = \mathbf{E}_0 e^{i(\mathbf{k} \cdot \mathbf{r} - \omega t)}$ and

$\mathbf{B} = \mathbf{B}_0 e^{i(\mathbf{k} \cdot \mathbf{r} - \omega t)}$ imply

$$\nabla \times \mathbf{E} = \nabla \left(e^{i(\mathbf{k} \cdot \mathbf{r} - \omega t)} \right) \times \mathbf{E}_0 + e^{i(\mathbf{k} \cdot \mathbf{r} - \omega t)} \nabla \times \mathbf{E}_0,$$

$$\nabla \times \mathbf{B} = \nabla \left(e^{i(\mathbf{k} \cdot \mathbf{r} - \omega t)} \right) \times \mathbf{B}_0 + e^{i(\mathbf{k} \cdot \mathbf{r} - \omega t)} \nabla \times \mathbf{B}_0.$$

Now, $\nabla \times \mathbf{E}_0 = 0$ since $\mathbf{E}_0$ is a constant vector, i.e. it doesn't vary spatially. The same is true for $\mathbf{B}_0$. Therefore, since $\nabla \left(e^{i(\mathbf{k} \cdot \mathbf{r} - \omega t)} \right) = i\mathbf{k} e^{i(\mathbf{k} \cdot \mathbf{r} - \omega t)}$,

$$\nabla \times \mathbf{E} = i\mathbf{k} \times \mathbf{E}, \tag{5.442}$$

$$\nabla \times \mathbf{B} = i\mathbf{k} \times \mathbf{B}. \tag{5.443}$$

We can also compute the time derivatives:

$$\frac{\partial \mathbf{B}}{\partial t} = -i\omega \mathbf{B}, \tag{5.444}$$

$$\frac{\partial \mathbf{E}}{\partial t} = -i\omega \mathbf{E}, \tag{5.445}$$

given the $e^{i(\mathbf{k} \cdot \mathbf{r} - \omega t)}$ dependence of $\mathbf{E}$ and $\mathbf{B}$.

In free space, Maxwell's equations **M3** and **M4** are

$$\nabla \times \mathbf{E} = -\frac{\partial \mathbf{B}}{\partial t}, \tag{5.446}$$

$$\nabla \times \mathbf{B} = \varepsilon_0 \mu_0 \frac{\partial \mathbf{E}}{\partial t}. \tag{5.447}$$

Substituting $\nabla \times \mathbf{E} = i\mathbf{k} \times \mathbf{E}$, $\nabla \times \mathbf{B} = i\mathbf{k} \times \mathbf{B}$, $\frac{\partial \mathbf{B}}{\partial t} = -i\omega \mathbf{B}$ and $\frac{\partial \mathbf{E}}{\partial t} = -i\omega \mathbf{E}$ yields

$$\mathbf{k} \times \mathbf{E} = \frac{\nabla \times \mathbf{E}}{i} = -i \left(-\frac{\partial \mathbf{B}}{\partial t} \right) \tag{5.448}$$

$$= -i \left(i\omega \mathbf{B} \right) \tag{5.449}$$

$$= \omega \mathbf{B}, \tag{5.450}$$

$$\mathbf{k} \times \mathbf{B} = \frac{\nabla \times \mathbf{B}}{i} \tag{5.451}$$

$$= -i \left(\varepsilon_0 \mu_0 \frac{\partial \mathbf{E}}{\partial t} \right) \tag{5.452}$$

$$= -i \left(-i \varepsilon_0 \mu_0 \omega \mathbf{E} \right) \tag{5.453}$$

$$= -\varepsilon_0 \mu_0 \omega \mathbf{E} \tag{5.454}$$

$$= -\frac{\omega}{c^2} \mathbf{E}. \tag{5.455}$$

So, $\mathbf{k}$, $\mathbf{E}$ and $\mathbf{B}$ must all be *mutually perpendicular*, as illustrated in Fig. 5.35. EM waves in free space need only be specified by a wavevector $\mathbf{k} = \frac{2\pi}{\lambda} \hat{\mathbf{k}}$, where $\hat{\mathbf{k}}$ is a unit vector in the direction of wave propagation, and electric field $\mathbf{E}$. Since

$\omega = ck$, the magnetic flux density $\mathbf{B}$ is

$$\mathbf{B} = \frac{\mathbf{k} \times \mathbf{E}}{ck}. \tag{5.456}$$

$$\therefore \mathbf{B} = \hat{\mathbf{k}} \times \tfrac{1}{c}\mathbf{E}. \tag{5.457}$$

Note that $\hat{\mathbf{k}}$, $\tfrac{1}{c}\mathbf{E}$ and $\mathbf{B}$ almost (!) form what is known as a 'right-handed set' of mutually perpendicular vectors. Cartesian unit vectors $\{\hat{\mathbf{x}}, \hat{\mathbf{y}}, \hat{\mathbf{z}}\}$ are a right-handed set since $\hat{\mathbf{x}} \times \hat{\mathbf{y}} = \hat{\mathbf{z}}$, $\hat{\mathbf{y}} \times \hat{\mathbf{z}} = \hat{\mathbf{x}}$ and $\hat{\mathbf{z}} \times \hat{\mathbf{x}} = \hat{\mathbf{y}}$, i.e. the cross-products cyclically permute. Now, $\mathbf{B} = \hat{\mathbf{k}} \times \tfrac{1}{c}\mathbf{E}$ and, since $\mathbf{k} \times \mathbf{B} = -\tfrac{\omega}{c^2}\mathbf{E}$ and $\omega = ck$, this implies $\mathbf{B} \times \hat{\mathbf{k}} = \tfrac{1}{c}\mathbf{E}$. Unfortunately, $\tfrac{1}{c}\mathbf{E} \times \mathbf{B} \neq \hat{\mathbf{k}}$, although the result is indeed parallel to $\hat{\mathbf{k}}$. The quantity $\mathbf{E} \times \mathbf{B}$ is related to the *Poynting flux* $\mathbf{N} = \mathbf{E} \times \mathbf{H}$, with the energy flux (i.e. power per unit area) of the EM radiation leaving a closed surface S that encloses volume V. In our free-space scenario, $\mathbf{H} = \tfrac{1}{\mu_0}\mathbf{B}$, so this means $\mathbf{E} \times \mathbf{B} = \mu_0 \mathbf{N}$. The Poynting flux formula $\mathbf{N} = \mathbf{E} \times \mathbf{H}$ is derived at the end of this chapter.

Since $\tfrac{1}{c}\mathbf{E} = \mathbf{B} \times \hat{\mathbf{k}}$, $\tfrac{1}{c}\mathbf{E} \times \mathbf{B} = \left(\mathbf{B} \times \hat{\mathbf{k}}\right) \times \mathbf{B}$. Using one of the cross-product identities defined earlier in this section,

$$\left(\mathbf{B} \times \hat{\mathbf{k}}\right) \times \mathbf{B} = (\mathbf{B} \cdot \mathbf{B})\hat{\mathbf{k}} - \left(\hat{\mathbf{k}} \cdot \mathbf{B}\right)\mathbf{B}.$$

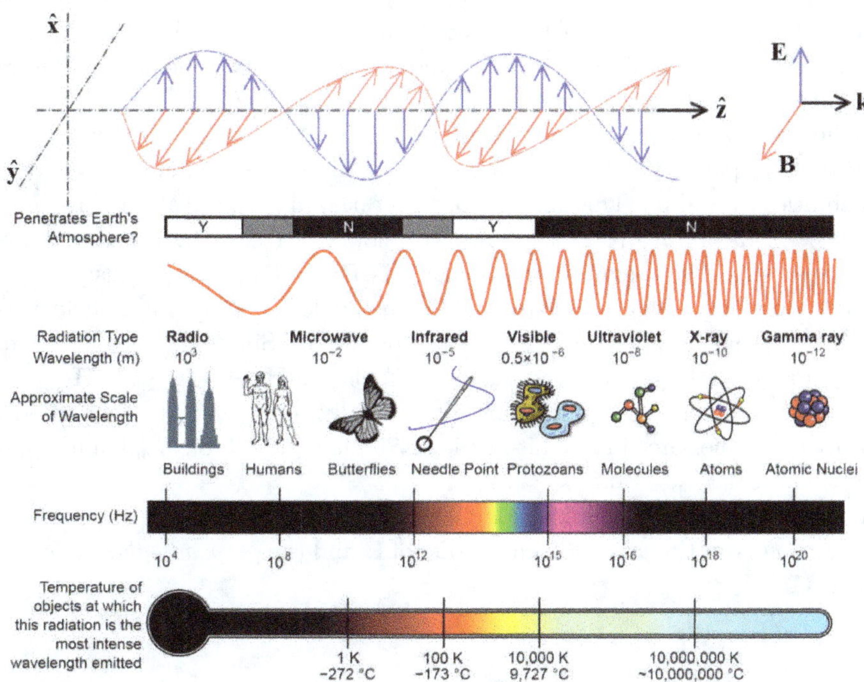

Fig. 5.35. Mutually perpendicular electric field $\mathbf{E}$, magnetic flux density $\mathbf{B}$ and wavevector $\mathbf{k}$ of an electromagnetic wave. The *electromagnetic spectrum* describes the range of wavelengths and frequencies. (Wikipedia, https://en.wikipedia.org/wiki/Electromagnetic_spectrum#/media/File:EM_Spectrum_Properties_edit.svg).

Since $\hat{\mathbf{k}}$ is perpendicular to $\mathbf{B}$, $\hat{\mathbf{k}} \cdot \mathbf{B} = \mathbf{0}$.

$$\therefore \tfrac{1}{c}\mathbf{E} \times \mathbf{B} = |\mathbf{B}|^2\,\hat{\mathbf{k}}.$$

Note that we could substitute $\mathbf{B} = \hat{\mathbf{k}} \times \tfrac{1}{c}\mathbf{E}$ instead and write

$$\tfrac{1}{c}\mathbf{E} \times \mathbf{B} = \tfrac{1}{c}\mathbf{E} \times \left(\hat{\mathbf{k}} \times \tfrac{1}{c}\mathbf{E}\right)$$

$$= \hat{\mathbf{k}}\tfrac{1}{c^2}(\mathbf{E} \cdot \mathbf{E}) - \left(\tfrac{1}{c}\mathbf{E} \cdot \hat{\mathbf{k}}\right)\tfrac{1}{c}\mathbf{E}$$

$$= \tfrac{1}{c^2}|\mathbf{E}|^2\,\hat{\mathbf{k}}.$$

5.13.5   *Continuity conditions for* $\mathbf{B}$ *and* $\mathbf{E}$

In the sections on the *toroidal inductor* and the *Fresnel equations*, it was asserted that *Maxwell's equations* tell us:

'At an interface between media of differing refractive index, vector components of total magnetic flux density $\mathbf{B}$ perpendicular to the interface surface must be *continuous* across the boundary. Also, components of the $\mathbf{E}$ and $\mathbf{H}$ fields which are *parallel* to the surface must be continuous across the boundary, as long as there is no surface charge and surface current.' Let's use Maxwell's equations to prove these results. The 'shrinking pill-box argument' that follows is adapted from Longair [22, pp. 126–135] and Bleaney & Bleaney [2, pp. 19 and 111].

Consider a pillbox volume at the boundary between two media, as illustrated in Fig. 5.36. The pillbox is deemed to be microscopically small,[57] so at its location we might assume a local isotropic approximation and characterise the media as having relative permittivities ε_1 and ε_2 either side of the boundary and relative permeabilities μ_1 and μ_2. This is the approach taken in the derivation of the Fresnel equations. For a non-isotropic medium, you could model it by allowing ε and μ to vary spatially. The magnetic flux density is $\mathbf{B}_1 = \mathbf{B}_{1,\|} + \mathbf{B}_{1,\perp}$ just above the interface boundary and $\mathbf{B}_2 = \mathbf{B}_{2,\|} + \mathbf{B}_{2,\perp}$ just below, i.e. a sum of field vectors parallel to ($\|$) and perpendicular to ($\perp$) the boundary. Similarly, the electric fields are $\mathbf{E}_{1,2} = \mathbf{E}_{1,2,\|} + \mathbf{E}_{1,2,\perp}$ and the magnetic fields are $\mathbf{H}_{1,2} = \mathbf{H}_{1,2,\|} + \mathbf{H}_{1,2,\perp}$. The top and bottom areas of the pillbox are differentially small, and the top and bottom *vector areas* are the same in magnitude, i.e. $dS = |d\mathbf{S}_{\text{top}}| = |d\mathbf{S}_{\text{bottom}}|$, and opposite in direction, i.e. $d\mathbf{S}_{\text{top}} = -d\mathbf{S}_{\text{bottom}}$.

Note that this idea of shrinking the height of the pillbox provides no constraints on the continuity of the *parallel* components of $\mathbf{B}$ and the *perpendicular* components of $\mathbf{E}$ and $\mathbf{H}$.

---

[57] In fact, we are going to take the limit that the pillbox height (and hence the total volume) tends to zero.

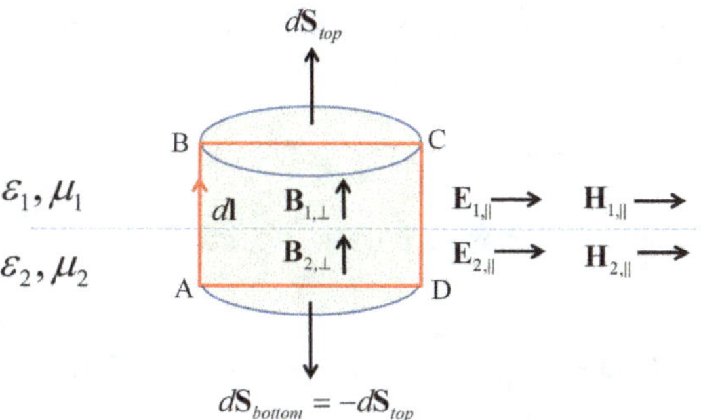

Fig. 5.36. A microscopic pillbox volume straddles the interface between two media of relative permittivities ε_1 and ε_2 either side of the boundary and relative permeabilities μ_1 and μ_2. This can be analysed using Maxwell's equations to prove that 'at an interface between media, vector components of total magnetic flux density **B** perpendicular ($\perp$) to the interface surface must be *continuous* across the boundary. Also, the components of the **E** and **H** fields which are *parallel* ($\parallel$) to the surface must be continuous across the boundary'.

5.13.5.1 *Continuity of* $\mathbf{B}_\perp$

M2 states the net flux of **B** through a closed surface S is zero, i.e. $\int_S \mathbf{B} \cdot d\mathbf{S} = 0$. If we take the limit where the pillbox height $\to 0$, then the distances AB and CD in Fig. 5.36 tend to zero, as do the areas of the sides of the pillbox. This is a valid thing to do since we are interested in the electric and magnetic fields at the boundary, and this limiting process allows us to approach this condition. So, as we get closer to the boundary by shrinking our pillbox,

$$\int_S \mathbf{B} \cdot d\mathbf{S} \to \mathbf{B}_{1,\perp} \cdot d\mathbf{S}_{\text{top}} + \mathbf{B}_{2,\perp} \cdot d\mathbf{S}_{\text{bottom}}. \tag{5.458}$$

Since the top and bottom pillbox vector areas are related by

$$d\mathbf{S}_{\text{top}} = -d\mathbf{S}_{\text{bottom}}, \tag{5.459}$$

then

$$\int_S \mathbf{B} \cdot d\mathbf{S} = (\mathbf{B}_{1,\perp} - \mathbf{B}_{2,\perp}) \cdot d\mathbf{S}_{\text{top}}. \tag{5.460}$$

Since $\int_S \mathbf{B} \cdot d\mathbf{S} = 0$ (Maxwell Equation **M2**),

$$\therefore \mathbf{B}_{1,\perp} = \mathbf{B}_{2,\perp}, \tag{5.461}$$

i.e. the vector components of total magnetic flux density **B** perpendicular to the interface surface must be *continuous* across the boundary.

5.13.5.2 *Continuity of* $\mathbf{D}_\perp$

The volume of the pillbox shrinks to zero in our boundary-approaching limit, so **M1**, i.e. $\int_S \mathbf{D} \cdot d\mathbf{S} = \int_V \rho \, dV$, implies that

$$\int_S \mathbf{D} \cdot d\mathbf{S} \to \sigma dS, \tag{5.462}$$

where σ is the local surface charge density (i.e. in units of coulombs/m^2). Since the sides of the pillbox shrink to zero area,

$$\int_S \mathbf{D} \cdot d\mathbf{S} \to (\mathbf{D}_{1,\perp} - \mathbf{D}_{2,\perp}) \cdot d\mathbf{S}_{\text{top}}. \tag{5.463}$$

If we define a unit normal vector $\hat{\mathbf{n}}$ parallel to $d\mathbf{S}_{\text{top}}$, this means $d\mathbf{S}_{\text{top}} = \hat{\mathbf{n}} dS$. Hence,

$$\int_S \mathbf{D} \cdot d\mathbf{S} = \sigma dS = (\mathbf{D}_{1,\perp} - \mathbf{D}_{2,\perp}) \cdot \hat{\mathbf{n}} dS. \tag{5.464}$$

Therefore,

$$(\mathbf{D}_{1,\perp} - \mathbf{D}_{2,\perp}) \cdot \hat{\mathbf{n}} = \sigma. \tag{5.465}$$

If there is no surface charge density, then $\sigma = 0$ and therefore

$$\mathbf{D}_{1,\perp} = \mathbf{D}_{2,\perp}, \tag{5.466}$$

i.e. the perpendicular components of $\mathbf{D}$ are continuous across a boundary, *as long as there is no surface charge density.*

5.13.5.3   *Continuity of* $\mathbf{E}_{\parallel}$

Now, consider the rectangular loop ABCD around our pillbox on the plane of the illustration in Fig. 5.36. **M3** states that $\oint_C \mathbf{E} \cdot d\mathbf{l} = -\frac{\partial}{\partial t} \int_S \mathbf{B} \cdot d\mathbf{S}$. In our limit of zero-height pillbox, the vector area of this loop must $\to 0$. Note that $d\mathbf{S}$, in this case, is perpendicular to the plane of the paper and points *into* the page (i.e. a right-handed rule with respect to the clockwise loop direction). $d\mathbf{S} \to 0$ implies $-\frac{\partial}{\partial t} \int_S \mathbf{B} \cdot d\mathbf{S} \to 0$. This means

$$\oint_C \mathbf{E} \cdot d\mathbf{l} = \mathbf{E}_{1,\parallel} \cdot \overrightarrow{BC} + \mathbf{E}_{2,\parallel} \cdot \overrightarrow{DA} = 0. \tag{5.467}$$

Since $\overrightarrow{BC} = -\overrightarrow{DA}$,

$$(\mathbf{E}_{1,\parallel} - \mathbf{E}_{2,\parallel}) \cdot \overrightarrow{BC} = 0, \tag{5.468}$$

which implies

$$\mathbf{E}_{1,\parallel} = \mathbf{E}_{2,\parallel}. \tag{5.469}$$

So, the components of the $\mathbf{E}$ fields which are *parallel* to the surface must be continuous across the boundary.

5.13.5.4   *Continuity of* $\mathbf{H}_{\parallel}$

Again, consider the rectangular loop ABCD around our pillbox on the plane of the illustration. **M4** states that $\oint_C \mathbf{H} \cdot d\mathbf{l} = \int_S \mathbf{J} \cdot d\mathbf{S} + \frac{\partial}{\partial t} \int_S \mathbf{D} \cdot d\mathbf{S}$. As we shrink our pillbox, the loop area $\to 0$, so the second term $\frac{\partial}{\partial t} \int_S \mathbf{D} \cdot d\mathbf{S} \to 0$. If there are any surface currents at the interface of media 1 and 2, then we might write $\int_S \mathbf{J} \cdot d\mathbf{S} = \hat{\mathbf{N}} \cdot \mathbf{J}_s \left| \overrightarrow{BC} \right|$, where $\hat{\mathbf{N}}$ is a unit normal parallel to $d\mathbf{S}$ (and into the plane of the page) and $\mathbf{J}_s$ is the current *per unit length* flowing along the interface. So, as our pillbox shrinks, **M4** becomes

$$\oint_C \mathbf{H} \cdot d\mathbf{l} \to \mathbf{H}_{1,\parallel} \cdot \overrightarrow{BC} + \mathbf{H}_{2,\parallel} \cdot \overrightarrow{DA} = \hat{\mathbf{N}} \cdot \mathbf{J}_s \left| \overrightarrow{BC} \right|. \tag{5.470}$$

Since $\overrightarrow{BC} = -\overrightarrow{DA}$,

$$(\mathbf{H}_{1,\parallel} - \mathbf{H}_{2,\parallel}) \cdot \overrightarrow{BC} = \hat{\mathbf{N}} \cdot \mathbf{J}_s \left| \overrightarrow{BC} \right|. \tag{5.471}$$

Now, since vectors $\hat{\mathbf{n}}$ (upwards on the plane of the page), $\hat{\mathbf{N}}$ (into the plane of the page) and $\overrightarrow{BC}$ (left to right on the plane of the page) are all mutually orthogonal:

$$\hat{\mathbf{N}} \times \hat{\mathbf{n}} = \frac{\overrightarrow{BC}}{\left| \overrightarrow{BC} \right|}. \tag{5.472}$$

Therefore,

$$(\mathbf{H}_{1,\parallel} - \mathbf{H}_{2,\parallel}) \cdot \left(\hat{\mathbf{N}} \times \hat{\mathbf{n}} \right) \left| \overrightarrow{BC} \right| = \hat{\mathbf{N}} \cdot \mathbf{J}_s \left| \overrightarrow{BC} \right|, \tag{5.473}$$

$$(\mathbf{H}_{1,\parallel} - \mathbf{H}_{2,\parallel}) \cdot \left(\hat{\mathbf{N}} \times \hat{\mathbf{n}} \right) = \hat{\mathbf{N}} \cdot \mathbf{J}_s. \tag{5.474}$$

The vector triple products of the form $\mathbf{a} \cdot (\mathbf{b} \times \mathbf{c})$ remain the same after a circular permutation of vectors $\mathbf{a}, \mathbf{b}, \mathbf{c}$. i.e.

$$\mathbf{a} \cdot (\mathbf{b} \times \mathbf{c}) = \mathbf{b} \cdot (\mathbf{c} \times \mathbf{a}) = \mathbf{c} \cdot (\mathbf{a} \times \mathbf{b}). \tag{5.475}$$

This means we can write

$$(\mathbf{H}_{1,\parallel} - \mathbf{H}_{2,\parallel}) \cdot \left(\hat{\mathbf{N}} \times \hat{\mathbf{n}} \right) = \hat{\mathbf{N}} \cdot (\hat{\mathbf{n}} \times (\mathbf{H}_{1,\parallel} - \mathbf{H}_{2,\parallel})). \tag{5.476}$$

Since $(\mathbf{H}_{1,\parallel} - \mathbf{H}_{2,\parallel}) \cdot (\hat{\mathbf{N}} \times \hat{\mathbf{n}}) = \hat{\mathbf{N}} \cdot \mathbf{J}s$,

$$\hat{\mathbf{N}} \cdot (\hat{\mathbf{n}} \times (\mathbf{H}_{1,\parallel} - \mathbf{H}_{2,\parallel})) = \hat{\mathbf{N}} \cdot \mathbf{J}_s. \tag{5.477}$$

Therefore,

$$\hat{\mathbf{n}} \times (\mathbf{H}_{1,\parallel} - \mathbf{H}_{2,\parallel}) = \mathbf{J}_s. \tag{5.478}$$

If there are no surface currents $\mathbf{J}_s = \mathbf{0}$, which then implies

$$\mathbf{H}_{1,\parallel} = \mathbf{H}_{2,\parallel}. \tag{5.479}$$

In summary, the components of the $\mathbf{H}$ fields which are *parallel* to the surface must be continuous across the boundary, *as long as there are no surface currents.*

5.13.6 *Energy and energy flow associated with electric and magnetic fields*

The goal of this section is to derive *Poynting's theorem*. The energy flux (i.e. power per unit area) of EM radiation, defined by electric field $\mathbf{E}$ and magnetic field $\mathbf{H}$ is

$$\mathbf{N} = \mathbf{E} \times \mathbf{H}.$$

We'll do it for the special case of *isotropic media*, although the result holds in general.[58]

Define energy density u to be the energy per unit volume associated with the electric and magnetic fields. The total energy U in the EM fields with volume V is

$$U = \int_V u\, dV. \tag{5.480}$$

Define $\mathbf{N}$ as the energy flux of EM radiation *leaving* a closed surface S that encloses volume V. The rate of change of internal energy U within V (a negative quantity) is the surface integral of $\mathbf{N}$ plus the work done on charges within V by the fields within V:

$$-\frac{\partial U}{\partial t} = -\frac{\partial}{\partial t}\int_V u\, dV = \int_S \mathbf{N} \cdot d\mathbf{S} + \int_V (\mathbf{f} \cdot \mathbf{v})\, dV. \tag{5.481}$$

Define $\mathbf{f}$ to be the force per unit volume so that the rate of work done per unit volume of the EM field space is

$$\mathbf{f} \cdot \mathbf{v} = \rho\,(\mathbf{E} + \mathbf{v} \times \mathbf{B}) \cdot \mathbf{v}, \tag{5.482}$$

where $\mathbf{v}$ is the velocity of charges of local charge density ρ. The vector triple product can be permuted, and since $\mathbf{v} \times \mathbf{v} = \mathbf{0}$,

$$(\mathbf{v} \times \mathbf{B}) \cdot \mathbf{v} = (\mathbf{v} \times \mathbf{v}) \cdot \mathbf{B} = 0. \tag{5.483}$$

Therefore,

$$\mathbf{f} \cdot \mathbf{v} = (\mathbf{E} \cdot \rho\mathbf{v}) = \mathbf{E} \cdot \mathbf{J}, \tag{5.484}$$

where $\mathbf{J}$ is the current per unit volume. Now, we can rewrite $\int_S \mathbf{N} \cdot d\mathbf{S} = \int_V (\nabla \cdot \mathbf{N})\, dV$ using the *divergence theorem*. Hence,

$$-\frac{\partial}{\partial t}\int_V u\, dV = \int_V (\nabla \cdot \mathbf{N})\, dV + \int_V (\mathbf{E} \cdot \mathbf{J})\, dV, \tag{5.485}$$

i.e.

$$-\frac{\partial u}{\partial t} = \nabla \cdot \mathbf{N} + \mathbf{E} \cdot \mathbf{J}. \tag{5.486}$$

---

[58] For a derivation of Poynting's theorem in non-isotropic media, see (for example) https://phys.libretexts.org/Bookshelves/Electricity_and_Magnetism/Electromagnetics_II_(Ellingson)/03%3A_Wave_Propagation_in_General_Media/3.01%3A_Poynting%E2%80%99s_Theorem (Accessed 18 June 2024).

Now, it can be shown that (e.g. Jackson [18] or Bleaney & Bleaney [2]) the energy density associated with electric and magnetic fields is

$$u = \tfrac{1}{2} \left(\mathbf{E} \cdot \mathbf{D} + \mathbf{B} \cdot \mathbf{H} \right). \tag{5.487}$$

If a medium is isotropic, $\mathbf{D} = \varepsilon\varepsilon_0 \mathbf{E}$ and $\mathbf{B} = \mu\mu_0 \mathbf{H}$.

$$\therefore u = \tfrac{1}{2} \left(\varepsilon\varepsilon_0 \mathbf{E} \cdot \mathbf{E} + \mu\mu_0 \mathbf{H} \cdot \mathbf{H} \right), \tag{5.488}$$

which means the time derivative of u is

$$\frac{\partial u}{\partial t} = \varepsilon\varepsilon_0 \mathbf{E} \cdot \frac{\partial \mathbf{E}}{\partial t} + \mu\mu_0 \mathbf{H} \cdot \frac{\partial \mathbf{H}}{\partial t}. \tag{5.489}$$

Maxwell's equations (**M3** and **M4**) for isotropic media imply

$$\nabla \times \mathbf{E} = -\mu\mu_0 \frac{\partial \mathbf{H}}{\partial t}, \tag{5.490}$$

$$\nabla \times \mathbf{H} = \mathbf{J} + \varepsilon\varepsilon_0 \frac{\partial \mathbf{E}}{\partial t}. \tag{5.491}$$

Therefore,

$$-\left(\nabla \times \mathbf{E} \right) \cdot \mathbf{H} = \mu\mu_0 \frac{\partial \mathbf{H}}{\partial t} \cdot \mathbf{H}, \tag{5.492}$$

$$\left(\nabla \times \mathbf{H} \right) \cdot \mathbf{E} - \mathbf{J} \cdot \mathbf{E} = \varepsilon\varepsilon_0 \mathbf{E} \cdot \frac{\partial \mathbf{E}}{\partial t}. \tag{5.493}$$

Hence, since $\frac{\partial u}{\partial t} = \varepsilon\varepsilon_0 \mathbf{E} \cdot \frac{\partial \mathbf{E}}{\partial t} + \mu\mu_0 \mathbf{H} \cdot \frac{\partial \mathbf{H}}{\partial t}$,

$$\frac{\partial u}{\partial t} = \left(\nabla \times \mathbf{H} \right) \cdot \mathbf{E} - \mathbf{J} \cdot \mathbf{E} - \left(\nabla \times \mathbf{E} \right) \cdot \mathbf{H}. \tag{5.494}$$

Now, the divergence of a cross-product[59] can be expanded as follows:

$$\nabla \cdot \left(\mathbf{E} \times \mathbf{H} \right) = \left(\nabla \times \mathbf{E} \right) \cdot \mathbf{H} - \left(\nabla \times \mathbf{H} \right) \cdot \mathbf{E}. \tag{5.495}$$

Therefore,

$$-\frac{\partial u}{\partial t} = \nabla \cdot \left(\mathbf{E} \times \mathbf{H} \right) + \mathbf{J} \cdot \mathbf{E}. \tag{5.496}$$

Equating this with $-\frac{\partial u}{\partial t} = \nabla \cdot \mathbf{N} + \mathbf{J} \cdot \mathbf{E}$,

$$\nabla \cdot \left(\mathbf{E} \times \mathbf{H} \right) + \mathbf{J} \cdot \mathbf{E} = \nabla \cdot \mathbf{N} + \mathbf{J} \cdot \mathbf{E}, \tag{5.497}$$

which implies

$$\mathbf{N} = \mathbf{E} \times \mathbf{H}. \tag{5.498}$$

This is called *Poynting's theorem*[60] and is a useful mechanism for calculating the power radiated in the form of EM waves from dipoles and other, more sophisticated antennae. It can also be used to calculate *radiation pressure p_{rad}* from the equation

$$p_{\text{rad}} = \frac{|\mathbf{N}|}{c}. \tag{5.499}$$

---

[59] http://www.eclecticon.info/index_htm_files/Calculus%20-%20Calculus%20toolbox%20summary.pdf.

[60] John Henry Poynting (1852–1914).

The strange idea that *light carries momentum* (since pressure is the rate of change of momentum per unit area) *but without mass* shall be explored in both the chapters on *special relativity* and *quantum mechanics* in the third volume of *Science by Simulation*.

5.14   A Selection of Electromagnetism Books

There are a vast number of books on electricity and magnetism. Here are a small selection that I have used over the years, and many have already been referenced directly in this chapter. My standard textbooks are the two volumes of *Electricity and Magnetism* by Bleaney & Bleaney [2] and, indeed, Jackson's *Classical Electrodynamics* [18] and *The Feynman Lectures on Physics* [6]. Longair's *Theoretical Concepts in Physics* [22] has a whole chapter on electromagnetism and its historical development. I own a copy of Tooley's *Electronic Circuits, Fundamentals and Applications* [35] and badly need to obtain a copy of Hammond's *Electromagnetism for Engineers* [14]. For physics undergraduates reading this, it is always a good idea to look at engineering standard texts as well as those in the physics section! Fleisch's *A Student's Guide to Maxwell's Equations* [8] is a superb primer (and mercifully slim). Read this before delving into the standard textbooks. Prior to teaching, my professional association with electromagnetism was as a radar systems engineer. So, I would certainly recommend Skolnik's *Introduction to Radar Systems* [34] and Richard's *Fundamentals of Radar Signal Processing* [31]. For the behaviour of electrons and condensed matter physics, my standard textbook is Kittel's *Introduction to Solid State Physics* [20], and a must-read is Feynman's *QED: The Strange Theory of Light and Matter* [7]. In terms of online resources, there is a huge amount that can be Googled, but I would personally start with R. Nave's amazing *Hyperphysics* (http://hyperphysics.phy-astr.gsu.edu/hbase/hframe.html). Much of the material presented here forms the basis of notes on my website http://www.eclecticon.info/physics_notes_em_elec.htm, which includes downloads of the electric fields simulations, Ising models and much more. I also have quite a significant number of videos of electromagnetism laboratory demonstrations at http://www.eclecticon.info/physics_notes_em_elec_videos.htm. I would highly recommend the online simulations at https://phet.colorado.edu/, especially the circuit design tools. For problems at the pre-university level, see http://www.eclecticon.info/physics_questions.htm, *Mastering Pre-University Physics* [23], *Isaac Physics* at https://isaacphysics.org/ and the website of the British Physics Olympiad https://www.bpho.org.uk/.

$$\text{Appendix A}$$

The Calculus of Variations

A large number of problems in applied mathematics involve finding the shape of a curve, $x(t)$, such that an integral, S, involving x (and possibly $\dot{x} = \frac{dx}{dt}$ and t also), is maximised (or minimised) with respect to small variations in x. Note that the variable x does not have to be displacement and variable t does not have to be time—the *calculus of variations* can be applied to a wider variety of systems involving parameterised curves than those associated with kinematics. To generalise, let the integrand L of integral S be a function of a family of variables, $\{x_i\}$, and associated derivatives:

$$S = \int_{t_a}^{t_b} L\left(x_1, x_2, \ldots, x_N, \dot{x}_1, \dot{x}_2, \ldots, \dot{x}_N, t\right). \tag{A.1}$$

Consider a small perturbation ε such that $X_i = x_i + \varepsilon \eta_i$, and $\eta_i(t_a) = \eta_i(t_b) = 0$. Define

$$S_\varepsilon = \int_{t_a}^{t_b} L\left(X_1, X_2, \ldots, X_N, \dot{X}_1, \dot{X}_2, \ldots, \dot{X}_N, t\right). \tag{A.2}$$

The goal is to find an equation for $x_i(t)$ such that $\frac{dS_\varepsilon}{d\varepsilon}\big|_{\varepsilon=0} = 0$, which is another way of saying S is *extremised* with respect to these perturbations in x_i since, when $\varepsilon = 0$, $X_i = x_i$ and therefore $S_\varepsilon = S$. The derivative $\frac{dS_\varepsilon}{d\varepsilon}$ can be written as

$$\frac{dS_\varepsilon}{d\varepsilon} = \int_{t_a}^{t_b} \frac{dL}{d\varepsilon} dt. \tag{A.3}$$

By the chain rule,

$$\frac{dL}{d\varepsilon} = \frac{\partial L}{\partial t} \frac{dt}{d\varepsilon} + \sum_{i=1}^{N} \left(\frac{\partial L}{\partial X_i} \frac{dX_i}{d\varepsilon} + \frac{\partial L}{\partial \dot{X}_i} \frac{d\dot{X}_i}{d\varepsilon} \right), \tag{A.4}$$

and since $X_i = x_i + \varepsilon \eta_i$ and $\dot{X}_i = \dot{x}_i + \varepsilon \dot{\eta}_i$,

$$\frac{dt}{d\varepsilon} = 0, \tag{A.5}$$

$$\frac{dX_i}{d\varepsilon} = \eta_i, \tag{A.6}$$

$$\frac{d\dot{X}_i}{d\varepsilon} = \dot{\eta}_i. \tag{A.7}$$

Hence,

$$\frac{dS_\varepsilon}{d\varepsilon} = \int_{t_a}^{t_b} \sum_{i=1}^{N} \left(\frac{\partial L}{\partial X_i} \eta_i + \frac{\partial L}{\partial \dot{X}_i} \dot{\eta}_i \right) dt \tag{A.8}$$

$$= \sum_{i=1}^{N} \left\{ \int_{t_a}^{t_b} \frac{\partial L}{\partial X_i} \eta_i dt + \int_{t_a}^{t_b} \frac{\partial L}{\partial \dot{X}_i} \dot{\eta}_i dt \right\}. \tag{A.9}$$

Using integration by parts,

$$\int_{t_a}^{t_b} \frac{\partial L}{\partial \dot{X}_i} \dot{\eta}_i dt = \left[\eta_i \frac{\partial L}{\partial \dot{X}_i} \right]_{t_a}^{t_b} - \int_{t_a}^{t_b} \eta_i \frac{d}{dt} \left(\frac{\partial L}{\partial \dot{X}_i} \right) dt. \tag{A.10}$$

Since $\eta_i(t_a) = \eta_i(t_b) = 0$,

$$\int_{t_a}^{t_b} \frac{\partial L}{\partial \dot{X}_i} \dot{\eta}_i dt = - \int_{t_a}^{t_b} \eta_i \frac{d}{dt} \left(\frac{\partial L}{\partial \dot{X}_i} \right) dt. \tag{A.11}$$

Therefore,

$$\frac{dS_\varepsilon}{d\varepsilon} = \sum_{i=1}^{N} \left\{ \int_{t_a}^{t_b} \frac{\partial L}{\partial X_i} \eta_i dt - \int_{t_a}^{t_b} \eta_i \frac{d}{dt} \left(\frac{\partial L}{\partial \dot{X}_i} \right) dt \right\} \tag{A.12}$$

$$= \sum_{i=1}^{N} \int_{t_a}^{t_b} \left\{ \frac{\partial L}{\partial X_i} - \frac{d}{dt} \left(\frac{\partial L}{\partial \dot{X}_i} \right) \right\} \eta_i dt. \tag{A.13}$$

If $\frac{dS_\varepsilon}{d\varepsilon} = 0$, when $X_i = x_i$, then this means $\frac{\partial L}{\partial x_i} - \frac{d}{dt} \left(\frac{\partial L}{\partial \dot{x}_i} \right) = 0$.

$$\therefore \frac{d}{dt} \left(\frac{\partial L}{\partial \dot{x}_i} \right) = \frac{\partial L}{\partial x_i}. \tag{A.14}$$

This rather elegant expression is the *Euler–Lagrange (EL) equation*, the solution of which will yield $x_i(t)$. Note that there is a *distinct equation for each variable* $x_i(t)$. This feature enables a complicated multi-variable calculation to be separated into its component parts, (i.e. one EL equation for each variable x_i), which will prove invaluable when we apply these ideas to mechanical systems, such as the two sliding blocks problem in Chapter 1. Several more interesting examples are worked through in http://www.eclecticon.info/index_htm_files/Mechanics%20-%20Lagrangians.pdf, many of which are inspired by problems in Morin's beautiful and inappropriately named *Introductory Mechanics* [26]. A large number of the problems are devilishly hard, so goodness knows what *Advanced Mechanics* would contain. Perhaps this will be a book aimed at an 'AI *minds*-only'[1] readership in the coming decades!

---

[1]See Ian M. Banks' *Culture* science fiction novels. Humans (and other sentient lifeforms in a futuristic pan-galactic society) are basically looked after and indulged by (mostly) benign sentient and self-evolving artificial intelligences, which Banks calls 'Minds'.

A.1   The Beltrami Identity

It is worth noting a special case when L does not explicitly include the integration variable t, i.e. $L = L(x, \dot{x})$ and, therefore, $\frac{\partial L}{\partial t} = 0$. By the chain rule,

$$\frac{dL}{dt} = \sum_{i=1}^{N} \left\{ \frac{\partial L}{\partial x_i} \dot{x}_i + \frac{\partial L}{\partial \dot{x}_i} \frac{d\dot{x}_i}{dt} \right\}. \tag{A.15}$$

From the EL equation,

$$\frac{d}{dt} \left(\frac{\partial L}{\partial \dot{x}_i} \right) = \frac{\partial L}{\partial x_i}. \tag{A.16}$$

Therefore,

$$\frac{dL}{dt} = \sum_{i=1}^{N} \left\{ \dot{x}_i \frac{d}{dt} \left(\frac{\partial L}{\partial \dot{x}_i} \right) + \frac{\partial L}{\partial \dot{x}_i} \frac{d\dot{x}_i}{dt} \right\} \tag{A.17}$$

$$= \sum_{i=1}^{N} \left\{ \frac{d}{dt} \left(\dot{x}_i \frac{\partial L}{\partial \dot{x}_i} \right) \right\}. \tag{A.18}$$

Hence,

$$\frac{d}{dt} \left(L - \sum_{i=1}^{N} \left(\dot{x}_i \frac{\partial L}{\partial \dot{x}_i} \right) \right) = 0 \tag{A.19}$$

and therefore

$$L - \sum_{i=1}^{N} \dot{x}_i \frac{\partial L}{\partial \dot{x}_i} = \text{constant}, \tag{A.20}$$

which is know as the *Beltrami*[2] *identity*.

A.2   Lagrangian Mechanics

Consider a particle of mass m moving in x, y, z directions under the influence of a force $\mathbf{f} = -\nabla V$, where $V(x, y, z)$ is an expression of potential energy.[3] The kinetic energy is

$$T = \tfrac{1}{2} m \left(\dot{x}^2 + \dot{y}^2 + \dot{z}^2 \right). \tag{A.21}$$

---

[2]Eugenio Beltrami (1835–1900).

[3]Both gravitational and electric fields give rise to forces, which can be expressed by $\mathbf{f} = -\nabla V$, i.e. force is a (negative) potential gradient.

Consider the quantity

$$L = T - V, \qquad (A.22)$$

where L is called the *Lagrangian*. Applying the EL equation(s) derived in the previous sections of this appendix:

$$\frac{d}{dt}\left(\frac{\partial L}{\partial \dot{x}}\right) = \frac{\partial L}{\partial x}, \qquad (A.23)$$

$$\frac{d}{dt}\left(\frac{\partial L}{\partial \dot{y}}\right) = \frac{\partial L}{\partial y}, \qquad (A.24)$$

$$\frac{d}{dt}\left(\frac{\partial L}{\partial \dot{z}}\right) = \frac{\partial L}{\partial z}. \qquad (A.25)$$

Now, $\frac{\partial L}{\partial \dot{x}} = m\dot{x}$, $\frac{\partial L}{\partial \dot{y}} = m\dot{y}$, and $\frac{\partial L}{\partial \dot{z}} = m\dot{z}$ since the potential energy $V(x, y, z)$ is not functionally dependent on $\dot{x}$, $\dot{y}$ or $\dot{z}$. Similarly, the kinetic energy $T = T(\dot{x}, \dot{y}, \dot{z})$. $\therefore \frac{\partial L}{\partial x} = -\frac{\partial V}{\partial x}$, $\frac{\partial L}{\partial y} = -\frac{\partial V}{\partial y}$ and $\frac{\partial L}{\partial z} = -\frac{\partial V}{\partial z}$.

Hence, the EL equations reduce to

$$m\ddot{x} = -\frac{\partial V}{\partial x}, \qquad (A.26)$$

$$m\ddot{y} = -\frac{\partial V}{\partial x}, \qquad (A.27)$$

$$m\ddot{z} = -\frac{\partial V}{\partial x}, \qquad (A.28)$$

which can be written in vector form as

$$m\ddot{\mathbf{r}} = -\nabla V. \qquad (A.29)$$

This vector equation is, of course, a statement of Newton's second law: *mass × acceleration = vector sum of force*. The Lagrangian or calculus-of-variations approach, therefore, gives us an alternative point of departure for the solution of mechanics problems. Rather than begin from Newton's second law, we can instead start from $L = T - V$ and then apply the EL equation for each variable of the system. For multi-variable systems (e.g. rigid bodies in contact, coupled pendulums), this can be a much more efficient scheme of analysis since L is a scalar quantity involving energy terms for a system, whereas Newton's second law requires us to carefully resolve the *vector* forces on each sub-component of the system. This may require several diagrams to correctly assign equal (and opposite) contact forces between components, which of course will add to the number of unknowns (and equations) that need to be solved simultaneously.

Appendix B

The Dispersion Relationship for Waves on the Interface Between Two Fluids

This appendix is largely inspired by Faber's *Fluid Dynamics for Physicists* [5]. Consider two *isotropic, incompressible* fluids which are *vorticity-free*, i.e. the fluid velocity $\mathbf{v}$ obeys $\nabla \cdot \mathbf{v} = 0$ (incompressible) and $\nabla \times \mathbf{v} = 0$ (vorticity-free). Since $\nabla \times \mathbf{v} = 0$, we can express fluid velocity in terms of a *scalar potential*, ϕ, such that

$$\mathbf{v} = \nabla\phi, \tag{B.1}$$

i.e. *'potential flow'* conditions.[1] Consider a disturbance $\Gamma(x, z, t)$ at an interface of two fluids, as illustrated in Fig. B.1. $z = 0$ is the interface when there is no disturbance. The upper fluid has density ρ_1, and the lower one has density ρ_2. The lower fluid has a finite depth D (i.e. when $z = -D$, there is a solid surface that the fluid cannot flow within), and $\Gamma \ll D$. Let the scalar potential $\phi_{1,2}(x, z, t)$ have the following wave-like form:

$$\phi_{1,2}(x, z, t) = f_{1,2}(z)e^{i(kx-\omega t)}, \tag{B.2}$$

i.e. we assert wave-like disturbances to propagate left to right along the interface with wavenumber $k = \frac{2\pi}{\lambda}$.

In potential flow conditions, *Bernoulli's equation*[2] is applicable:

$$\frac{p}{\rho} + \tfrac{1}{2}v^2 + gz + \frac{\partial\phi}{\partial t} = \text{constant}, \tag{B.3}$$

where p is the fluid pressure. If we apply Bernoulli's equation to either side of the interface, assuming a disturbance of $z = \Gamma$ and also that the fluid velocity is small enough that we can neglect the v^2 term,

$$p_1 + \rho_1 g\Gamma + \rho_1 \left.\frac{\partial\phi_1}{\partial t}\right|_{z=\Gamma} = p_2 + \rho_2 g\Gamma + \rho_2 \left.\frac{\partial\phi_2}{\partial t}\right|_{z=\Gamma}. \tag{B.4}$$

---

[1] $\nabla \times \nabla\phi = 0$, where ϕ is any scalar function. So, if $\nabla \times \mathbf{v} = 0$, then we can write $\mathbf{v} = \nabla\phi$.

[2] Daniel Bernoulli (1700–1782).

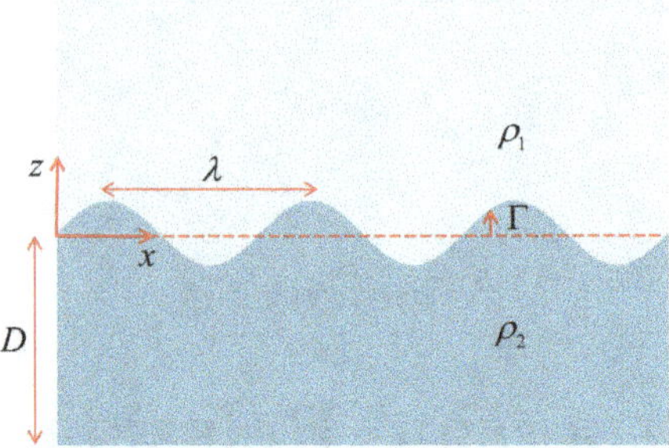

Fig. B.1. Waves of wavelength λ on an interface between fluids of density ρ_2 and ρ_1, where the lower fluid (of density ρ_2) is at depth D, obey the *dispersion relationship* $\omega^2 = \frac{\sigma k^3 + (\rho_2 - \rho_1)gk}{\rho_1 + \rho_2 \cotanh(kD)}$. The quantity σ is the *surface tension* of the interface.

Define *surface tension* σ such that the pressure difference across the interface is

$$p_2 - p_1 = -\sigma \frac{\partial^2 \Gamma}{\partial x^2}. \tag{B.5}$$

Now, if we assume

$$\frac{\partial \Gamma}{\partial t} = \mathbf{v} \cdot \hat{\mathbf{z}}, \tag{B.6}$$

i.e. the interface speed is the projection of fluid velocity in the z direction, and since $\mathbf{v} \cdot \hat{\mathbf{z}} = \nabla \phi \cdot \hat{\mathbf{z}} = \frac{\partial \phi}{\partial z}$,

$$\frac{\partial \Gamma}{\partial t} = \frac{\partial \phi}{\partial z} = \left. \frac{df_{1,2}}{dz} \right|_\Gamma e^{i(kx - \omega t)}. \tag{B.7}$$

$$\therefore \; \Gamma = \frac{1}{-i\omega} \left. \frac{df_{1,2}}{dz} \right|_\Gamma e^{i(kx - \omega t)} + \text{constant}. \tag{B.8}$$

Since there is no reason for Γ to have a fixed displacement about which there are oscillations, let us set this constant to be zero. Also, $\frac{1}{i} = -i \therefore \frac{1}{-i\omega} = \frac{i}{\omega}$.

$$\therefore \Gamma = \frac{i}{\omega} \left. \frac{df_{1,2}}{dz} \right|_\Gamma e^{i(kx - \omega t)}, \tag{B.9}$$

which means

$$e^{i(kx - \omega t)} = \frac{-i\omega \Gamma}{\left. \frac{df_{1,2}}{dz} \right|_\Gamma}. \tag{B.10}$$

The only x functional dependence of $\Gamma = \frac{i}{\omega} \left. \frac{df_{1,2}}{dz} \right|_\Gamma e^{i(kx - \omega t)}$ is e^{ikx}, which means

$$\frac{\partial^2 \Gamma}{\partial x^2} = -k^2 \Gamma. \tag{B.11}$$

Since $p_2 - p_1 = -\sigma \frac{\partial^2 \Gamma}{\partial x^2}$, the pressure difference across the interface can therefore be written as

$$p_2 - p_1 = \sigma k^2 \Gamma. \tag{B.12}$$

Now, since $\phi_{1,2}(x, z, t) = f_{1,2}(z) e^{i(kx - \omega t)}$,

$$\left. \frac{\partial \phi_{1,2}}{\partial t} \right|_{z=\Gamma} = -i\omega f_{1,2}(\Gamma) e^{i(kx - \omega t)}. \tag{B.13}$$

So, using $e^{i(kx-\omega t)} = \frac{-i\omega\Gamma}{\frac{df_{1,2}}{dz}}$, it implies that

$$\left.\frac{\partial \phi_{1,2}}{\partial t}\right|_{z=\Gamma} = -\omega^2 \frac{f_{1,2}(\Gamma)}{\left.\frac{df_{1,2}}{dz}\right|_{\Gamma}}\Gamma. \tag{B.14}$$

Substituting $p_2 - p_1 = \sigma k^2 \Gamma$ and $\left.\frac{\partial \phi_{1,2}}{\partial t}\right|_{z=\Gamma} = -\omega^2 \frac{f_{1,2}(\Gamma)}{\left.\frac{df_{1,2}}{dz}\right|_{\Gamma}}\Gamma$ into Bernoulli's

equation $p_1 + \rho_1 g\Gamma + \rho_1 \left.\frac{\partial \phi_1}{\partial t}\right|_{z=\Gamma} = p_2 + \rho_2 g\Gamma + \rho_2 \left.\frac{\partial \phi_2}{\partial t}\right|_{z=\Gamma}$ yields

$$\rho_1 \left.\frac{\partial \phi_1}{\partial t}\right|_{z=\Gamma} - \rho_2 \left.\frac{\partial \phi_2}{\partial t}\right|_{z=\Gamma} = p_2 - p_1 + (\rho_2 - \rho_1)\, g\Gamma. \tag{B.15}$$

$$\therefore -\omega^2 \left(\rho_1 \frac{f_1}{\frac{df_1}{dz}} - \rho_2 \frac{f_2}{\frac{df_2}{dz}} \right) \Gamma = \sigma k^2 \Gamma + (\rho_2 - \rho_1)\, g\Gamma,$$

which yields a *dispersion relationship*:

$$\omega^2 = \frac{\sigma k^2 + (\rho_2 - \rho_1)\, g}{\rho_2 \frac{f_2(\Gamma)}{\left.\frac{df_2}{dz}\right|_{\Gamma}} - \rho_1 \frac{f_1(\Gamma)}{\left.\frac{df_1}{dz}\right|_{\Gamma}}}. \tag{B.16}$$

Now, for the fluids to be incompressible, $\nabla \cdot \mathbf{v} = 0$, which means

$$\nabla^2 \phi = 0 \tag{B.17}$$

$$\frac{\partial^2 \phi_{1,2}}{\partial x^2} + \frac{\partial^2 \phi_{1,2}}{\partial z^2} = 0 \tag{B.18}$$

$$-k^2 f_{1,2} e^{i(kx-\omega t)} + \frac{d^2 f_{1,2}}{dz^2} e^{i(kx-\omega t)} = 0. \tag{B.19}$$

$$\therefore \frac{d^2 f_{1,2}}{dz^2} = k^2 f_{1,2}. \tag{B.20}$$

This second-order ordinary differential equation has solutions of the form

$$f_{1,2}(z) = A_{1,2} e^{kz} + B_{1,2} e^{-kz}. \tag{B.21}$$

Therefore,

$$\phi_{1,2}(x, z, t) = \left(A_{1,2} e^{kz} + B_{1,2} e^{-kz} \right) e^{i(kx-\omega t)}.$$

In our scenario, the boundary conditions for fluid velocity are

$$\mathbf{v}(z = -D) \cdot \hat{\mathbf{z}} = 0, \tag{B.22}$$

$$\mathbf{v}(z = \infty) \cdot \hat{\mathbf{z}} = 0. \tag{B.23}$$

Since $\mathbf{v} = \nabla \phi$,

$$\left.\frac{\partial \phi_1}{\partial z}\right|_{z=\infty} = 0, \tag{B.24}$$

$$\left.\frac{\partial \phi_2}{\partial z}\right|_{z=-D} = 0. \tag{B.25}$$

For the upper fluid,

$$\frac{\partial \phi_1}{\partial z} = \left(kA_1 e^{kz} - kB_1 e^{-kz}\right) e^{i(kx - \omega t)}. \tag{B.26}$$

So, if $\left.\frac{\partial \phi_1}{\partial z}\right|_{z=\infty} = 0$, then this means $A_1 = 0$, to remove the exponentially increasing term e^{kz} as $z \to \infty$. Hence,

$$f_1(z) = B_1 e^{-kz}. \tag{B.27}$$

For the lower fluid,

$$\frac{\partial \phi_2}{\partial z} = \left(kA_2 e^{kz} - kB_2 e^{-kz}\right) e^{i(kx - \omega t)}. \tag{B.28}$$

So, if $\left.\frac{\partial \phi_2}{\partial z}\right|_{z=-D} = 0$, then this means

$$0 = kA_2 e^{-kD} - kB_2 e^{kD}. \tag{B.29}$$

$$\therefore A_2 = B_2 e^{2kD}. \tag{B.30}$$

Hence,

$$f_2(z) = B_2 \left(e^{2kD} e^{kz} + e^{-kz}\right) \tag{B.31}$$

$$= B_2 \left(e^{kD} e^{kz+kD} + e^{kD} e^{-kD} e^{-kz}\right) \tag{B.32}$$

$$= 2B_2 e^{kD} \tfrac{1}{2} \left(e^{k(z+D)} + e^{-k(z+D)}\right) \tag{B.33}$$

$$= 2B_2 e^{kD} \cosh\left(k(z + D)\right). \tag{B.34}$$

With $f_1(z) = B_1 e^{-kz}$ and $f_2(z) = 2B_2 e^{kD} \cosh\left(k(z + D)\right)$, we can now replace the derivatives in the dispersion relationship:

$$\omega^2 = \frac{\sigma k^2 + (\rho_2 - \rho_1)\,g}{\rho_2 \frac{f_2(\Gamma)}{\frac{df_2}{dz}\big|_\Gamma} - \rho_1 \frac{f_1(\Gamma)}{\frac{df_1}{dz}\big|_\Gamma}} \tag{B.35}$$

$$= \frac{\sigma k^2 + (\rho_2 - \rho_1)\,g}{\rho_2 \frac{2B_2 e^{kD} \cosh(k(\Gamma+D))}{2B_2 e^{kD} k \sinh(k(\Gamma+D))} - \frac{B_1 e^{-k\Gamma}}{-kB_1 e^{-k\Gamma}} \rho_1} \tag{B.36}$$

$$= \frac{\sigma k^3 + (\rho_2 - \rho_1)\,gk}{\rho_2 \cotanh\left(k(\Gamma + D)\right) + \rho_1}. \tag{B.37}$$

Finally, since we have asserted a small perturbation approximation $\Gamma \ll D$,

$$\omega^2 = \frac{\sigma k^3 + (\rho_2 - \rho_1)\,gk}{\rho_1 + \rho_2 \cotanh\left(kD\right)}. \tag{B.38}$$

Afterword

Like any recipe book, *Science by Simulation* is best read actively. This means the mathematical recipes should serve as inspiration to construct spreadsheets and write computer code yourself. And if you have access to some basic laboratory equipment, do the experiments too! Consider joining or setting up a physics club so you have sufficient time to nurture your laboratory skills and work collaboratively in practical work, theoretical model-building and problem-solving. If you haven't already done so, sign up to *Isaac Physics* (https://isaacphysics.org/) and register for the *British Physics Olympiad* (https://www.bpho.org.uk/) (or equivalent in your country if you are not from the UK). *Science by Simulation* should be well thumbed and contain numerous annotations, chocolate sauce and coffee stains, although perhaps not if your copy has been borrowed from a library. If you feel adding a gastronomic marker is necessary for reasons of personalisation and comfort, try not to obscure the mathematics. World Scientific will release an electronic version of this book, but if you wish to peruse the majority of the MATLAB codes and Excel sheets referred to in this publication, do please download the files from my website: http://www.eclecticon.info/scibysim_vol2.htm. *The Eclecticon* also contains a fairly comprehensive collection of notes and resources that cover the majority of pre-university physics and mathematics, as well as a wide range of scientific computing examples. And a few other more arty things too.

Bibliography

[1] Akrill, T., Bennet, G., and Millar, C. *Practice in Physics*, 4th edn. Hodder Education, 2011.

[2] Bleaney, B.I. and Bleaney, B. *Electricity and Magnetism (Volumes 1 and 2)*, 3rd edn. Oxford Science Publications, 1976.

[3] Chadha, G., Bone, G., and Saunders, N. *A Level Physics for OCR A*. Oxford University Press, 2015.

[4] Cullerne, J.P. and Machacek, A. *The Language of Physics*. Oxford University Press, 2008.

[5] Faber, T.E. *Fluid Dynamics for Physicists*. Cambridge University Press, 1995.

[6] Feynman, R.P., Leighton, R.B., and Sands M. *The Feynman Lectures on Physics*, Vols. I–III, New Millennium edn. Basic Books, 2010.

[7] Feynman, R.P. *QED: The Strange Theory of Light and Matter*. Princeton University Press (USA), 1985.

[8] Fleisch, D. *A Student's Guide to Maxwell's Equations*. Cambridge University Press, 2008. http://www.danfleisch.com/maxwell/index.shtml.

[9] French, A. *Science by Simulation – Volume 1: A Mezze of Mathematical Models*. World Scientific, 2022. http://www.eclecticon.info/scibysim_vol1.htm.

[10] French, A., Cullerne J.P., and Kanchanasakdichai O. Numerical methods as an introduction to calculus. *Phys. Educ.* **54**, 045009 (2019).

[11] French, A. www.eclecticon.info.

[12] Fowler, C.N.R. *The Solid Earth*. Cambridge University Press, 1990.

[13] Goldstein, H., Poole, C., and Safko, J. *Classical Mechanics*, 3rd edn. Pearson Education International, 2002.

[14] Hammond, P. *Electromagnetism for Engineers*, 4th edn. Oxford Science Publications, 1997.

[15] Hand, L.N. and Finch, J.D. *Analytical Mechanics*. Cambridge University Press, 1998.

[16] Hanselman, D. and Littlefield, B., *Mastering MATLAB*, International edn. Pearson, 2012.

[17] Hecht, E. *Optics*, 3rd edn. Addison-Wesley Longman, 1998.

[18] Jackson, J.D. *Classical Electrodynamics*, 3rd edn. John Wiley & Sons, Inc., 1998.

[19] Kirk, T. *Physics for the IB Diploma. Standard and Higher Level*. Oxford University Press, 2007.

[20] Kittel, C. *Introduction to Solid State Physics*, 7th edn. John Wiley & Sons, Inc., 1996.

[21] Kleppner, D. and Kolenkow, R. *An Introduction to Mechanics*, 2nd edn. Cambridge University Press, 2014.

[22] Longair, M. *Theoretical Concepts in Physics*, 3rd edn. Cambridge University Press, 2020.

[23] Machacek, A.C. and Crowter, J.J. *Mastering Pre-University Physics*, 2nd edn. Periphyseos Press, Cambridge, 2016.

[24] Mandl, F. *Statistical Physics*, 2nd edn. John Wiley & Sons, 1988.

[25] MacKay, D.J.C. *Sustainable Energy – Without the Hot Air*. UIT Cambridge Ltd., 2009. www.withouthotair.com.

[26] Morin, D. *Introduction to Classical Mechanics with Problems and Solutions*. Cambridge University Press, 2007.

[27] Povey, T. *Professor Povey's Perplexing Problems*. Oneworld Publications, 2015.

[28] Press, W.H., Teukolsky, S.A., Vetterling, W.T., and Flannery, B.P. *Numerical Recipes in C++. The Art of Scientific Computing*, 2nd edn. Cambridge University Press, 2002.

[29] Quadling, D. *Mechanics 3 & 4*, 4th printing. Cambridge Advanced Mathematics Series. Cambridge University Press, 2010.

[30] Rees, W.G. *Physics by Example. 200 Problems and Solutions*. Cambridge University Press, 1994.

[31] Richards, M.A. *Fundamentals of Radar Signal Processing*. McGraw-Hill Electronic Engineering, 2005.

[32] Halpern, A. *3,000 Solved Problems in Physics*. Schaum's Outlines. McGraw-Hill, 2011.

[33] Shearer, P.M. *Introduction to Seismology*. Cambridge University Press, 1999.

[34] Skolnik, M.I. *Introduction to Radar Systems*, 3rd edn. (International edn.) McGraw-Hill, 2001.

[35] Tooley, M. *Electronic Circuits. Fundamentals and Applications*, 2nd edn. Newnes (An Imprint of Elsevier), 2004.

[36] Woan, G. *The Cambridge Handbook of Physics Formulas*. Cambridge University Press, 2000.

Index